New Frontiers in Crustacean Biology:
Proceedings of the TCS Summer Meeting, Tokyo,
20-24 September 2009

# New Frontiers in Crustacean Biology: Proceedings of the TCS Summer Meeting, Tokyo, 20-24 September 2009

*By*

Akira Asakura, Kobe University, Japan
(Editor in chief)

Raymond T. Bauer, University of Louisiana at Lafayette, U.S.A.
Anson H. Hines, Smithsonian Environmental Research Center, U.S.A.
Martin Thiel, University Católica del Norte, Chile
Keiji Wada, Nara Women's University, Japan
Toshiyuki Yamaguchi, Chiba University, Japan
Christoph Held, Alfred Wegener Institute of Polar and Marine Research, Germany
Christoph Schubart, University of Regensburg, Germany
James M. Furse, Griffith University, Queensland, Australia
Jason Coughran, Croaking Environment Resources, Australia
Tadashi Kawai, Hokkaido Fisheries Experiment Station, Japan
Susumu Ohtsuka, Hiroshima University, Japan
Miguel V. Archdale, Kagoshima University, Japan
Antonio Baeza, Smithsonian Marine Station at Fort Pierce, Florida, U.S.A.
Mikio Moriyasu, Gulf Fisheries Centre, Canada
(Editorial board)

CRUSTACEANA MONOGRAPHS, 15

BRILL

LEIDEN • BOSTON

This book is printed on acid-free paper.

**Library of Congress Cataloging-in-Publication Data**

The Library of Congress Cataloging-in-Publication Data is available from the Publisher.

ISBN: 378-90-04-17425-2

PRINTED BY DRUKKERIJ WILCO B.V. - AMERSFOORT, THE NETHERLANDS

CONTENTS



   

## PREFACE

The Crustacean Society Summer Meeting jointly held with the 47th Annual Meeting of the Carcinological Society of Japan, was held 20-24 September 2009 at Shinagawa Campus, Tokyo University of Marine Science and Technology, Tokyo, Japan. This meeting was a truly landmark event in the history of both the CSJ and the TCS, as it was the first time that TCS held its annual meeting in the Asian region and was likened to a happy marriage of the two crustacean societies.

Around 350 participants from 25 countries of all over the world came to Tokyo. About 260 papers were presented, including keynote addresses, symposium presentations, and symposium related and general contributed papers of both oral and poster presentations. The symposia held included: Life history migrations of freshwater shrimps – ecological and adaptive significance; Phylogeography and population genetics in decapod Crustacea; Speciation and biogeography in non-decapod crustaceans; Biology of Anomura III; Crustacean chemoreception – identification of cues and their applications; Integrative biology – crustaceans as model systems; Ecology and behavior of peracarids – progress and prospects; Reproductive behavior of decapod crustaceans; The new perspective on barnacle research; Symbiosis in crustaceans – diversity and evolutionary trends; Current status of fisheries and biological knowledge of snow and tanner crabs genus Chionoecetes in the world; Diversity and ecology of thalassinidean shrimps; Impacts of human exploitation on large decapod resources; Conservation biology of freshwater crayfishes – new challenges from Japan, Eastern Asia.

On behalf of the organizing committee (Keiji Baba, Akira Asakura, Katsuyuki Hamasaki, Wataru Doi, Seiichi Watanabe, Keiji Wada, Kooichi Konishi), I am grateful to all participants, the sponsors (below) and supporting organizations (below), staffs and students for making this meeting a great success. My special gratitude must go to TCS Past and Current Presidents, Drs. Frederick R. Schram, Jens Høeg, Gary C. B. Poore, Trisha Spears, Jeffrey D. Shields, and Rafael Lemaitre, as well as the board of TCS for their help, efforts and encouragement to the council of Carcinological Society of Japan and myself to make this meeting possible.

I am convinced that this volume of Crustacean Monographs Series, containing 29 contributions based on the presentations at the meeting, reflects the high standard of the meeting and will contribute to further advancement.

The Tokyo Meeting was kindly sponsored by: Carcinological Society of Japan, Toshimitsu Odawara Memorial Fund for Crustacean Research, The Crustacean Society, Tokyo University of Marine Science and Technology, Inoue Foundation for Science, Tokyo University of Marine Science and Technology, Kiwi Breeze, Taiheiyo Synthetic Consultant Co., Ltd., Zukosha Co., Ltd., Seibutsu Kenkyusha Co., Ltd., and The Zoological Society of Japan.

Furthermore, the Tokyo Meeting was kindly supported by: Society of Evolutionary Studies, Japan, The Genetic Society of Japan, The Plankton Society of Japan, The Japanese Society of Fisheries Science, The Sessile Organisms Society of Japan, Japanese Society for Aquaculture Research, The Biological Society of Okinawa, The Zoological Society of Japan, Ecology and Civil Engineering Society, Japanese Association of Benthology, The Biogeographical Society of Japan, Ecological Society of Japan, The Society of Population Ecology, Japanese Coral Reef Society, Japan Ethological Society, Japanese Society of Biological Scientists, The Japanese Society of Soil Zoology.

AKIRA ASAKURA
*Editor-in-Chief, for the present volume*
*New Frontiers in Crustacean Biology*
Secretary General, Organizing Committee, TCS Summer Meeting in Tokyo
President, The Crustacean Society

# SYMBIOSIS OF PLANKTONIC COPEPODS AND MYSIDS WITH EPIBIONTS AND PARASITES IN THE NORTH PACIFIC: DIVERSITY AND INTERACTIONS

BY


SUSUMU OHTSUKA[1,8]), TAKEO HORIGUCHI[2]), YUKIO HANAMURA[3]),
ATSUSHI YAMAGUCHI[4]), MICHITAKA SHIMOMURA[5]), TOSHINOBU SUZAKI[6]),
KIMIAKI ISHIGURO[2]), HIDEO HANAOKA[1]), KAYOKO YAMADA[1])
and SHUJI OHTANI[7])

[1]) Takehara Marine Science Station, Setouchi Field Center, Graduate School of Biosphere Science, Hiroshima University, 5-8-1 Minato-machi, Takehara, Hiroshima 725-0024, Japan
[2]) Department of Natural History Sciences, Faculty of Science, Hokkaido University, Sapporo 060-0810, Japan
[3]) National Research Institute of Fisheries Science, 2-12-4, Fukuura, Kanazawa, Yokohama, Kanagawa, 236-8648, Japan
[4]) Laboratory of Marine Biology, Graduate School of Fisheries Sciences, Hokkaido University, 3-1-1 Minatomachi, Hakodate, Hokkaido 041-8611, Japan
[5]) The Kitakyushu Museum of Natural History and Human History, 2-4-1 Higashida, Yahatahigashi-ku, Kita-Kyushu, Fukuoka 805-0071, Japan
[6]) Department of Biology, Faculty of Science, Kobe University, 1-1 Rokkodai-cho, Nada-ku, Kobe 657-8501, Japan
[7]) Faculty of Education, Shimane University, 1060 Nishikawatsu-cho, Matsue, Shimane 690-8504, Japan


## ABSTRACT


Planktonic crustaceans such as copepods and mysids are two of the most abundant components of the marine zooplankton community and, although they harbor a diversity of symbionts, their real interactions have been poorly understood. We have been investigating planktonic symbiosis and briefly review the biology of symbionts on planktonic crustaceans based mainly on our research conducted in the North Pacific.

Symbiotic histophagous apostome ciliates probably have a significant negative impact on their coastal copepod hosts in view of their high prevalence and their worldwide distributions in the coastal ecosystems. Such symbionts are also likely to impact the populations of the copepod's predators such as chaetognaths. In contrast, symbiosis between copepods and epibionts such as diatoms and suctorian ciliates may be more or less harmless to the host.


---


[8]) Corresponding author; e-mail: ohtsuka@hiroshima-u.ac.jp

Various endoparasitic alveolates have been discovered infecting from copepods, some of which could have evolved as parasitoids.

Epibiont peritrichians found on the body of gastrosaccid mysids are generally regarded as a commensal, and showed a remarkably high host-specificity to intertidal species as well as a distinct geographic cline with a preference for boreal waters. The sand-burrowing behavior of the mysids, coupled with the diversity and abundance of their preys possibly contributes substantially to the establishment of the symbiotic association with epibionts. A dajid isopod and a nicothoid copepod compete for the space and possibly, food within the marsupium of the mysid host *Siriella okadai*. The annual egg production of the host *S. okadai* seems to be significantly suppressed by these two parasites. Prior to the appearance of mature adults of each of these parasites within the host marsupium, immature individuals occupy particular microhabitats within the host dependent upon the state of maturity of the host. It is important to pay more attention to parasitoid protists on zooplankters in order to better understand the aquatic ecosystem.

## INTRODUCTION

The study of marine plankton has paid more attention to prey-predator relationships than to symbiosis, in part because the impact of the latter had been improperly underestimated so that symbiosis was considered to play only a minor role in the ecological interactions structuring pelagic communities (Ohtsuka et al., 2007). Recent investigations have, however, clearly revealed that symbionts have more complex and significant impacts on the population dynamics of their host zooplankters. For example, alveolate parasitoids sometimes lead to mass mortalities of host zooplankters including tintinnids, copepods, and euphausiids (Cachon & Cachon, 1987; Coats & Heisler, 1989; Kimmerer & McKinnon, 1990; Capriulo et al., 1991; Gómez-Gutiérrez et al., 2003, 2006, 2009; Ohtsuka et al., 2004, 2007; Skovgaard & Saiz, 2006). Their interactions broadly range from phoresy, to mutualism through to commensalism and parasitism to parasitoidism (Bush et al., 2001; Rhode, 2005).

The present paper briefly reviews the symbiotic relationship of copepods and mysids with a variety of microscopic symbionts based mainly on our recent investigations carried out in Japanese waters. Symbiosis is generally defined as an association between two different organisms living together, and usually with a gradient of beneficial or deleterious consequences for at least one of them (Bush et al., 2001). However, we redefine this term considering the interspecific relationships in which usually large-sized "hosts" are infested or infected by symbionts.

# COPEPOD HOSTS

## Apostome ciliates

Apostomes are symbiotic ciliates that mainly infest planktonic and benthic crustaceans at least during one phase of their complex life cycles, which typically include four functionally different morphs: resting phoront, feeding trophont, divisional tomont, and infective tomite stages (Chatton & Lwoff, 1935).

*Vampyrophrya pelagica* Chatton & Lwoff, 1930 is well investigated in terms of its morphology, cytology and ecology, and most likely plays a pivotal role in the brackish to coastal pelagic ecosystems in the world oceans due to its high prevalence and harmful impact both directly on planktonic copepods and indirectly on their invertebrate predators such as chaetognaths (Chatton & Lwoff, 1935; Grimes & Bradbury, 1992; Ohtsuka et al., 2004). Zooplanktonic invertebrate predators that prey upon parasitized copepods obtain fewer nutrients from their preys, because infected copepods are at least partly consumed by the histophagous apostome. The ciliates can increase mortality in the copepod population, if the copepod is injured by any other means, thus reducing the prey availability in the pelagic ecosystem. Thus, the trophic behavior of the histophagous apostome ciliates could have two adverse affects on higher trophic levels in the zooplankton food web: increasing copepod mortality and depleting the resources available to carnivorous zooplanktonic predators.

The life cycle of *V. pelagica* is briefly summarized below based on Ohtsuka et al. (2004), and on our unpublished data from studies carried out in the Seto Inland Sea, Japan. An oval, encysted phoront, within which one cell is enclosed, is typically attached to the ventral side of the prosome and/or to the prosomal appendages of copepods (fig. 1A, B). Host-specificity was expressed: calanoid and "poecilostomatoid" copepods (see Boxshall & Halsey, 2004), were preferred hosts in comparison of other copepod taxonomic groups, irrespective of body sizes and/or behavior, while some species of the cyclopoid *Oithona* were not selected. The infective tomite of *Vampyrophrya* was observed swimming rapidly around the body of *Oithona*, but finally moving away without successful settlement, from which we inferred some kind of physio-chemical interaction between the tomite and host body surface. High incidence of this apostome was observed between August and January, especially in later summer and fall, when prevalence was nearly 100% in the numerically dominant calanoid *Paracalanus parvus* (Claus, 1863) s.l. Its intensity could exceed 40 cells per host. This stage was characterized by a

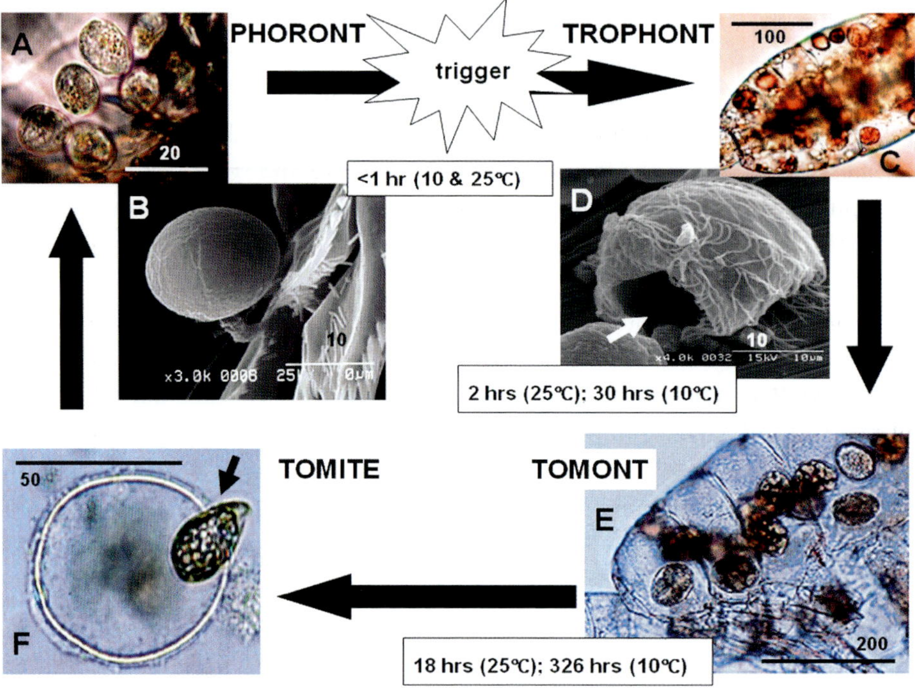


Fig. 1. Life cycle of the apostome ciliate *Vampyrophrya pelagica* infecting planktonic copepods. A, B, phoronts; C, D, trophonts (cytostome indicated by white arrow in D); E, tomonts; F, tomite (arrowed) releasing from tomont cyst. The phoront excyst having a metamorphosis to trophont stage in about 1 hr after the copepods were crashed by needles (trigger stimulus). Approximate stage durations of trophonts and tomonts at 10 and 25°C are indicated in D and E panels, respectively. Scales are in $\mu$m. (After Ohtsuka et al., 2004 with permission from Inter-Research.)

specialized intracellular structure, numerous lamellae of ca. 0.04 $\mu$m thick, which was identified as a precursor of the food vacuole membrane of the trophont.

A trophont (fig. 1C, D) excysts from a phoront when the parasitized copepods are: (1) fed upon by invertebrate predators such as chaetognaths or medusae, that break the copepod body allowing the ciliates to excyst, (2) physically damaged, and presumably, (3) unsuccessful in molting (Grimes & Bradbury, 1992; Ohtsuka et al., 2004). In any case, body fluids leaking from the copepod or physically damaged copepods trigger the ciliate metamorphosis. Feeding of fish larvae of fish such as *Plecoglossus altivelis altivelis* (Temminck & Schlegel, 1846) on parasitized copepods didn't result in hatching of phoronts in the laboratory. Trophonts enter via fissures in the copepod exoskeleton, and commence to consume copepod tissues by a large cytostome (fig. 1D). The cell volume eventually increases to about 30 times that of the initial trophont cell. This increase in trophont size is enabled by the intracellular precursor material of the phoront. Fully-grown trophonts metamorphose into encysted tomonts inside the empty body of the consumed copepod (fig. 1E). Up to 20 tomites cells are released from a tomont. These infective cells search for a new copepod host (fig. 1F) and then metamorphose into phoronts again.

In the cold-water season they almost completely disappeare from the hosts except in large-sized species such as *Calanus sinicus* Brodsky, 1965 (~3 mm long) with sufficient long longevity to harbor phoronts, which seem to attach just before the phase of intensive production of tomites finished. This phenomenon can be explained by the relationship between seasonal fluctuation in water temperature and duration of development of apostome life stages. In the laboratory low temperature clearly causes delay in developmental duration of each stage, in particular, of the tomont stages. Completion of the divisional stage took about 330 h for 50% cells at 10°C, about 20 times longer than at 25°C. Production of the infesting tomites appears to be suppressed by low temperature.

<br>

Suctorian and peritrich ciliates

Symbiotic relationships between suctorian ciliates and pelagic copepods have been intensively investigated in the subarctic waters of the North Pacific (Yamaguchi, 2006, unpubl.). The attachment of four epibiont suctorian genera, *Paracineta*, *Pelagacineta*, *Tokophrya* and *Trophogemma* on copepods was observed exclusively on the urosome of 10 relatively large-sized, mesopelagic species of six calanoid copepod genera. *Tokophrya* and *Trophogemma* exhibited high host-specificity on *Paraeuchaeta birostrata* Brodsky, 1950 and *P. elongata* (Esterly, 1913), respectively. The attachment of suctorians to copepods seems to be species- and stage-specific. Carnivorous hosts and adult females had higher infestation rates, suggesting higher host-specificity on hosts of larger size, greater longevity, and/or higher escape ability from predators. This means that suctorians are less impacted by being eaten by predators of the copepod hosts.

*Paracineta* (fig. 2F, G) and *Pelagacineta* suctorian ciliates exclusively infest adult females of the particle-feeding calanoid *Metridia pacifica* Brodsky, 1950. Prevalence was relatively higher in the Bering Sea and at higher latitudes in the North Pacific in summer and fall with an average of 9.4% (range 0-70%) (table I). A geographical gradient was also observed, with higher attachment in cold waters (<10°C) (table I).

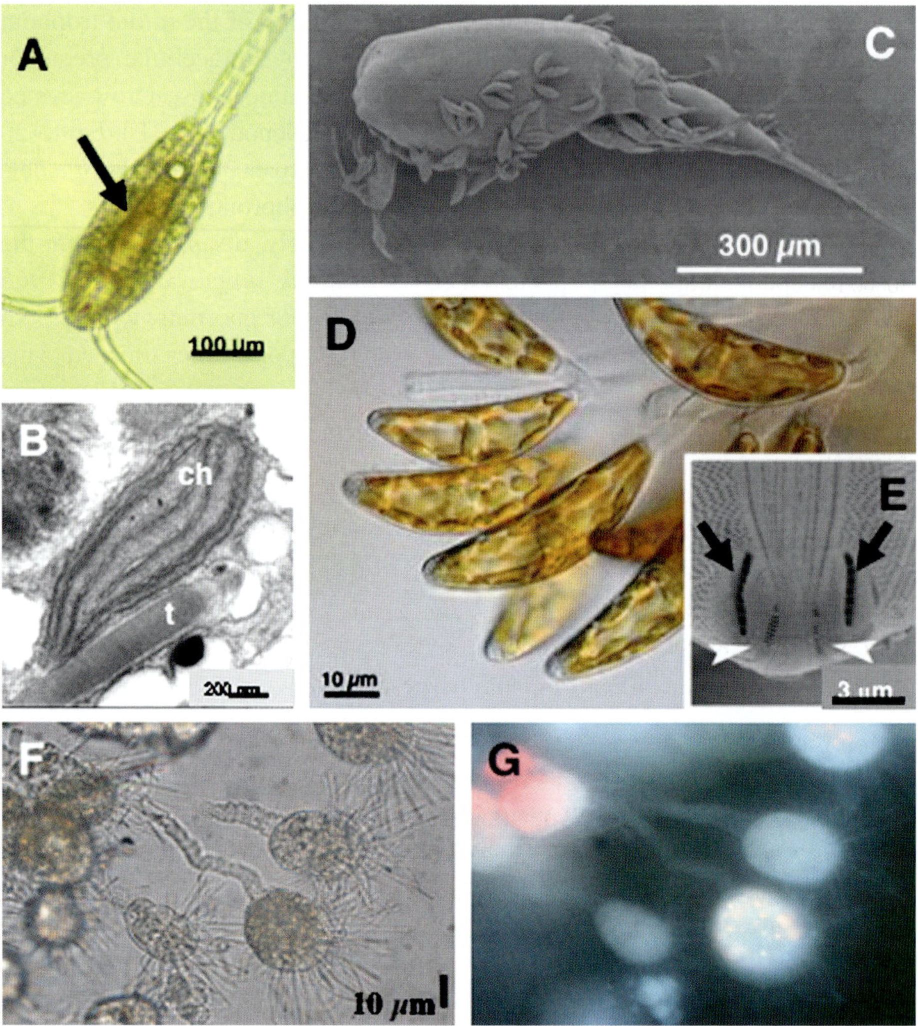

Fig. 2. A, *Blastodinium oviforme* in gut of *Oithona* sp. (arrowed); B, TEM microphoto of trophont cell of *B. oviforme*, ch = chloroplast, t = trichocyst; C, *Corycaeus affinis* infected by the diatom *Pseudohimantidium pacificum*; D, in-situ *P. pacificum* with stalks; E, two pairs of pores (attachment secretory pores indicated white arrowhead; stalk-substance releasing pores by black arrows); F, G, *Paracineta* sp., epibiont on urosome of adult female of *Metridia pacifica* (G in blue excitation with fluorescent microscope).

Adult females of *M. pacifica* infested by the suctorian ciliates contained 2.0-5.4 times higher concentration of fluorescent pigment than uninfested copepods, suggesting that the epibionts are also particle feeders (fig. 2G). Thus, copepods infested with suctorians are nutritionally more enriched for

TABLE I

Latitudinal changes in prevalence of suctorian ciliates of adult females of *Meteridia pacifica* collected during June to August 2004 in the North Pacific (after Yamaguchi, 2006)

| Location | Temperature (°C) | Density (indiv. m$^{-3}$) | Prevalence (%) |
|---|---|---|---|
| Bering Sea (177°W) | | | |
| 53°30'N | 8.1 | 63.0 | 67.6 |
| Northeast Pacific (165°W) | | | |
| 50°00'N | 7.9 | 33.1 | 48.1 |
| 49°00'N | 8.2 | 108.4 | 23.3 |
| 48°00'N | 8.0 | 4.0 | 4.0 |
| 47°00'N | 9.2 | 5.9 | 4.8 |
| 45°30'N | 10.4 | 9.2 | 0.0 |
| 44°00'N | 10.8 | 19.7 | 0.0 |
| 42°30'N | 11.0 | 1.7 | 0.0 |
| 41°00'N | 13.8 | 0.5 | 0.0 |

the copepod predators of copepods. The abundance and biomass of suctorians on copepods were estimated at $1.7 \times 10^5$ cells m$^{-2}$ and 816 $\mu$g C m$^{-2}$, corresponding about 0.1-0.3% and 0.3-0.4% of those of free-living ciliates, respectively.

## Dinoflagellates and their closely related alveolates

Symbiotic dinoflagellates sensu lato comprise a phylogenetically miscellaneous assemblage, according to recent genetic analyses (López-Garcia et al., 2001; Silberman et al., 2004; Skovgaard et al., 2005; Dolvin et al., 2007; Harada et al., 2007). Several dinoflagellates are extracellular parasites, while others are intracellular. Their effects on hosts vary from almost harmless (*Blastodinium*) to fatal (*Atelodinium, Syndinium*, etc.) (Ianora et al., 1987; Kimmerer & McKinnon, 1990; Shields, 1994; Horiguchi et al., 2006; Skovgaard & Saiz, 2006).

*Blastodinium* is an endoparasitic dinoflagellate found in the gut of a variety of copepods (fig. 2A). *Blastodinium* infection causes reduced survival of starved adults and sterility of infected females of the poecilostomatoid *Oncaea*, although it is almost harmless to other copepods (Skovgaard, 2005). Our observations of the ultrastructure of the trophont of *Blastodinium oviforme* Chatton, 1912 that infects *Oithona* sp. revealed the presence of chloroplasts and trichocysts in the cell (see fig. 2B), implying photosynthetic activity within the copepod gut (Pasternak et al., 1984; Ianora et al., 1987; Skovgaard, 2005). Skovgaard & Saiz (2006) reported the remarkable annual variation in occurrence of *Blastodinium* spp. parasitizing planktonic copepods in the north-western Mediterranean Sea. They found that the highest incidence coincided with the greatest abundance of the hosts. They estimated that sterility was in the range of 0.05 to 0.16 $d^{-1}$ for *Oncaea scottodicarloi* Heron & Bradford-Grieve, 1995. Skovgaard & Saiz (2006) also estimated that mortality rate caused by infection by *Syndinium turbo* Chatton, 1910 ranged from 0.08 to 0.15 $d^{-1}$ for *Paracalanus parvus* females. Other endoparasites belonging to *Atelodinium* can be fatal and have been reported to cause mass mortalities of copepods in the Mediterranean Sea (Ianora et al., 1987) and Australia (Kimmerer & McKinnon, 1990). *Atelodinium* sp. parasitic on *Paracalanus indicus* Wolfenden, 1905 is reported to kill the females at the maximum rate of 41% per day (Kimmerer & McKinnon, 1990).

Diatoms

A variety of pennate diatoms (fig. 2C-E) are associated with planktonic copepods as epibionts and they show high host-specificity (Hiromi et al., 1985). Symbiosis between the diatom *Pseudohimantidium pacificum* Hustedt & Krasske, 1941 and the members of the poecilostomatoid family Corycaeidae has been investigated (Russel & Norris, 1971; Hanaoka et al., un-publ.). During the copepod mating, the diatoms can move from one to another host, using viscous secretion released from tips of the cells (white arrowheads in fig. 2E) and this transfer takes about 10 minutes. After settlement, stalk sub-stance is released from another pair of pores on both sides of the tip (black arrows in fig. 2E). The diatoms can proliferate asexually on the body surface of the host, although the life cycle is not as yet elucidated.

Generally, these epibiont diatoms might be able to derive the following benefits from the symbiotic association (cf. Hiromi et al., 1985): (1) enhanced photosynthesis due to the near-surface distribution of copepods during the daytime; (2) replenishment of nutrient supply due to the host movement during its diel vertical migration; (3) avoidance of particle-feeders; (4) utilization of nutrients released from host bodies and/or from captured prey animals. In contrast, copepods may suffer from negative impacts due to the diatom attachment: (1) loss of energy due to increased drag in swimming and feeding; (2) greater susceptibility to visual predators due to increasing apparent volume; (3) interference with host mating.

Some ecological characteristics of copepods are related to the attachment of epibiont diatoms. Carnivorous copepods such the calanoid families Candaci-idae and Pontellidae are preferred by some epibiont diatoms as hosts. However, *Sceptronema orientale* Takano, 1983 exhibits an extremely high host-specificity to the harpacticoid *Euterpina acutifrons* (Dana, 1848) which is possibly a particle-feeder (Skovgaard & Saiz, 2006; Hanaoka et al., unpubl.).

## MYSID HOSTS

### Ciliates

Approximately ten or more ciliate species belonging to subclasses Chonotrichia, Suctoria, and Peritrichia are known to be epibionts of mysids (Hanamura & Nagasaki, 1996; Fernandez-Leborans & Tato-Porta, 2000a, b; Fernandez-Leborans, 2001; Hanamura, 2000, 2004; Ohtsuka et al., 2006). Some species are inferred to have negative impacts on their hosts due to interference with visual perception, swimming and respiration, and to competition for food particles, while others show no discernible effects on them (Hanamura, 2000; Ohtsuka et al., 2006).

Epibiont peritrich ciliates belonging to the Vorticellidae and Epistylididae exhibit high host-specificity and show a distinct geographical cline in prevalence on the genus *Archaeomysis*, a sand-burrowing member of the Gastrosaccinae (Hanamura & Nagasaki, 1996, Hanamura, 2000; Ohtsuka et al., 2006). The ciliate infestation is restricted to intertidal boreal species, but never occur on the southern temperate inhabitants or infralittoral species or non-sand-burrowing mysids (Hanamura & Nagasaki, 1996; Ohtsuka et al., 2006). These regional infection patterns seem to be due partly to remarkable behavior of intertidal *Archaeomysis*, which swims in the water in high tide and burrows into sediments as soon as the water recedes. It is possible that high food availability may contribute to high incidence of the ciliates in northern waters. In Ishikari Bay, northern Japan, the prevalence of the peritrich ciliates on *Archaeomysis articulata* Hanamura, 1997 annually reached on average 92%, suggesting that the ciliates are capable of colonizing the new integument of the host shortly after molting takes place (Hanamura, 2000). Suctorian ciliates on the oceanic calanoid copepods show similar geographical clines, reflecting higher prevalence in cold waters (see table I).

### Copepods and isopods

Dajid isopods and nicothoid copepods are common ectoparasites on the body surface or within the marsupium of mysids (Hansen, 1897; Mauchline, 1980; Rhode, 2005; Ohtsuka et al., 2005, 2006, 2007). Recently we have found

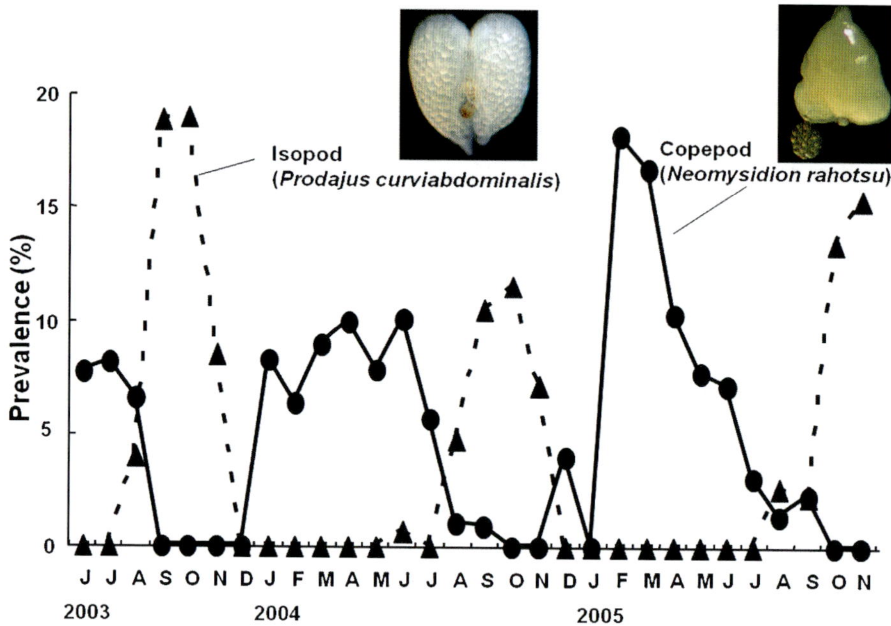


Fig. 3. Seasonal segregation of two crustacean parasites (dajid isopod *Prodajus curviabdominalis* and nicothoid copepod *Neomysidion rahotsu*) parasitizing within marsupium of the mysid *Siriella okadai*. (After Ohtsuka et al., 2007 with permission from Inter-Research.)

new taxa of these crustaceans within the marsupium of *Siriella okadai* Ii, 1964 in the Seto Inland Sea, Japan: the isopod *Prodajus curviabdominalis* Shinomura, Ohtsuka & Naito, 2005 and the copepod *Neomysidion rahotsu* Ohtsuka, Boxshall & Harada, 2005 (Ohtsuka et al., 2005, 2006, 2007; Shimomura et al., 2005). The adult copepod and, possibly, the isopod feed voraciously on host eggs. The parasites alternatively occupy the marsupium of the host (fig. 3): the presence of the parasitic isopod was restricted to the middle summer through late fall (water temperature marked >20°C), whereas the parasitic copepod was present from the mid winter to summer (<20°C) (Ohtsuka et al., 2007). Their coexistence on the same host individual was extremely rare. The isopod dominance may be caused because the females have larger body size, more motility, and possibly higher feeding rates in comparison with those of the copepod. During the occupation of the host marsupium by the isopod, the copepod may exhibit an unusual behavior that allows it to avoid predation from the isopod. The isopod life cycle probably includes a pelagic phase in which larvae utilize intermediate hosts such as planktonic copepods while adults are absent from the mysids (Ohtsuka et al., 2007).

Interestingly, immature females of these two parasites show a specific behavior to enter the host before they arrive at their final microhabitat, the marsupial lumen. The isopod penetrates the space between the carapace and dorsal tergites, while the copepod embeds itself in the host body tissue. However, the infection pathways from this initial infection site in the host body to the marsupium are still unknown. This unique parasitic behavior is shown only in the mysid females. Presumably, the infective stage of the Dajidae male directly enters the mysid female marsupium where they subsequently metamorphose into dwarf adult. The infection behavior of the Dajidae female allows for temporal coupling until the mysid host becomes mature with a fully developed marsupium, and also lowers the chance of detachment from the host. In contrast, the parasitic copepods seem to grow within the host body tissue, feeding on it.

The prevalence of females of the isopod and the copepod in the marsupium of the host mysid usually averaged 9 and 7%, respectively, with a maximum of about 20% for both parasitic species (Ohtsuka et al., 2007). Adult females of the nicothoid copepods fed on host eggs at a rate of 7 to 10 eggs female$^{-1}$ d$^{-1}$ (Ohtsuka et al., 2007). Considering the high prevalence and intensive feeding on the mysid host eggs, it is likely that these two parasites impact the dynamics of the mysid population. A closely related nicothoid copepod *Hansenulus trebax* Heron & Damkaer, 1986 showed high prevalence, up to 52% within on the marsupium of *Neomysis mercedis* Holmes, 1896 (Daly & Damkaer, 1986). This level of parasitism was suggested as having an adverse influence on higher trophic levels such as fish.

## PERSPECTIVES

Marine zooplankters provide a diverse and dynamic substrate in the vast extent of the water column for epibionts, and can serve as food sources for parasites and parasitoids. However, precise interactions between zooplankton hosts and their symbionts are rarely studied and thus little understood. The host specificity and the life cycles of these symbionts, including information on infective stages, deserve further investigations. In particular, symbiotic alveolate protists have significant negative impacts on the population dynamics of their hosts. Therefore symbionts should be regarded as important components of aquatic food webs. We must also pay an attention to relationships between global climate change and symbiosis in the marine ecosystem (cf. Kutz et al., 2005).

## ACKNOWLEDGEMENTS

We would like to express our sincere thanks to Drs. Geoffrey A. Boxshall and Rony Huys for reviewing the draft. This study was partly supported by grants from the Japan Society for the Promotion of Sciences (Nos. 20380110 awarded to SO; 21770100 to MS).

## REFERENCES

BOXSHALL, G. A. & S. H. HALSAY, 2004. An introduction to copepod diversity: 1-966. (The Ray Society, London).

BUSH, A. O., J. C. FERNÁNDEZ, G. W. ESCHE & J. R. SEED, 2001. Parasitism. The diversity and ecology of animal parasites: i-ix, 1-566. (Cambridge University Press, Cambridge).

CACHON, J. & M. CACHON, 1987. Parasitic dinoflagellates. Bot. Monogr., **21**: 571-610.

CAPRIULO, G. M., M. J. REDONE & E. B. SMALL, 1991. High apostome ciliate endoparasite infection rates found in Bering Sea euphausiid *Thysanoessa inermis*. Mar. Ecol. Prog. Ser., **72**: 203-204.

CHATTON, É. & A. LWOFF, 1935. Les Ciliés Apostomes. 1. Aperçu historique et général; étude mongraphique des genres et des espèces. Arch. Zool. Exp. Gén., **77**: 1-453.

COATS, D. W. & J. J. HEISLER, 1989. Spatial and temporal occurrence of the parasitic dinoflagellate *Duboscquella cachoni* and its tintinnine host *Eutintinnus pectinis* in Chesapeake Bay. Mar. Biol., **101**: 401-409.

DALY, K. L. & D. M. DAMKAER, 1986. Population dynamics and distribution of *Neomysis mercedis* and *Alienacanthomysis macropsis* (Crustacea: Mysidacea) in relation to the parasite copepod *Hansenulus trebax* in the Columbia River estuary. J. crust. Biol., **6**: 840-857.

DOLVEN, J. K., C. LINDQVIST, V. A. ALBERT, K. R. BJØRKLUND, T. YUASA, O. TAKAHASHI & S. MAYAMA, 2007. Molecular diversity of alveolates associated with neritic North Atlantic radiolarians. Protist, **158**: 65-76.

FERNANDEZ-LEBORAN, G., 2001. A review of the species of protozoan epibionts on crustaceans. III. Chonotrich ciliates. Crustaceana, **74**: 581-607.

FERNANDEZ-LEBORAN, G. & M. L. TATO-PORTA, 2000a. A review of the species of protozoan epibionts on crustaceans. I. Peritrich ciliates. Crustaceana, **73**: 643-683.

— — & — —, 2000b. A review of the species of protozoan epibionts on crustaceans. II. Suctorian ciliates. Crustaceana, **73**: 1205-1237.

GÓMEZ, F., P. LÓPEZ-GARCÍA, A. NOWACZYK & D. MOREIRA, 2009. Epibiotic suctorians and enigmatic ecto- and endoparasitoid dinoflagellates of euphausiid eggs (Euphausiacea) off Oregon, USA. J. Plankton Res., **31**: 777-786.

GÓMEZ-GUTIÉRREZ, J., W. T. PETERSON & J. F. MORADA, 2006. Discovery of a ciliate parasitoid of euphausiids off Oregon, USA: *Collinia oregonensis* n. sp. (Apostomatida: Colliniidae). Dis. aquat. Org., **71**: 33-49.

GÓMEZ-GUTIÉRREZ, J., W. T. PETERSON, A. D. ROBERTIS & R. D. BRODEUR, 2003. Mass mortality of krill caused by parasitoid ciliates. Science, **301**: 339.

GRIMES, B. H. & P. C. BRADBURY, 1992. The biology of *Vampyrophrya pelagica* (Chatton & Lwoff, 1930), a histophagous apostome ciliate associated with marine calanoid copepods. J. Protozool., **39**: 65-79.

HANAMURA, Y., 2000. Seasonality and infestation pattern of epibiosis in the beach mysid *Archaeomysis articulata*. Hydrobiologia, **427**: 121-127.

——, 2004. Ciliate-mysid epibiontic association on sandy beaches of Japan. In: K. NAGASAWA (ed.), Aquaparasitology in a field of Japan: 108-122. (Tokai University Press, Tokyo). [In Japanese.]

HANAMURA, Y. & K. NAGASAKI, 1996. Occurrence of the sandy beach mysids *Archaeomysis* spp. (Mysidacea) infested by epibiontic peritrich ciliates (Protozoa). Crust. Res., **25**: 25-33.

HANSEN, H. J., 1897. The Choniostomatidae, a family of Copepoda, parasites on Crustacea Malacostraca: 1-206. (Host and Son, Copenhagen).

HARADA, A., S. OHTSUKA & T. HORIGUCHI, 2007. Species of the parasitic genus *Duboscquella* are members of the enigmatic Marine Alveolate Group I. Protist, **158**: 337-347.

HIROMI, J., S. KADOTA & H. TAKANO, 1985. Diatom infestation of marine copepods (review). Bull. Tokai reg. Fish. Res. Lab., **117**: 37-46.

HORIGUCHI, T., A. HARADA & S. OHTSUKA, 2006. Taxonomic studies on parasitic dinoflagellates in Japan. Bull. Plankton Soc. Japan, **53**: 21-29. [In Japanese with English abstract.]

IANORA, A., M. G. MAZZOCCHI & B. S. CARLO, 1987. Impact of parasitism and intersexuality on Mediterranean populations of *Paracalanus parvus* (Copepoda: Calanoida). Dis. aquat. Org., **3**: 29-36.

KIMMERER, W. J. & A. D. MCKINNON, 1990. High mortality in a copepod population caused by a parasitic dinoflagellate. Mar. Biol., **107**: 449-452.

KUTZ, S. J., E. P. HOBERG, L. POLLEY & E. J. JENKINS, 2005. Global warming is changing the dynamics of Arctic host-parasite systems. Proc. Roy. Soc., (B) **272**: 2571-2576.

LÓPEZ-GARCIA, P., F. RODRIGUEZ-VALERA, C. PEDRÓS-ALIÓ & D. MOREIRA, 2001. Unexpected diversity of small eukaryotes in deep-sea Antarctic plankton. Nature, **401**: 603-607.

MAUCHLINE, J., 1980. The biology of mysids and euphausiids. Adv. mar. Biol., **18**: 1-680.

OHTSUKA, S., G. A. BOXSHALL & S. HARADA, 2005. A new genus and species of nicothoid copepod (Crustacea: Copepoda: Siphonostomatoida) parasitic on the mysid *Siriella okadai* Ii from off Japan. Syst. Parasitol., **62**: 65-81.

OHTSUKA, S., Y. HANAMURA, S. HARADA & M. SHIMOMURA, 2006. Recent advances in studies of parasites on mysid crustaceans. Bull. Plankton Soc. Japan, **53**: 37-44. [In Japanese with English abstract.]

OHTSUKA, S., S. HARADA, M. SHIMOMURA, G. A. BOXSHALL, R. YOSHIZAKI, D. UENO, Y. NITTA, S. IWASAKI, H. OKAWACHI & T. SAKAKIHARA, 2007. Temporal partitioning: the dynamics of alternating occupancy of a host microhabitat by two different crustacean parasites. Mar. Ecol. Prog. Ser., **348**: 261-272.

OHTSUKA, S., M. HORA, T. SUZAKI, M. ARIKAWA, G. OMURA & K. YAMADA, 2004. Morphology and host-specificity of the apostome ciliate *Vampyrophrya pelagica* infecting pelagic copepods in the Seto Inland Sea, Japan. Mar. Ecol. Prog. Ser., **282**: 129-142.

PASTERNAK, A. F., Y. G. ARASHKEVICH & Y. S. SOROKIN, 1984. The role of the parasitic algal genus *Blastodinium* in the ecology of planktonic copepods. Oceanology, **24**: 748-751.

RHODE, K. (ed.), 2005. Marine parasitology: 1-565. (CSIRO Publishing, Wallingford).

RUSSEL, D. J. & R. E. NORRIS, 1971. Ecology and taxonomy of an epizoic diatom. Pacific Sci., **25**: 357-367.

SHIELDS, J. D., 1994. The parasitic dinoflagellates of marine crustaceans. Ann. Rev. Fish Dis., **4**: 241-271.

SHIMOMURA, M., S. OHTSUKA & K. NAITO, 2005. *Prodajus curviabdominalis* n. sp. (Isopoda: Epicaridea: Dajidae), an ectoparasite of mysids, with notes on morphological changes, behaviour and life-cycle. Syst. Parasitol., **60**: 39-57.

SILBERMAN, J. D., A. G. COLLINS, L. A. GERSHWIN, P. J. JOHANSON & A. J. ROGER, 2004. Ellobiopsids of the genus *Thalassomyces* are alveolates. J. eukaryot. Microbiol., **51**: 246-252.

SKOVGAARD, A., 2005. Infection with the dinoflagellate parasite *Blastodinium* spp. in two Mediteranean copepods. Aquat. microb. Ecol., **38**: 93-101.

SKOVGAARD, A., R. MASSANA, V. BALABUÉ & E. SAIZ, 2005. Phylogenetic position of the copepod-infesting parasite *Syndinium turbo* (Dinoflagellata, Syndinea). Protist, **156**: 413-423.

SKOVGAARD, A. & E. SAIZ, 2006. Seasonal occurrence and role of protistan parasites in coastal marine zooplankton. Mar. Ecol. Prog. Ser., **327**: 37-49.

YAMAGUCHI, A., 2006. Suctorian ciliate epibionts on calanoid copepods in the subarctic Pacific. Bull. Plankton Soc. Japan, **53**: 29-36. [In Japanese with English abstract.]

First received 2 November 2009.
Final version accepted 11 January 2010.

# THE BIOLOGY OF *ARGULUS* SPP. (BRANCHIURA, ARGULIDAE) IN JAPAN: A REVIEW

BY

KAZUYA NAGASAWA[1])

Graduate School of Biosphere Science, Hiroshima University, 1-4-4 Kagamiyama, Higashi-Hiroshima, 739-8528 Japan

## ABSTRACT

Branchiurans of the genus *Argulus* are ectoparasites of freshwater and marine fishes. A total of nine species of this genus has been reported from Japan: four freshwater species (*A. americanus*, *A. coregoni*, *A. japonicus*, and *A. lepidostei*) and five marine species (*A. caecus*, *A. kusafugu*, *A. matuii*, *A. onodai*, and *A. scutiformis*). This paper reviews various aspects of the biology of these nine *Argulus* species, particularly *A. japonicus* and *A. coregoni*, in Japan. *A. japonicus* is usually found on cyprinid fishes, while *A. coregoni* prefers salmonid fishes. These species usually overwinter as eggs, and after hatching in spring, they abundantly infect their hosts from spring to fall. These species can cause disease problems in fish farms.

## INTRODUCTION

Branchiurans of the genus *Argulus* Müller, 1785 (Arguloidea: Argulidae) are ectoparasites of freshwater and marine fishes (Yamaguti, 1963). This genus comprises more than 120 species, which accounts for about 85% of the known species in the subclass Branchiura (Kabata, 1988). Argulid branchiurans can cause disease problems and mortality of fishes in aquaculture and aquaria. This paper reviews various aspects of the biology of *Argulus* spp. in Japan, especially *A. japonicus* and *A. coregoni* that have been well studied there. Due to limited restrictions, many papers cannot be not cited in this review, and as such, Nagasawa (2009) should be consulted for further information on the literature. The fish names used here are those recommended by Nakabo (2002).

---

[1]) e-mail: ornatus@hiroshima-u.ac.jp

## *ARGULUS* SPP. FOUND IN JAPAN

Nine species of the genus *Argulus* are known to occur in Japan (Nagasawa, 2009): four freshwater species (*A. americanus* Wilson, *A. coregoni* Thorell, *A. japonicus* Thiele, and *A. lepidostei* Kellicott) and five marine species (*A. caecus* Wilson, *A. kusafugu* Yamaguti & Yamasu, *A. matuii* Sikama, *A. onodai* Tokioka, and *A. scutiformis* Thiele). Of these, *A. americanus* and *A. lepidostei* are not native to Japan, as they were introduced from North America by ornamental fish trade (Tsutsumi, 1968; Shimura & Asai, 1984). These two species were found only in aquaria. Unspecified *Argulus* have been also reported from freshwater and marine fishes in Japan (Nagasawa, 2009).

## BIOLOGY OF *ARGULUS JAPONICUS* IN JAPAN

*Argulus japonicus* was originally but poorly described by Thiele (1900) based on a female specimen collected in Tokyo (no host information was available). This species was soon described in details by the same author (Thiele, 1904) using specimens from Yokohama. Some other authors also described the species from Japan (see Nagasawa, 2009 for the literature). Cyprinid fishes (family Cyprinidae) are its preferred hosts, i.e., common carp and koi carp *Cyprinus carpio*, goldfish *Carassius auratus*, silver crucian carp *C. auratus langsdorfi*, Japanese crucian carp *C. cuvieri*, unspecified crucian carp *Carassius* sp., silver carp *Hypophthalmichthys molitrix*, and bighead carp *Aristichthys nobilis* (Nagasawa, 2009). As the species was also found on "many other freshwater fishes" (Tokioka, 1936a), it can perhaps infect fishes of other families as well. Nonetheless, this parasite may prefer certain cyprinid species, because experiments have shown that it more heavily infects the common carp and the Japanese crucian carp as compared to the silver carp (Kimura, 1970). The attachment site of the species is the body surface of its hosts. It can swim in the water to transfer to other hosts, and as such, free-living individuals have been collected in plankton samples (Kimura, 1970; Nagasawa et al., 2009).

*Argulus japonicus* is native to East or Southeast Asia, but it has been introduced to many regions of the world, except Antarctica (Poly, 2008). In Japan, it has been reported from Hokkaido and central and western Honshu only (Nagasawa, 2009). As cyprinids and other freshwater fishes occur in southern Japan, such as Shikoku, Kyushu and the Ryukyu Islands, the species is likely to be present there as well.

There is a definite seasonal change in occurrence of *A. japonicus* in fish farms. Based on a monthly survey of free-swimming individuals collected by net tows in fish ponds, this species was found to be abundant from April to October, with peaks in May and October, but abruptly declined in number from November to December (Kimura, 1970). A few individuals were collected from late December to February. The observed spring peak in abundance was attributed to both hatching of overwintered eggs and spawning of overwintered adults. Most of fish-infecting and free-swimming individuals were dead in late fall to early winter (Kimura, 1970).

In fish ponds, female *A. japonicus* deposits egg masses during the evening on the deeper areas of the concrete enclosure (Kimura, 1970). The number of eggs per mass varies widely, ranging from 24-534. A single female spawns one to 10 times and deposits a total of 403-2,393 eggs during her lifetime. The number of days for hatching depends on water temperature: eggs hatch, for example, in 10.7 and 44 days on average at 30 and 16.1°C in mean water temperature, respectively. After hatching, female individuals make the first egg deposition in 20.4 and 49.7 days at 28.3 and 16.6°C, respectively (Kimura, 1970). This species typically has seven larval stages and a final adult stage (Tokioka, 1936b) although nine larval stages were also reported (Stammer, 1959). The morphology of individuals for each stage was described by Tokioka (1936b).

*Argulus japonicus* causes a disease problem in goldfish farms (Nakazawa, 1914). This species is also known to induce a secondary bacterial infection in koi carp (Miyazaki et al., 1976). Several chemicals were tested to treat and control the parasite (e.g., Kimura, 1960, 1966).


## BIOLOGY OF *ARGULUS COREGONI* IN JAPAN


*Argulus coregoni* is primarily a parasite of salmonid fishes (family Salmonidae) in the Palaearctic region. In Japan, this species has been reported from various salmonids reared in farms and fisheries research institutions and from wild host individuals caught in rivers and a lake. The known hosts are brook trout *Salvelinus fontinalis*, gogi charr *S. leucomaenis imbrius*, yamato charr *S. leucomaenis japonicus*, brown trout *Salmo trutta*, amago salmon *Oncorhynchus masou ishikawae*, cherry salmon *O. masou masou*, and rainbow trout *O. mykiss* (cf. Nagasawa et al., 1987; Nagasawa, 2009). *Argulus coregoni* also infects wild and reared ayu *Plecoglossus altivelis altivelis* (family Plecoglossidae) (Hoshina, 1955; Nagasawa & Ohya, 1996).

There is a record of *A. coregoni* from a cyprinid *Acheilognathus melanogaster* (cf. Tokioka, 1936a), indicating that its host specificity is not entirely strict. Yamaguti (1937) described *Argulus plecoglossi* from ayu in Japan, which has, however, been relegated to a junior synonym of *A. coregoni* (Shimura, 1981). The attachment sites of this species are the body surface and fins of its hosts.

It was previously suggested that *A. coregoni* had been introduced with unspecified fishes from Europe into Japan (Tokioka, 1965), but it is currently considered to be a native species in this country (Nagasawa & Kawai, 2008). It occurs in central and western Honshu but has not been reported so far from the northern region, including Tohoku and Hokkaido (Nagasawa, 2009). Since this species usually infects salmonids and is regarded as a cold-water species, the reason for its absence in northern Japan is unknown.

The egg deposition, larval development, generation alternation, and attachment sites of *A. coregoni* infecting mainly the cherry salmon and the rainbow trout were studied in a Tokyo fish hatchery (Shimura & Egusa, 1980; Shimura, 1981, 1983). This species deposits egg clusters at night, predominantly near the bottom of fish ponds. It overwinters as eggs and produces one or two generations within a year. Individuals that hatch from overwintered eggs are abundant in May to July and lay eggs in August. Larvae of the second generation mainly hatch in September and adults deposit eggs in October to November. Nine larval stages are recognized. Small individuals are found on almost all parts of the skin and fins, whereas large ones mainly on the skin near the pectoral and pelvic fins.

Little information is available on the ecology of *A. coregoni* in wild fish populations. In a river in central Japan, this species infects the amago salmon and possibly the ayu but does not occur on cyprinids (Takegami, 1984). It is abundantly found on the skin near the pectoral fins.

The level of secondary infection by pathogenic bacteria (e.g., *Aeromonas salmonicida*) in salmonids increases after parasitization by *A. coregoni* (Shimura et al., 1983a). Salmonids infected by this parasite showed some hematological changes: for example, erythrocyte and leucocyte counts, hemoglobin concentration, and hematocrit value decreased 10 days after infection (Shimura et al., 1983b). Pesticidal effects of trichlorfon on *A. coregoni* were examined by Inoue et al. (1980).


## OTHER FRESHWATER SPECIES OF *ARGULUS* IN JAPAN

*Argulus americanus* and *A. lepidostei* were introduced with bowfin *Amia calva* (family Amiidae) and spotted gar *Lepisosteus oculatus* (family Lepisosteidae), respectively, from North America into aquaria in Japan (Tsutsumi, 1968; Shimura & Asai, 1984). Although argulids from the spotted gar were initially identified as *A. foliaceus* Linnaeus by Tsutsumi (1968), they were later suggested to be *A. lepidostei* or a morphologically similar species (Shimura & Asai, 1984). In the laboratory, hatched larvae of *A. americanus* infected Mozambique tilapia *Oreochromis mossambicus* and grew to adults (Shimura & Asai, 1984).

## MARINE SPECIES OF *ARGULUS* IN JAPAN

Five marine species of *Argulus* have been reported from Japan, but their biological information is quite limited (Nagasawa, 2009). *Argulus caecus* was originally described by Wilson (1922) and later redescribed by Tokioka (1936a). Its known hosts are pufferfishes (as "*Spheroides* spp.", Tokioka, 1936a). *Argulus onodai* and *A. kusafugu* infect fine-patterned puffer *Takifugu poecilonotus* (see Nagasawa, 2009 for the nomenclature) and grass puffer *T. niphobles* (cf. Tokioka, 1936a; Yamaguti & Yamasu, 1959), respectively. *Argulus scutiformis* and *A. matuii*, which were originally described by Thiele (1900) and Sikama (1938) from wild fishes, are known to occur on farmed Japanese pufferfish *T. rubripes* and bastard halibut *Paralichthys olivaceus*, respectively (Egusa, 1978; Nagasawa & Fukuda, 2009). Based on the above host information, Japanese marine species of *Argulus* appear to prefer tetraodontiform fishes as hosts.

## FUTURE WORK

Since our knowledge on two freshwater species (*A. japonicus* and *A. coregoni*) was chiefly based on research conducted in fish ponds and fish hatcheries, we need more information on the ecology (i.e., host utilization, spawning, growth, seasonal occurrence) and pathogenicity (i.e., impact on the host populations) of these species infecting wild fishes.

As for the marine species of *Argulus* in Japanese waters, a redescription is necessary for each of them, because their past descriptions were incomplete. A survey on their host range (especially tetraodontiform fishes) and geographical distribution from coastal waters is also needed.

## ACKNOWLEDGEMENTS

I thank Dr. Danny Tang, Hiroshima University, for his review of the manuscript.

## REFERENCES

EGUSA, S., 1978. The infectious diseases of fishes: 1-554. (Koseisha Koseikaku, Tokyo). [In Japanese.]

HOSHINA, T., 1950. Über eine *Argulus*-Art im Salmonidenteiche. Bull. Japan. Soc. Sci. Fish., **16**: 239-243.

INOUE, K., S. SHIMURA, M. SAITO & K. NISHIMURA, 1980. The use of trichlorfon in the control of *Argulus coregoni*. Fish Pathol., **15**: 37-42. [In Japanese with English abstract.]

KABATA, Z., 1988. Copepoda and Branchiura. In: L. MARGOLIS & Z. KABATA (eds.), Guide to the parasites of fishes of Canada. Part II. Crustacea. Can. Spec. Publ. Fish. Aquat. Sci., **101**: 1-184.

KIMURA, S., 1960. Control of the fish lice, *Argulus japonicus* Thiele, with dipterex. Suisan-zoshoku, **8**: 141-150. [In Japanese.]

— —, 1966. Control of the fish lice, *Argulus japonicus* Thiele, with ciodrin: 1-12. (Freshwater Fisheries Research Laboratory, Hino). [In Japanese.]

— —, 1970. Notes on the reproduction of water lice (*Argulus japonicus* Thiele). Bull. Freshwater Fish. Res. Lab., **20**: 109-126. [In Japanese with English abstract.]

MIYAZAKI, T., S. S. KUBOTA & S. EGUSA, 1976. Histopathological studies of gliding-bacterial ulcer disease of the color carp (*Cyprinus carpio*). I. Infected ulcers in the body surface. Bull. Fac. Fish., Mie Univ., **3**: 49-58. [In Japanese with English abstract.]

NAGASAWA, K., 2009. Synopsis of branchiurans of the genus *Argulus* (Crustacea, Argulidae), ectoparasites of freshwater and marine fishes, in Japan (1900-2009). Bull. Biogeogr. Soc. Japan, **64**: 135-148. [In Japanese with English abstract.]

NAGASAWA, K. & Y. FUKUDA, 2009. A record of a crustacean parasite *Argulus matuii* (Branchiura: Argulidae) in finfish mariculture in Japan. Biosp. Sci., **48**: 37-41.

NAGASAWA, K. & K. KAWAI, 2008. New host record for *Argulus coregoni* (Crustacea: Branchiura: Argulidae), with discussion on its natural distribution in Japan. Biosp. Sci., **47**: 23-28.

NAGASAWA, K. & S. OHYA, 1996. Infection of *Argulus coregoni* (Crustacea: Brachiura) on *Plecoglossus altivelis* reared in central Hunshu, Japan. Bull. Fish. Lab., Kinki Univ., **5**: 89-92.

NAGASAWA, K., S. URAWA & T. AWAKURA, 1987. A checklist and bibliography of parasites of salmonids of Japan. Sci. Rep. Hokkaido Salmon Hatchery, **41**: 1-75.

NAGASAWA, K., D. UYENO & T. TOCHIMOTO, 2009. *Argulus japonicus* Thiele and *A. coregoni* Thorell caught in western Honshu, Japan. Biosp. Sci., **48**: 43-47. [In Japanese with English abstract.]

NAKABO, T. (ed.), 2002. Fishes of Japan with pictorial keys to the species (English edition): 1-1749. (Tokai University Press, Tokyo).

NAKAZAWA, K., 1914, Studies on the fish lice, *Argulus japonicus*, on goldfish. J. Imp. Fish. Inst., **9**: 306-316. [In Japanese.]

POLY, W. J., 2008. Global diversity of fishlice (Crustacea: Branchiura: Argulidae) in freshwater. Hydrobiologia, **595**: 209-212.

SHIMURA, S., 1981. The larval development of *Argulus coregoni* Thorell (Crustacea: Branchiura). J. Nat. History, **15**: 331-348.

— —, 1983. Seasonal occurrence, sex ratio and site preference of *Argulus coregoni* Thorell (Crustacea: Branchiura) parasitic on cultured freshwater salmonids in Japan. Parasitology, **86**: 537-552.

SHIMURA, S. & M. ASAI, 1984. *Argulus americanus* (Crustacea: Branchiura) parasitic on the bowfin, *Amia calva*, imported from North America. Fish Pathol., **18**: 199-203. [In Japanese with English abstract.]

SHIMURA, S. & S. EGUSA, 1980. Some ecological notes on the egg deposition of *Argulus coregoni* Thorell (Crustacea, Branchiura). Fish Pathol., **15**: 43-47. [In Japanese with English abstract.]

SHIMURA, S., K. INOUE, M. KUDO & S. EGUSA, 1983a. Studies on effects of parasitism of *Argulus coregoni* (Crustacea: Branchiura) on furunculosis of *Oncorhynchus masou* (Salmonidae). Fish Pathol., **18**: 37-40. [In Japanese with English abstract.]

SHIMURA, S., K. INOUE, K. KASAI & M. SAITO, 1983b. Hematological changes of *Oncorhynchus masou* (Salmonidae) caused by the infection of *Argulus coregoni* (Crustacea: Branchiura). Fish Pathol., **18**: 157-162. [In Japanese with English abstract.]

SIKAMA, Y., 1938. On a new species of *Argulus* found in [sic] a marine fish in Japan. J. Shanghai Sci. Inst., (III) **4**: 129-134.

STAMMER, H. J., 1959. Beiträge zur Morphologie, Biologie und Bekämpfung de Karpfenläuse. Z. Parasitenk., **19**: 135-208.

TAKEGAMI, T., 1984. On *Argulus coregoni* parasitic on *Salmo* (*Oncorhynchus*) *masou macrostomus* in Hiki River. Nankiseibutu, **26**: 45-50. [In Japanese.]

THIELE, J., 1900. Diagnoses neuer Arguliden-Arten. Zool. Anz., **23**: 46-48.

— —, 1904. Beiträge zur Morphologie der Arguliden. Mitteil. Zool. Mus. Berlin, **2**: 1-51, pls. 6-9.

TOKIOKA, T., 1936a. Preliminary report on Argulidae in Japan. Annot. Zool. Japon., **15**: 334-343.

— —, 1936b. Larval development and metamorphosis of *Argulus japonicus*. Mem. Coll. Sci., Kyoto Imp. Univ., (B), **12**: 93-114.

— —, 1965. Branchiura. In: Y. OKADA, S. UCHIDA & T. UCHIDA (eds.), New illustrated encyclopedia of the fauna of Japan (II): 503-504. (Hokuryu-kan, Tokyo). [In Japanese.]

TSUTSUMI, T., 1968. Fish lice, *Argulus foliaceus* Linne found in [sic] the imported spotted gar fish, *Lepisosteus productus*. Sci. Rep. Keikyu Aburatsubo Mar. Park Aquarium, **1**: 13-15. [In Japanese with English abstract.]

WILSON, C. B., 1922. Parasitic copepods from Japan, including five new species. Ark. Zool., **14** (10): 1-16, pls. 1-4.

YAMAGUTI, S., 1937. On two species of *Argulus* from Japan. In: R. E. S. SHULZ & M. P. GNYEDINA (eds.), Papers on helminthology published in commemoration of the 30 year jubileum of the scientific, educational and social activities of the honoured worker of science K. J. Skrjabin, M. Ac. Sci. and of 15th anniversary of All-Union Institute of Helminthology: 781-784. (Moscow).

— —, 1963. Parasitic Copepoda and Branchiura of fishes: 1-1103. (Interscience Publishers, New York).

YAMAGUTI, S. & T. YAMASU, 1959. On two species of *Argulus* (Branchiura, Crustacea) from Japanese fishes. Biol. J. Okayama Univ., **5**: 167-175.

First received 1 November 2009.
Final version accepted 17 December 2009.

# TWO NEW SPECIES OF ECTOPARASITIC ISOPODS (ISOPODA, DAJIDAE) FROM MYSIDS IN JAPAN

BY


MICHITAKA SHIMOMURA[1,3]) and SUSUMU OHTSUKA[2])

[1]) Kitakyushu Museum of Natural History and Human History, 2-4-1 Higashida, Yahatahigashi-ku, Kitakyushu, Fukuoka 805-0071, Japan

[2]) Takehara Marine Science Station, Graduate School of Biosphere Science, Hiroshima University, 5-8-1 Minato-machi, Takehara, Hiroshima 725-0024, Japan



## ABSTRACT

Two new species of ectoparasitic isopods are described from mysid hosts collected in Japan. *Notophryxus ocellatus* n. sp. was obtained from the ventral surface of the pleon of *Rhopalophthalmus orientalis* O. S. Tattersall. The present species appears most closely related to the Indian congener *N. lobatus* Pillai, 1963 in having well-developed lateral plates on the pereon in the female and distinctly segmented pleomeres in the male, but differs from the latter primarily by having a pair of small eyes on the cephalon in the female, rounded anterior margin of the female cephalon and 2 transverse ridges on the pereomeres in the male. *Aspidophryxus japonicus* n. sp. infected the upper surface of the thorax of *Holmesiella affinis* Ii, and can be distinguished from its congeners in having a short frontal part of the female cephalon and uniramous uropods in the male. Both genera are recorded from Japan for the first time.


## INTRODUCTION

The Dajidae, a family of the suborder Cymothoida, consists of about 50 species belonging to 19 genera, all of which are exclusively ectoparasites of mysid, euphausiid and decapod crustaceans (Schotte et al., 1995). Four dajid species have so far been recorded from Japan: *Prodajus bilobatus* Shiino, 1943 from the mysid *Anisomysis ijimai* Nakazawa, 1910, *P. curviabdominalis* Shimomura, Ohtsuka & Naito, 2005 from the mysid *Siriella okadai* Ii, 1964, *Holophryxus fusiformis* Shiino, 1937 from the sergestid *Sergestes prehensilis* Bate, 1881 and *Heterophryxus appendiculatus* G. O. Sars, 1885 from the

---


[3]) e-mail: shimomura@kmnh.jp

euphausiid *Euphausia recurva* Hansen, 1905 (Shiino, 1937, 1943; Shimomura et al., 2005; Shimomura & Ohtsuka, 2008).

Our parasitological surveys of planktonic invertebrates in Japanese waters have yielded two new species of *Notophryxus* and *Aspidophryxus* as the first occurrences of both genera from Japan.

Total length as indicated in "Material examined" was measured from the tip of the cephalon to the end of pleotelson. The terminology follows Shimomura et al. (2005). The type specimens are deposited in the Kitakyushu Museum of Natural History and Human History (KMNH).

## SYSTEMATICS

### Dajidae Giard & Bonnier, 1887

**Notophryxus** G. O. Sars, 1882

**Notophryxus ocellatus** n. sp. (fig. 1)

Material examined. — Holotype, ovigerous ♀, 2.1 mm (KMNH IvR 500,434), obtained from ventral surface of the pleonite 3 of *Rhopalophthalmus orientalis* O. S. Tattersall, 1957, St-9, 34°23.38′N 135°07.91′E, 21-20 m depth, off Misaki-cho, Osaka Bay, Seto Inland Sea, Japan, 5 October 2008, sledge-net, towed by TR/V "Toyoshio-Maru". Paratypes: 5 ovigerous ♀♀, 2.2 mm (KMNH IvR 500,435), 1.9 mm (KMNH IvR 500,436), 1.9 mm (KMNH IvR 500,437), 1.8 mm (KMNH IvR 500,438), 1.7 mm (KMNH IvR 500,439), 6 ♂♂, 1.1 mm (KMNH IvR 500,440), obtained from the holotype, 1.1 mm (KMNH IvR 500,441), obtained from the female (KMNH IvR 500,435), 1.0 mm (KMNH IvR 500,442), obtained from the female (KMNH IvR 500,436), 1.1 mm (KMNH IvR 500,443), obtained from the female (KMNH IvR 500,437), 1.0 mm (KMNH IvR 500,444), obtained from the female (KMNH IvR 500,438), 1.0 mm (KMNH IvR 500,445), obtained from the female (KMNH IvR 500,439), same data as holotype; 7 ovigerous ♀♀, 2.0 mm (KMNH IvR 500,446), 1.9 mm (KMNH IvR 500,447), 1.8 mm (KMNH IvR 500,448), 1.8 mm (KMNH IvR 500,449), 1.5 mm (KMNH IvR 500,450), 1.5 mm (KMNH IvR 500,451), 1.4 mm (KMNH IvR 500,452), 6 ♂♂, 1.1 mm (KMNH IvR 500,453), obtained from the female (KMNH IvR 500,446), 1.1 mm (KMNH IvR 500,454), obtained from the female (KMNH IvR 500,447), 1.1 mm (KMNH IvR 500,455), obtained from the female (KMNH IvR 500,448), 1.0 mm (KMNH IvR 500,456), obtained from the female (KMNH IvR 500,449), 1.1 mm (KMNH IvR 500,457), obtained from the female (KMNH IvR 500,450), 1.1 mm (KMNH IvR 500,458), obtained from the female (KMNH IvR 500,451), St-12, 34°34.48′N 134°49.76′E, 25-35 m depth, Shika-no-se, Harima-nada, Seto Inland Sea, Japan, 7 October 2008, sledge-net, towed by TR/V "Toyoshio-Maru".

Description of female. — Body (fig. 1A–C) elongate ovate, approximately 1.5 times as long as maximum width (including lateral lamellae), highly vaulted dorsally, with pair of broad lateral lamellae filled with many eggs. Egg diameter ranged from 69.3 to 83.7 $\mu$m (N = 20; average, 76.7; SD, 4.9).

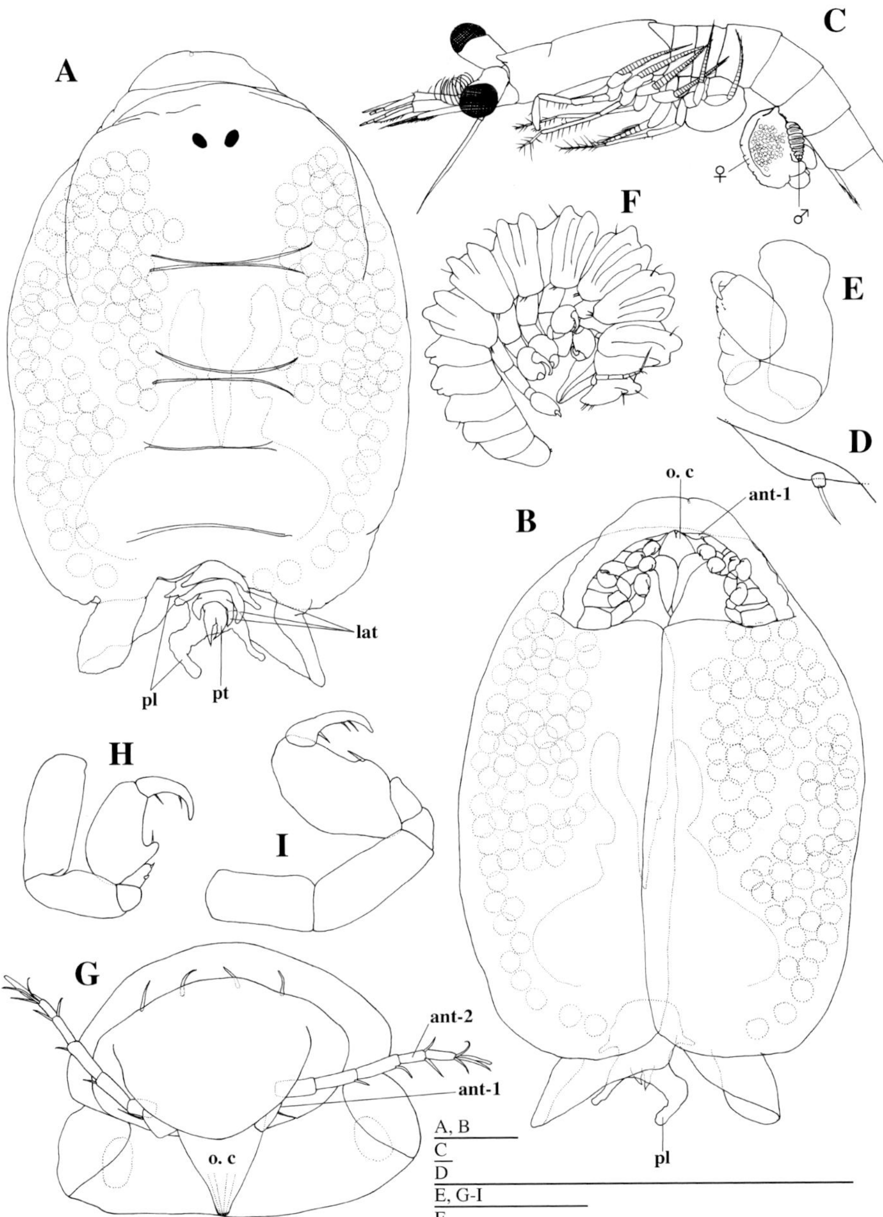



Fig. 1. *Notophryxus ocellatus* n. sp., A–E, holotype female (KMNH IvR 500,434); F–I, paratype male (KMNH IvR 500,440). A, habitus, dorsal; B, habitus, ventral; C, position on the host, lateral; D, left antenna 1, ventral; E, left pereopod 5, lateral; F, habitus, lateral; G, cephalon and pereomeres 1-2, ventral; H, right pereopod 4, lateral; I, right pereopod 6, lateral. Scales: 300 $\mu$m. Abbreviations: lat, lateral plate; pt, pleotelson; pl, pleopods; o. c, oral cone; ant-1, antenna 1; ant-2, antenna 2.

Lateral lamellae prolonged into two large lobes extending beyond pleon. Cephalon (fig. 1A, B) ventrally located, with pair of small, oval-shaped eyes; anterior margin rounded; posterior margin imperfectly defined. Pereomeres (fig. 1A) indistinctly defined in dorsal view, with trace of segmentation. Pleomeres (fig. 1A, B) distinctly separated in dorsal view, but indistinct in ventral view, with 2 pairs of pleopods ventrally: pleomeres 1-3 each with lateral plate. Pleotelson (fig. 1A) distinct, with pair of projections posteriorly, with terminal, open-slit anus. Antenna 1 (fig. 1B, D) composed of broad article and minute article bearing long seta terminally. Antenna 2 lacking. Oral cone (fig. 1B) conical, surpassing cephalon ventrally. Pereopods (fig. 1B, E); pereopods 1-5 similar in shape; basis longest, without setae; ischium about half as long as basis, without setae; merus partly fused with carpus; carpus with few short setae ventrally; propodus ovate, with few short setae ventrally; dactylus smallest, with minute claw.

Description of male. — Body (fig. 1F) curved ventrally, scattered setae dorsally and laterally. Cephalon (fig. 1F, G) without eyes, partly fused with pereomere 1, with transverse dorsal ridge; anterior margin convex. Pereomeres 1-7 (fig. 1F) each with 2 transverse dorsal ridges. Pleon 6-segmented, without uropods; pleomeres 1 and 2 each with 2 transverse dorsal ridges. Antenna 1 (fig. 1G) composed of broad article and long seta. Antenna 2 (fig. 1E) composed of 5 articles: articles 2 and 3 subequal in length, each with simple seta medially; article 4 shorter than article 3, with 2 simple setae distally; article 5 with 3 simple setae and aesthetasc apically. Oral cone (fig. 1G) with pair of mandibular gnathobases protruding from mouth opening. Pereopods 1-4 (fig. 1H) similar in shape; carpus with short setae ventrally; propodus with long projection and long simple seta proximoventrally; dactylus curved inward, with 2 simple setae ventrally. Pereopods 5-7 (fig. 1I) similar in shape: carpus without setae; projection of propodus short and blunt; dactylus curved inward, with 3 simple setae ventrally.

Remarks. — *Notophryxus* contains 9 species, all of which infest mysid and euphausiid crustaceans (Koehler, 1911; Schotte et al., 1995).

*Notophryxus ocellatus* n. sp. is most similar to *N. lobatus* Pillai, 1963 obtained from *Rhopalophthalmus tattersallae* Pillai, 1961 of the Indian Ocean in having pleomeres 1-3 each with well-developed lateral plates in female and distinctly segmented pleomeres in male. The new species is distinguished from *N. lobatus* by the following features (those of *N. lobatus* in parentheses): female cephalon with pair of small eyes (without eyes); anterior margin of female cephalon rounded (subtruncate); female pereomeres imperfectly defined in dorsal view, with trace of segmentation (without any trace of segmentation); male cephalon with transverse dorsal ridge (without dorsal ridge); male pereomeres 1-7 and pleomeres 1-2 each with 2 transverse dorsal ridges (each with 1 dorsal ridge); and male antenna 2 composed of 5 articles (6 articles).

Etymology. — The species name is derived from the Latin word *ocellatus*, which means "small eye".

## Aspidophryxus G. O. Sars, 1882

### Aspidophryxus japonicus n. sp. (fig. 2)

Material examined. — Holotype, ovigerous ♀, 2.2 mm (KMNH IvR 500,460), obtained from upper surface of the thorax of *Holmesiella affinis* Ii, 1937, St-18, 31°26.40′N 131°28.70′E, 129 m depth, off Cape Toi-misaki, Japan, 31 May 2000, sledge-net, towed by TR/V "Toyoshio-Maru". Paratypes: 2 ovigerous ♀♀, 2.5 mm (KMNH IvR 500,461), 2.2 mm (KMNH IvR 500,462), 3 ♂♂, 1.4 mm (KMNH IvR 500,463), obtained from the holotype, 1.4 mm (KMNH IvR 500,464), obtained from the female (KMNH IvR 500,461), 1.3 mm (KMNH IvR 500,465), obtained from the female (KMNH IvR 500,462), same data as holotype.

Description of female. — Body (fig. 2A–C) elongate ovate, approximately 1.3 times as long as maximum width (including lateral lamellae), highly vaulted dorsally, with pair of broad lateral lamellae filled with many eggs; lateral lamellae not reaching beyond frontal margin of cephalon. Egg diameter ranged from 58.2 to 75.2 $\mu$m (N = 20; average, 66.7; SD, 4.4). Cephalon (fig. 2A, B) ventrally located, without eyes; frontal part short, approximately 0.2 times as long as wide; anterior margin rounded; posterior margin imperfectly defined in dorsal view. Pereon (fig. 2A) with indistinct transversal folds as indication of segments. Pleon (fig. 2A, B) 3-segmented, without lateral plates and pleopods: pleomeres not visible in dorsal view; pleomere 1 small; pleomere 2 large, swollen ventrally; pleotelson partly visible in dorsal view, with pair of projections posteriorly, with terminal, open-slit anus. Antenna 1 (fig. 2B, D) composed of one article lacking setae. Antenna 2 (fig. 2B, D) composed of 5 articles; article 5 with long simple seta terminally. Oral cone (fig. 2B) conical, surpassing cephalon ventrally. Pereopods 1-5 similar in shape (fig. 2B, E); basis longest, without setae; ischium about half as long as basis, without setae; merus fused with carpus, with few short setae ventrally; propodus ovate; dactylus smallest, with minute claw.

Description of male. — Body (fig. 2F) curved ventrally, scattered setae laterally. Cephalon (fig. 2F, G) without eyes, partly fused with pereomere 1; anterior margin convex. Pereomeres 2-7 (fig. 2F) separated. Pleon (fig. 2F, H)

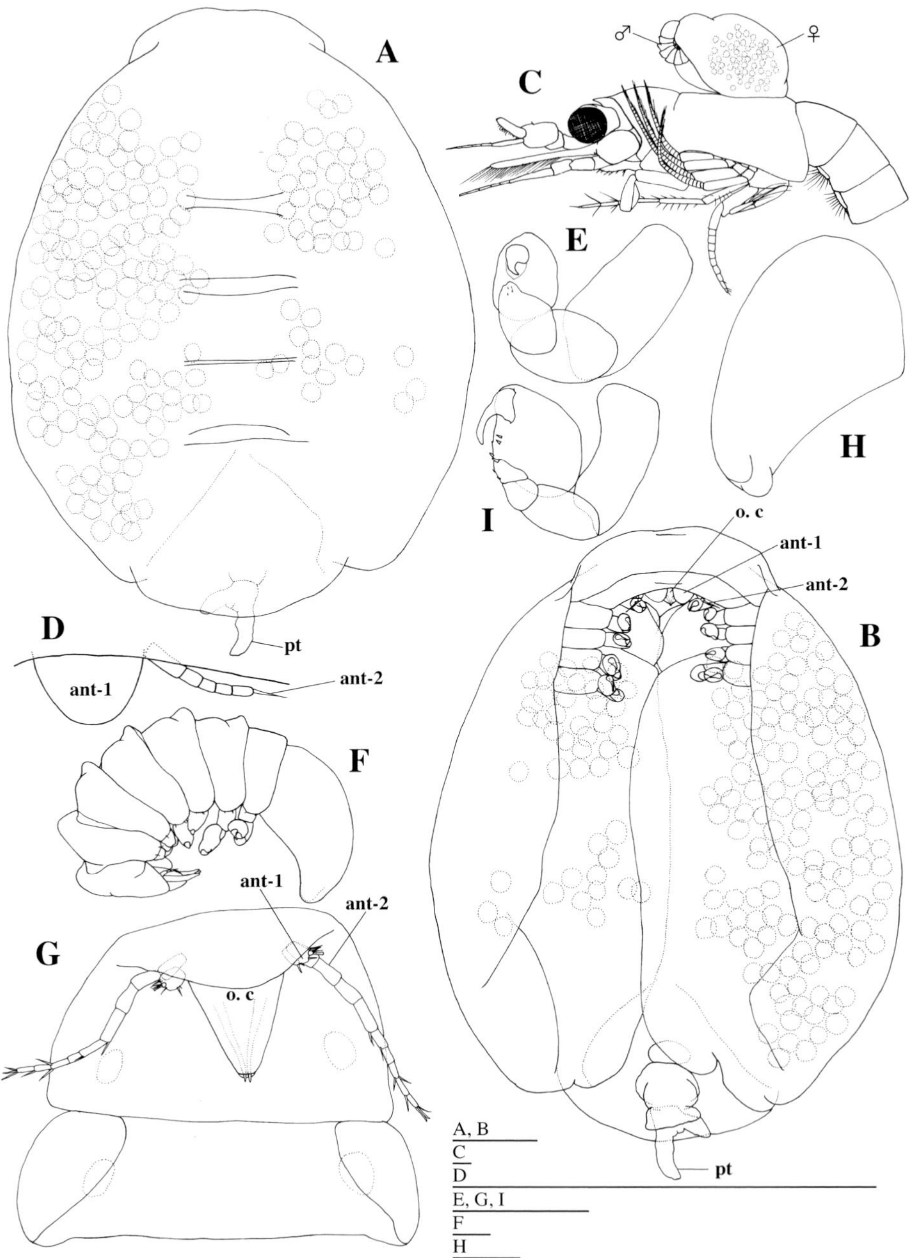

Fig. 2. *Aspidophryxus japonicus* n. sp. A–E, holotype female (KMNH IvR 500,460); F–I, paratype male (KMNH IvR 500,464). A, habitus, dorsal; B, habitus, ventral; C, position on the host, lateral; D, left antennae 1 and 2, ventral; E, left pereopod 5, lateral; F, habitus, lateral; G, cephalon and pereomeres 1-2, ventral; H, pleon, ventral; I, right pereopod 3, medial. Scales: 300 μm. Abbreviations: pt, pleotelson; o. c, oral cone; ant-1, antenna 1; ant-2, antenna 2.

uni-segmented, with pair of protrusions posteriorly and slit-like anus between uropodal protrusions. Uropod (fig. 2H) uniramous. Antenna 1 (fig. 2G) composed of broad basal article and smaller second article: article 1 with simple seta laterally and 3 simple setae medially; article 2 with robust and 2 simple setae apically. Antenna 2 (fig. 2G) composed of 9 articles: articles 1-3 subequal in length, without setae; article 4 slightly longer than article 3, with 1 simple seta medially; articles 5 and 8 each with simple seta medially; article 7 with 2 simple setae; article 9 with 3 simple setae apically. Oral cone (fig. 2E): pair of mandibular gnathobases protruding from mouth opening. Pereopods (fig. 2I) similar in shape; carpus with 3 short setae ventrally; propodus with 3 short setae ventrally; dactylus curved inward, without setae.

Remarks. — *Aspidophryxus* is a small genus in the Dajidae, now containing 3 species (including the new species). The two previously described species are *A. frontalis* Bonnier, 1887 infecting *Siriella norvegica* G. O. Sars, 1869 from off Morocco and *A. peltatus* G. O. Sars, 1882 infecting *Erythrops erythrophthalma* (Goes, 1864), *E. elegans* (G. O. Sars, 1863), *E. serrata* (G. O. Sars, 1863), *E. microps* (G. O. Sars, 1864), *Parerythrops obesa* (G. O. Sars, 1864) and *Mysidopsis didelphys* Norman, 1863 from Norway and Ireland (G. O. Sars, 1882; Bonnier, 1887; Giard & Bonnier, 1889; Koehler, 1911; Stephensen, 1912).

*Aspidophryxus japonicus* n. sp. is similar to *A. peltatus* G. O. Sars, 1882 in having indistinct transversal folds on pereon as an indication of segments. The two species, however, differ from one another in the following characters (those of *A. peltatus* in parentheses): female lateral lamellae not expanded anteriorly (well-expanded anteriorly); frontal part of female cephalon short, approximately 0.2 times as long as wide (long, approximately 1.1 times as long as wide); female pleon 3-segmented (unsegmented); male pleon completely unsegmented (divided into 5 not very sharply defined segments); and male uropod uniramous (biramous).

Etymology. — The specific name is derived from the country where the present new species was collected.


## ACKNOWLEDGEMENTS

We express our sincere thanks to the captain and crew of TR/V "Toyoshio-Maru", Hiroshima University, for their cooperation at sea. Thanks are also extended to the editor Dr. Akira Asakura of Natural History Museum and Institute, Chiba, and an anonymous reviewer, for improving this manuscript with their comments. This study was partly supported by grants from the Ministry of Education, Science, Sports and Culture to MS (No. 21770100) and the Japan Society of the Promotion of Sciences to SO (No. 20380110).

## REFERENCES

BONNIER, J., 1887. Catalogue des Crustacés Malacostracés recueilles dans la baie de Concarneau. Bull. Scient. France et Belgique, **18**: 199-262, 296-356, 361-422.

GIARD, A. & J. BONNIER, 1889. Sur les épicarides de la famille des Dajidae. Bull. Scient. France et Belgique, **20**: 252-372.

KOEHLER, R., 1911. Isopodes nouveaux de la famille Dajidés provenant des campagnes de la "Princesse-Alice". Bull. Inst. Océanogr. Monaco, **196**: 1-34.

SARS, G. O., 1882. Oversigt af Norges Crustaceer med foreløbige Bemaerkninger over de nye eller mindre bekjendte Arter. I. (Podophthalmata, Cumacea, Isopoda, Amphipoda). Forh. Vidensk.-selsk. Christ., **18**: 1-124.

SCHOTTE, M., B. F. KENSLEY & S. SHILLING, 1995. World list of Marine, Freshwater and Terrestrial Crustacea Isopoda. (National Museum of Natural History Smithsonian Institution, Washington, D.C., U.S.A.). Available at http://invertebrates.si.edu/isopod/ (accessed 2 November 2009).

SHIINO, S. M., 1937. *Holophryxus fusiformis*, a new species of the family Dajidae, Epicaridea. Annot. Zool. Japon., **16**: 188-192.

— —, 1943. On *Prodajus bilobatus*, a new species of the family Dajidae (Epicaridea, Isopoda). Jour. Shigenkagaku Kenkyusho, **1**: 115-118.

SHIMOMURA, M. & S. OHTSUKA, 2008. New record of a euphausiid ectoparasitic isopod, *Heterophryxus appendiculatus* G. O. Sars, 1885 (Crustacea: Dajidae) from Japan. Proc. Biol. Soc. Wash., **121**: 326-330.

SHIMOMURA, M., S. OHTSUKA & K. NAITO, 2005. *Prodajus curviabdominalis* n. sp. (Isopoda: Epicaridea: Dajidae), an ectoparasite of mysids, with notes on morphological changes, behaviour and life-cycle. Syst. Parasitol., **60**: 39-57.

STEPHENSEN, K., 1912. Report on the Malacostraca collected by the "Tjalfe" Expedition, under the direction of Ad. S. Jensen, especially at W. Greenland. Vidensk. Medd. Naturh. Foren., **64**: 57-134.

First received 14 November 2009.
Final version accepted 19 January 2010.

# RECENT ADVANCES IN THE BIOLOGY OF THE PARASITIC COPEPOD *PSEUDOCALIGUS FUGU* (SIPHONOSTOMATOIDA, CALIGIDAE), HOST SPECIFIC TO PUFFERFISHES OF THE GENUS *TAKIFUGU* (ACTINOPTERYGII, TETRAODONTIDAE)


BY

B. A. VENMATHI MARAN[1]), SUSUMU OHTSUKA[1,4]), IKUO TAKAMI[2]),
SHINYA OKABE[1]) and GEOFFREY A. BOXSHALL[3])

[1]) Takehara Marine Science Station, Graduate School of Biosphere Science, Hiroshima
University, 5-8-1 Minato-machi, Takehara 725 0024, Hiroshima, Japan
[2]) Nagasaki Prefectural Institute of Fisheries, Nagasaki, Japan
[3]) Department of Zoology, The Natural History Museum, Cromwell Road,
London SW7 5BD, U.K.



ABSTRACT

*Pseudocaligus fugu* Yamaguti, 1936, a sea louse belonging to the family Caligidae, causes serious economic loss in pufferfish aquaculture in Japan. This parasite is known only from East Asian countries. Recently, cultured tiger puffer has been suffering from serious infections with *P. fugu* in western Japan. In order to understand the importance of *P. fugu*, its biology and ecology has been briefly reviewed here. The seasonal cycles of prevalence and intensity of *P. fugu* infections were studied from April 2000 to May 2001 in the Seto Inland Sea, Japan. The post-embryonic stages of *P. fugu* have been described and comprise 2 naupliar, 1 copepodid, and 4 chalimus stages preceding the adult. A brief review is given on the structure of the frontal filament of caligid chalimi, which varies within the main fish-parasitic siphonostomatoid lineage. *Pseudocaligus fugu* is host-specific to pufferfishes, including *Takifugu niphobles*, *T. pardalis*, *T. poecilonotus*, and *T. rubripes*, all of which are known to produce tetrodotoxin (TTX) and related toxins. Our studies on this parasite of toxic puffer have revealed that, through feeding on the mucus and skin of its host, it accumulates TTX in all of its body tissues except for the reproductive organs, epicuticle, and part of the hindgut, indicating that TTX will not be transmitted to the parasite's offspring. Epibiontic bacteria on *P. fugu* isolated from *T. pardalis* appear to produce TTX. The possible origin of the TTX and information on accumulation are briefly reviewed.


---


[4]) Corresponding author; e-mail: ohtsuka@hiroshima-u.ac.jp

## INTRODUCTION

Parasitic copepods are common on cultured and wild marine finfish and members of the siphonostomatoid family Caligidae, known as sea lice, cause significant commercial losses in aquaculture worldwide (Johnson et al., 2004; Ho & Lin, 2004). With the development of semi-intensive and intensive brackish water and marine aquaculture, the importance of sea lice as disease-causing agents has become more evident (Boxshall & Defaye, 1993). The family Caligidae comprises 34 genera and more than 450 species (Ho & Lin, 2004; Boxshall & Halsey, 2004; Boxshall & Justine, 2005; Boxshall, 2008). Of the 34 genera, the genus *Caligus* Müller, 1785 is the most speciose, comprising more than 250 species, *Lepeophtheirus* von Nordmann, 1832 contains in excess of 110 species, and *Pseudocaligus* Scott, 1901 has 11 species (Boxshall & Halsey, 2004). Two species of *Pseudocaligus* are known from Japan: *Pseudocaligus fugu* Yamaguti, 1936 (fig. 1A, C) and *P. tenuicauda* Shiino, 1964 (= *P. fistulariae* Pillai, 1961; see Pillai, 1985). The elongate abdomen of the latter species distinguishes it from the former (Shiino, 1964).

In Japan five caligid species were recognized as pests in finfish aquaculture by Ho & Lin (2004): *Caligus lalandei* Barnard, 1948 on greater amberjack, *Seriola dumerili* (Risso, 1810); *C. orientalis* Gusev, 1951 on common carp, *Cyprinus carpio* Linnaeus, 1758, and rainbow trout *Oncorhynchus mykiss* (Walbaum, 1792); *C. spinosus* Yamaguti, 1939 on Japanese amberjack, *Seriola quinqueradiata* Temminck & Schlegel, 1845 and coho salmon, *Oncorhynchus kisutch* (Walbaum, 1792); *C. sclerotinosus* Roubal, Armitage & Rohde, 1983 on red seabream, *Pagrus major* (Temminck & Schlegel, 1843); and *Lepeophtheirus longiventralis* Yü & Wu, 1932 on spotted halibut, *Verasper variegatus* (Temminck & Schlegel, 1846) (cf. Froese & Pauly, 2009). Subsequently, Ohtsuka et al. (2009) recognized *Pseudocaligus fugu* parasitic on Japanese pufferfish (known also as tiger puffer), *Takifugu rubripes* (Temminck & Schlegel, 1850), as a pest, and since 2003 cultured tiger puffer has suffered serious infections (fig. 2) in western Japan, resulting in severe economic losses. Heavily infected puffers are irritated by skin lesions caused by sea lice and secondary bacterial infections can result in significant mortalities (Ohtsuka et al., 2009). In addition to *P. fugu*, another *Pseudocaligus* species, *P. apodus* Brian, 1924 has been reported as causing severe disease in mullet culture in the eastern Mediterranean (Paperna, 1975; Johnson et al., 2004). The host-specificity of *P. fugu* is high, as it is restricted to toxic puffers belonging to the genus *Takifugu* (cf. Ogawa, 2006). We have studied the presence of tetrodotoxin on *P. fugu* (Ikeda et al., 2006; Ito et al., 2006) and also the relations between *P. fugu*

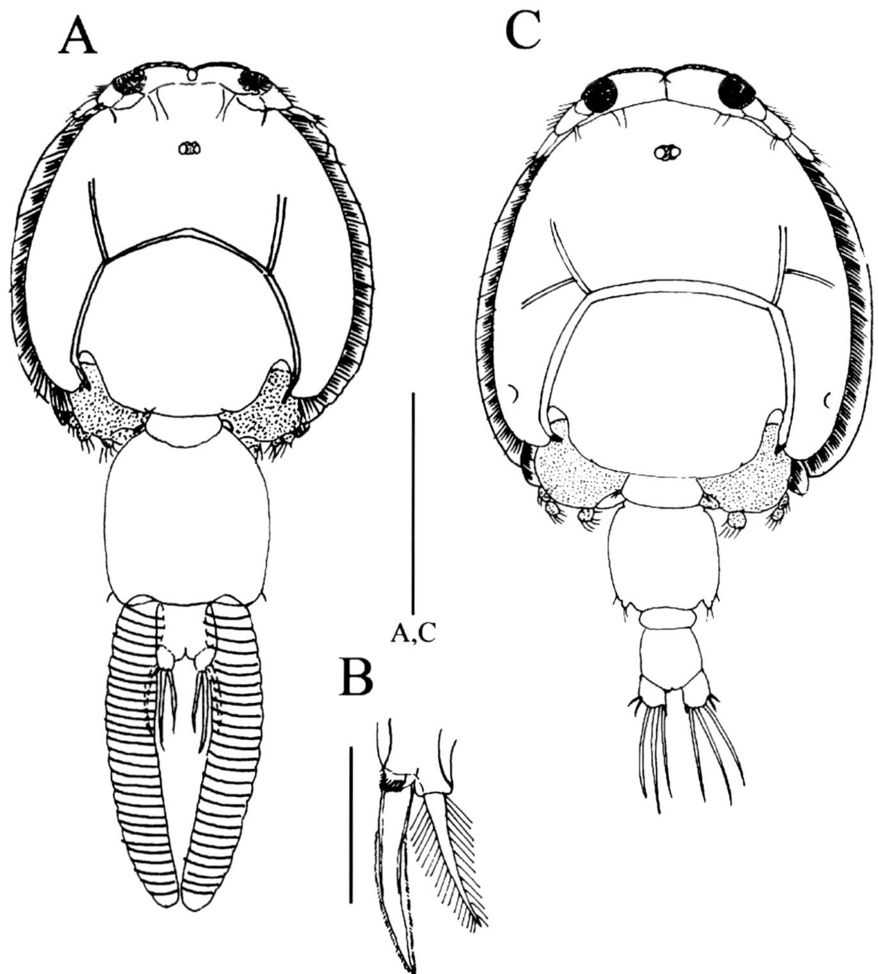

Fig. 1. *Pseudocaligus fugu* Yamaguti, 1936 parasitic on puffers. A, habitus female; B, leg 4; C, habitus male. Scale bars: A, C — 1 mm; B — 0.05 mm.

and epibiontic bacteria in order to reveal the dynamics of tetrodotoxin (Venmathi Maran et al., 2007). This paper briefly reviews the biology and ecology of *P. fugu* mainly based on our recent studies.


## DISTRIBUTION AND HOST SPECIFICITY

*Pseudocaligus fugu* from host puffers has been reported by several authors in Japan (Yamaguti, 1936; Shiino, 1963; Ogawa & Inouye, 1997; Okabe, 2003; Ogawa, 2006; Ohtsuka et al., 2009), and from Korea (Kim, 1998).

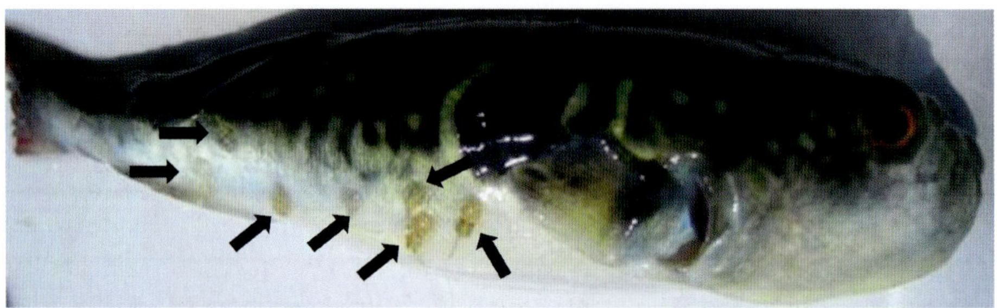

Fig. 2. Heavy infection of *Pseudocaligus fugu* Yamaguti, 1936 on tiger puffer *Takifugu rubripes* in western Japan (Ohtsuka et al., 2009, with permission from Taylor & Francis, U.K.).

The hosts of *P. fugu* are puffers belonging to the genus *Takifugu* Abe, 1949 (Actinopterygii: Tetraodontidae). In Japan four species of *Takifugu* are known as sea-lice hosts, the grass puffer, *Takifugu niphobles* (Jordan & Snyder, 1901); the panther puffer, *T. pardalis* (Temminck & Schlegel, 1850); fine patterned puffer, *T. poecilonotus* (Temminck & Schlegel, 1850); and Japanese pufferfish, *T. rubripes* (Yamaguti, 1936; Kanoh, 1988; Kim, 1998; Ikeda et al., 2006; Ito et al., 2006; Venmathi Maran et al., 2007; Ogawa, 2006; Ohtsuka et al., 2009). However, in Korea only *T. niphobles* has been recorded as host of *P. fugu* (Kim, 1998).

The host puffers are highly toxic, containing lethal amounts of the poison tetrodotoxin (TTX) and the distribution of TTX in their bodies appears to be species-specific (Noguchi et al., 2004, 2006). In marine pufferfish species, liver and ovary generally show the highest toxicity, followed by intestines and skin and it is suggested that the puffers accumulate TTX as a biological defense agent (Noguchi et al., 2006).

Differences in host-parasite relations were observed for *P. fugu* (cf. Takegami, 1986). Takegami (1986), while conducting rearing experiments on *P. fugu*, revealed that copepodids grew to maturity on *T. niphobles*, but, after initial infection by the copepodid larva, there was no further development on *T. poecilonotus* and *T. vermicularis*. We consider it is possible that in nature juveniles might be unable to develop successfully on these puffer species and that the presence of adults on these fishes might possibly result from migration of adults originally from *T. niphobles*, *T. pardalis*, and *T. rubripes*.

The toxicity of puffers varies from species to species. The Japanese puffer-fish, *T. rubripes*, is the least toxic (toxicity expressed in Mouse Unit gram$^{-1}$) among its congeners, its skin containing 61 MUg$^{-1}$ of TTX. In contrast, *T. poecilonotus* has 6100 MUg$^{-1}$ followed by *T. vermicularis* with 1300 MUg$^{-1}$, *T. niphobles* with 752 MUg$^{-1}$, and *T. pardalis* with 700 MUg$^{-1}$ (Harada &

Abe, 1993; Noguchi et al., 2006). *Pseudocaligus fugu* isolated from *T. niphobles* was screened for the presence of TTX and the toxicity level was 269.0-520.7 MUg$^{-1}$, similar to that of another parasitic copepod *Taeniacanthus* sp. (338.0-345.5 MUg$^{-1}$) found on the same puffer (Ito et al., 2006). Experimental studies on *P. fugu* were carried out mostly from the host puffers: *T. rubripes*, followed by *T. niphobles* and then *T. pardalis* (cf. Okabe, 2003; Ikeda et al., 2006; Ito et al., 2006; Venmathi Maran et al., 2007; Ohtsuka et al., 2009). According to Harada & Abe (1993) the toxicity in these three species is considerably lower than in *T. poecilonotous* and *T. vermicularis* and leads to the suggestion that the high toxicity of the skin might be linked to the failure of the parasite's development at the copepodid stage.

## REPRODUCTION, DEVELOPMENTAL STAGES, AND LIFE CYCLE

Some young adult females still attached by a frontal filament have mated, as indicated by spermatophores they carried. They can mate with many males (cf. fig. 9 in Ohtsuka et al., 2009). Mating behavior has never been observed in *P. fugu*. The clutch size varies between species in the Caligidae and with the physiological state of individual females. Ohtsuka et al. (2009) reported that adult female *P. fugu* carried a mean total of 138.7 eggs with a range from 67 to 178 in a pair of egg strings.

## CHARACTERISTICS OF DEVELOPMENTAL STAGES

During its development, *P. fugu* exhibits with some peculiar features relative to other caligids, such as:

1. Egg numbers: in comparison with those of other caligids, egg numbers are relatively high for *P. fugu*, with a mean of 69 eggs per string (Ohtsuka et al., 2009). However, the number of eggs per string ranges from 4 to 253 in species of caligids reported from Taiwanese and Japanese waters (Ho & Lin, 2004).

2. Duration of development: in *P. fugu* the minimum development time from copepodid to adult was 9 days at ca. 20°C (Ohtsuka et al., 2009). In comparison, it was 8 days in *C. epidemicus* Hewitt, 1971 at 24.5°C (Lin & Ho, 1993), and was 38 and 29 days in *Lepeophtheirus salmonis* (Krøyer, 1837) at 10°C, for females and males, respectively (Bjørn & Finstad, 1998). Such life history data are available for relatively few caligid species so generalizations,

other than that recognizing a relationship between temperature and generation time, are not possible at present.

3. Nauplius: the second naupliar stage has been found with a post-mandibular anlage, first appearing in this stage, as a pair of slender, posteriorly directed processes, which subsequently develops into the maxilliped (Piasecki, 1996; Ohtsuka et al., 2009).

4. Leg 4: reduction of leg 4 (fig. 1B) is the diagnostic feature of adult *Pseudocaligus* (cf. Yamaguti, 1936; Kim, 1998; Ohtsuka et al., 2009). The development of leg 4 in *P. fugu* starts only at the chalimus II stage and is represented by a lobe-like vestige armed with one or two setal elements (Ohtsuka et al., 2009). In contrast, in *Caligus* species, leg 4 first appears at chalimus I as an anlage, and develops to attain the near-adult condition by chalimus IV (Kim, 1993; Ho & Lin, 2004).

5. Frontal filament: it is an attachment organ, an evolutionary novelty exhibited by many siphonostomatoids parasitic on fishes (Piasecki & MacKinnon, 1993). It is produced by the infective copepodid after location and attachment to the host. The number of nodes on the frontal filament of *P. fugu* was consistent with other features of the chalimus stages and can be used as a character to distinguish between successive stages. However, in the family Caligidae the formation of the frontal filament and its fate during moulting follows two different patterns, the multi-node type (found in *Caligus* and *Pseudocaligus* species) or the single-node type as found in *Lepeophtheirus* species (Ohtsuka et al., 2009).

Multi-node pattern: In *P. fugu*, the pre-formed frontal filament is extruded and attached to the host prior to the moult to chalimus I. At the subsequent moult an additional bulb of material is secreted by the frontal organ and appended to the proximal end of the filament, likewise at each moult between successive chalimus stage a further bulb of material (an extension lobe) is secreted and added in the same manner. So chalimus II, III, and IV, have two, three, and four nodes, respectively, incorporated within the length of the filament at its proximal end. What is important in this pattern is that the filament formed by copepodid (and attached to host at this stage) is continuously used and modified by all subsequent chalimus stages. This pattern has also been observed in *C. elongatus* von Nordmann, 1832 (cf. Piasecki & MacKinnon, 1993) *C. punctatus* Shiino, 1955 (cf. Kim, 1993), *C. orientalis* (cf. Hwa, 1965), *C. pageti* Russel, 1925 (cf. Ben Hassine, 1983), and *C. rotundigenitalis* Yü, 1933 (cf. Lin et al., 1997, as *C. multispinosus* Shen, 1957).

Single-node pattern: In contrast, in *Lepeophtheirus* the frontal filament of all chalimus stages is simple and constitutes an integral part of the anterior cephalothorax of the chalimus, with its surface sheath being continuous with the integument of the chalimus (Pike et al., 1993). What is important: The filament is secreted de-novo by each successive chalimus stage and no extension lobes are secreted and incorporated at moults through the chalimus phase. This pattern is observed in *Lepeophtheirus pectoralis* (Müller, 1776) and *L. salmonis* (Boxshall, 1974; Johnson & Albright, 1991; Pike et al., 1993).

In the Caligidae, the life cycle had been elucidated for 15 species representing just two genera, *Caligus* (11 species) and *Lepeophtheirus* (4 species) (Ho & Lin, 2004). The life cycle of caligids was considered to comprise 2 free-living naupliar stages, 1 infective copepodid stage, 4-6 chalimus stages each tethered via a frontal filament to the host, zero to 2-preadult stages, and the adult stage (Johnson & Albright, 1991; Lin & Ho, 1993; Piasecki & MacKinnon, 1995; Lin et al., 1997; Ho & Lin, 2004; Ohtsuka et al., 2009). *Pseudocaligus* represents the third caligid genus (after *Caligus* and *Lepeophtheirus*) in which the full life cycle has been elucidated (Ohtsuka et al., 2009). It has direct life cycle consisting of 8 stages: 2 free-living nauplii, 1 infective copepodid, 4 chalimi, and the adult. Existing papers on the development of members of the genus *Lepeophtheirus* interpret the life cycle as comprising a total of 4 chalimus and 2 preadult stages, but by using changes in the setation patterns of the antennules as stage indicators through the post-naupliar phase of development of 4 species of *Caligus*, 2 species of *Lepeophtheirus*, and 1 each of *Alebion* and *Pseudocaligus*, Ohtsuka et al. (2009) concluded that the evidence supported the recognition of only 4 moult-separated stages between copepodid and sexually mature adult in the family Caligidae.


## SEASONAL OCCURRENCE OF *P. FUGU* PARASITIC ON *T. NIPHOBLES* IN THE SETO INLAND SEA, JAPAN

Grass puffer, *T. niphobles*, collected during April 2000–May 2001 from off Takehara City, Hiroshima, the Seto Inland Sea, western Japan, displayed a seasonal cycle of occurrence of *P. fugu* in prevalence, intensity and occurrence of ovigerous females (Okabe, 2003). Prevalence values of adults, immature adults with frontal filaments, and chalimi were high in May (68.8%), June (33.3%), and May (93.75%), but low in September (0%), September, November, January, February (0%), and December (5%), respectively (fig. 3A). Mean intensities of adults, immature adults with frontal filaments and chalimi were

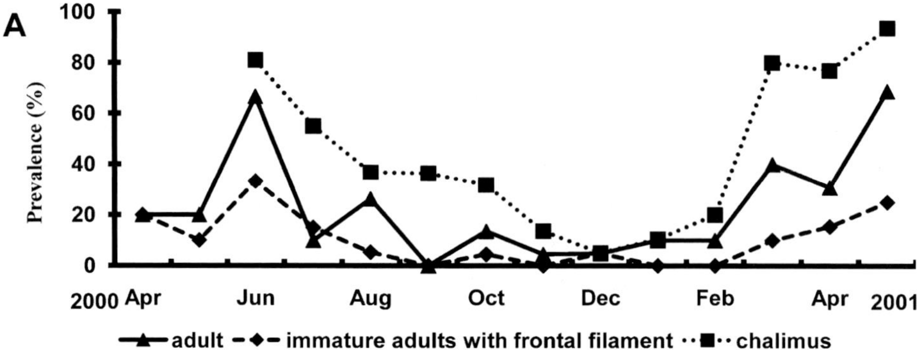


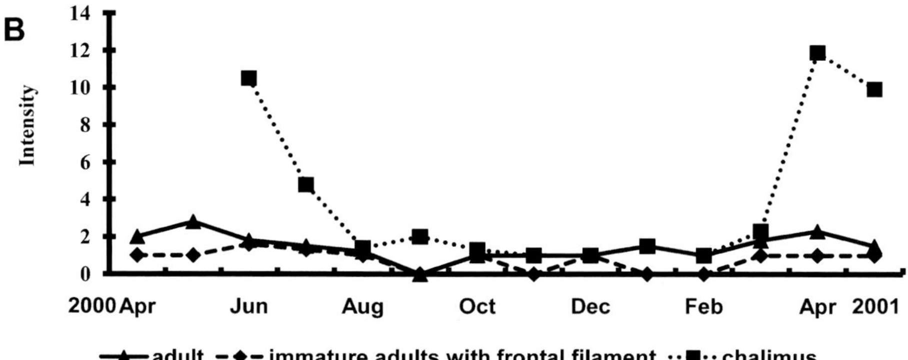


Fig. 3. Seasonal changes in prevalence (A) and intensity (B) of *Pseudocaligus fugu* Yamaguti, 1936 on grass puffer, *Takifugu niphobles*, in the Seto Inland Sea, Japan during April 2000–May 2001.

high in May (2.8%), June (1.6%) and April (11.9%), and low in September (0%), September, November, January, February (0%) and November, December, February (1%), respectively (fig. 3B). Ovigerous females were, however, found throughout the year (Okabe, 2003).

## IMPACT OF *P. FUGU* ON AQUACULTURE OF TIGER PUFFER

The annual wild catch of puffer has been relatively stable ranging from 7800 to 11 000 t per year from 1995 to 2002 in Japan (Kikuchi, 2006). According to Kikuchi (2006), aquaculture of Japanese pufferfish was started in the 1960s and commercial production became more extensive in the 1980s with the development of fingerling production and rearing techniques. Production increased during the past 2 decades to 5200 t in 2002 and puffer ranked 5th among marine cultured finfish in Japan, next to Japanese amberjack, *Seriola quinqueradiata*; red sea bream, *Pagrus major*; coho salmon, *Oncorhynchus kisutch*; and bastard halibut, *Paralichthys olivaceus* (Temminck & Schlegel, 1846) (Ogawa, 1996; Kikuchi, 2006).

Tiger puffer is considered to be one of the most attractive species for net cage aquaculture because of its high market price (Kikuchi, 2006). However, one of the most serious problems in the net cage aquaculture of tiger puffer is low productivity due to high mortality from outbreaks of parasitic diseases (Ogawa & Inouye, 1997). Survival during the production period is estimated to be less than 50% (Kikuchi, 2006). Although infection with *P. fugu* is one of the major problems faced in net cage aquaculture in Japan, precise estimates for economic losses due to *P. fugu* are not yet available.

## BIOLOGICAL CONTROL

Under intensive aquaculture, several parasites have emerged as serious pathogens (Matsuoka, 1995; Ogawa & Yokoyama, 1998). Economic losses due to parasitic diseases have been estimated to exceed US\$ 5.5 million annually in Japan (Hirazawa et al., 2001). However, no single method has proved effective in controlling such parasitic diseases (Ogawa & Yokoyama, 1998). Initial studies conducted to develop effective treatments to control *P. fugu* (Hirazawa et al., 2000, 2001; Tensha & Momoyama, 2006) had limited success.

Hirazawa et al. (2000, 2001) found that caprylic acid, a medium-chain fatty acid in coconut oil, butter and other edible oil, has an antiparasitic effect against the monogenean infecting tiger puffer, but no clear effect was found against the copepodids and adults of *P. fugu* during in vitro trials. Observations on juvenile tiger puffer heavily infected with *P. fugu* in Yamaguchi Prefecture, western Japan found that both adult and chalimi were detached from the host fish by bathing in 0.025% Marine Sour SP30 (Katayama Chemical Inc, Osaka) for 24 h, however, freshwater bathing for 5 h was not effective against *P. fugu* (Tensha & Momoyama, 2006). Effective methods are urgently needed to control these parasitic diseases.

## ACCUMULATION OF TTX ON *P. FUGU* PARASITIC ON *T. PARDALIS*

*Pseudocaligus fugu* is parasitic on the toxic skin of puffer fishes, which may indicate that it is endowed with some tolerance to TTX. Okabe et al. (2003) detected a small amount of TTX present in *P. fugu* that had been removed from the skin of *T. niphobles*. Ito et al. (2006) revealed the presence of TTX from *P. fugu* and *Taeniacanthus* sp. from *T. niphobles* using HPLC and GC-MS analysis. To clarify the origin of TTX in the caligid, the micro-distribution of TTX in *P. fugu* specimens collected from toxic wild (*T. pardalis*) and non-toxic cultured pufferfish (*T. rubripes*) (fig. 4A) was investigated using immunohistochemical techniques (Ikeda et al., 2006). TTX was observed in the body of *P. fugu* but varies from region to region (fig. 4B). The epicuticle, hindgut, ovary, oviduct, uterus, and eggs of *P. fugu* are not exposed to TTX (fig. 4a, b, c). The presence of TTX along the gut indicates that the caligids probably acquire TTX through feeding. They feed on the skin and mucus of toxic pufferfish, resulting in the dense localization of TTX in the gut tissues (Ikeda et al., 2006). Ito et al. (2006) reported that the accumulation of TTX in *P. fugu* is related to feeding on the mucus and skin of the host puffer fish, since these regions show high toxicity.

TTX accumulation has been studied only in the adult stage. Chalimus stages are also parasitic and the acquisition of TTX by these stages has yet to be studied. The reason why TTX is not localized in the ovary and eggs despite being distributed over most other tissues in *P. fugu* is unclear. However, it can be inferred that TTX is not transmitted vertically to its offspring.

In general, TTX is thought to be indispensable for the survival of TTX-bearing organisms. However, this may not be true of caligids: they appear to retain the TTX transferred from their host fish only temporarily, and may use it for a specific purpose such as defense against predators.


## EPIBIONTS ON *P. FUGU* PARASITIC ON *T. PARDALIS*

Using scanning electron microscopy, naturally-occurring rod-shaped bacteria were found on both dorsal and lateral surfaces of the cephalothorax of *P. fugu* (Venmathi Maran et al., 2007). In an experimental study carried out to find the adhesive affinity of bacteria on *P. fugu* using a shrimp carapace assey, fifty strains were isolated from the homogenates of *P. fugu*. These were grouped into six different colony types, two of which were found with high adhesive affinity to shrimp carapace. These two highly adhesive strains were identified through 16S rRNA sequencing as *Shewanella woodyi* and *Roseobacter* sp. (cf. Venmathi Maran et al., 2007).

The TTX-bearing animals are considered not to synthesize the toxin by themselves but to accumulate TTX through the food chain starting from marine bacteria that produce TTX (Noguchi et al., 2006). It is known that

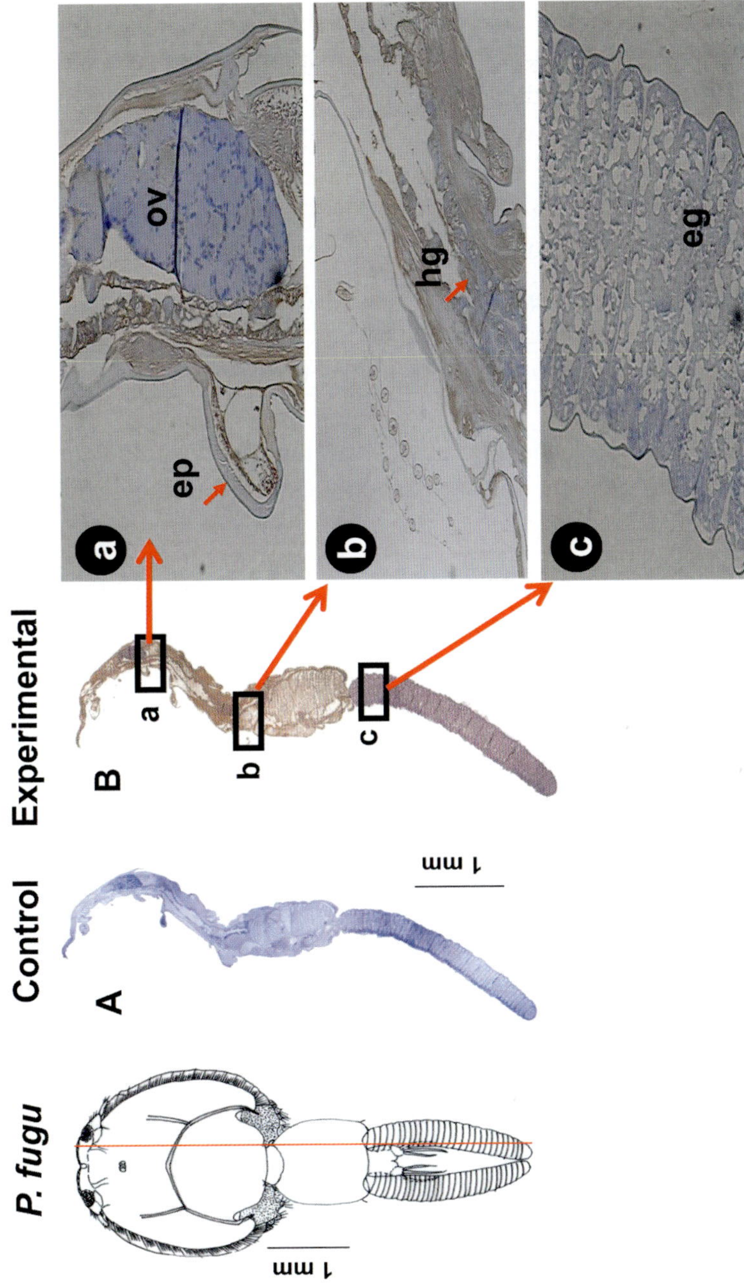


Fig. 4. Accumulation of tetrodotoxin (TTX) in female *Pseudocaligus fugu* Yamaguti, 1936; control (A); experimental (B); and enlarged sections (a, b, c); ep — epicuticle; ov — ovary; hg — hindgut; eg — eggs (Ikeda et al., 2006, with permission from Elsevier, The Netherlands).

marine bacteria like *Vibrio*, *Pseudomonas*, *Shewanella*, *Alteromonas*, and others isolated from many TTX-bearing organisms, produce TTX although the amount of TTX produced was very small (Noguchi et al., 2006). *Roseobacter* sp. isolated from *P. fugu* was shown to exhibit production of TTX and its derivatives, using LC-MS and GC-MS analyses. This was the first discovery for *Roseobacter* as a tetrodotoxin producer (Venmathi Maran et al., 2007).

## PERSPECTIVES

The sea louse *Pseudocaligus fugu* is an economically important parasite in marine aquaculture in Japan. Recent studies have revealed the life cycle and TTX-dynamics of *P. fugu*, but more studies on this species are needed to address the following issues:

1. Host specificity: the mechanism underlying the ability of *P. fugu* to successfully develop on some toxic puffer species but not others needs to be studied in detail.
2. The metabolic pathway of TTX in *P. fugu*: the mechanism whereby it tolerates TTX and why it is not accumulated in the epicuticle, the ovary, oviduct, uterus, and egg sacs remain to be elucidated.
3. The interaction between parasite, host, and epibiontic bacteria has to be further elaborated, in particular in relation to the TTX dynamics.
4. Molecular phylogenetic studies would give insight into the open question of the validity of the genus *Pseudocaligus* in the family Caligidae.

## ACKNOWLEDGEMENTS

The senior author is grateful to the Graduate School of Biosphere Science, Hiroshima University for providing the post doctoral fellowship. The present study was partially supported by a grant-in-aid from the Japan Society for the Promotion of Science to SO (No. 20380110). We would like to thank the reviewer for critical comments.

## REFERENCES

BEN HASSINE, O. K., 1983. Les Copépodes parasites de poissons Mugilidae en Méditerranée occidentale (côtes françaises et tunisiennes). Morphologie, bio-écologie, cycles évolutifs: i-vi, 1-452, bibliography 1-19. (Thèse de Doctorat d'Etat, Université des Sciences et Techniques du Languedoc, Montpellier).

BJØRN, P. A. & B. FINSTAD, 1998. The development of salmon lice (*Lepeophtheirus salmonis*) on artificially infected post smolts of sea trout (*Salmo trutta*). Can. J. Zool., **76**: 970-977.

BOXSHALL, G. A., 1974. The developmental stages of *Lepeophtheirus pectoralis* (Müller, 1776) (Copepoda: Caligidae). J. Nat. Hist., **8**: 681-700.

— —, 2008. A new genus of sea-louse (Copepoda: Siphonostomatoida: Caligidae) parasitic on the bluespine unicornfish (*Naso unicornis*). Folia Parasitol., **55**: 231-240.

BOXSHALL, G. A. & D. DEFAYE (eds.), 1993. Pathogens of wild and farmed fish: sea lice: 1-378. (Ellis Horwood, London).

BOXSHALL, G. A. & S. H. HALSEY, 2004. An introduction to copepod diversity: 1-966. (The Ray Society, London).

BOXSHALL. G. A. & J. L. JUSTINE, 2005. A new genus of parasitic copepod (Siphonostomatoida: Caligidae) from a Razorback Scabbardfish, *Assurger anzac* Alexander (Trichiuridae) off New Caledonia. Folia Parasitol., **52**: 349-358.

FROESE, R. & D. PAULY, 2009. FishBase. (World Wide Web electronic publication). Available at www.fishbase.org; accessed 10 Nov. 2009.

HARADA, Y. & T. ABE, 1993. Taxonomy and toxicity of imported pufferfishes in Japan: 1-130. (Kouseisha Kouseikaku Co. Ltd., Tokyo).

HIRAZAWA, N., S. OSHIMA & K. HATA, 2000. Challenge trials on the antihelminthic effect of drugs and natural agents against the monogenean *Heterobothrium okamotoi* in the tiger puffer *Takifugu rubripes*. Aquaculture, **188**: 1-13.

— —, — — & — —, 2001. In vitro assessment of the antiparasitic effect of caprylic acid against several fish parasites. Aquaculture, **200**: 251-258.

HO, J.-S. & C. L. LIN, 2004. Sea lice of Taiwan (Copepoda: Siphonostomatoida: Caligidae): 1-388. (The Sueichan Press, Keelung).

HWA, T.-K., 1965. Studies on the life history of a fish louse (*Caligus orientalis* Gusev). Acta Zool. Sinica., **17**: 48-57. [In Chinese.]

IKEDA, K., B. A. VENMATHI MARAN, S. HONDA, S. OHTSUKA, O. ARAKAWA, T. TAKATANI, M. ASAKAWA & G. A. BOXSHALL, 2006. Accumulation of tetrodotoxin (TTX) in *Pseudocaligus fugu*, a parasitic copepod from panther puffer *Takifugu pardalis*, but without vertical transmission — using an immunoenzymatic study. Toxicon, **41**: 291-295.

ITO, K., S. OKABE, M. ASAKAWA, K. BESSHO, S. TANIYAMA, Y. SHIDA & S. OHTSUKA, 2006. Detection of tetrodotoxin (TTX) from two copepods infecting the grass puffer *Takifugu niphobles*: TTX attracting the parasites? Toxicon, **48**: 620-626.

JOHNSON, S. C. & L. J. ALBRIGHT, 1991. The developmental stages of *Lepeophtheirus salmonis* (Krøyer, 1837) (Copepoda: Caligidae). Can. J. Zool., **69**: 929-950.

JOHNSON, S. C., J. W. TREASURER, S. BRAVO, K. NAGASAWA & Z. KABATA, 2004. A review of the impact of parasitic copepods on marine aquaculture. Zool. Stud., **43**: 229-243.

KANOH, S., 1988. Distribution of tetrodotoxin in vertebrates. In: K. HASHIMOTO (ed.), Recent advances in Tetrodotoxin: 32-44. (Kouseikaku-Koseisha, Tokyo). [In Japanese.]

KIKUCHI, K., 2006. Present status of research and production of tiger puffer *Takifugu rubripes* in Japan. Memorias del VIII Simposium Internacional de Nutrición Acuícola. 15 al 17 de Noviembre de 2006, Mazatlán, Sinaloa, México: 20-28.

KIM, I. H., 1993. Developmental stages of *Caligus punctatus* Shiino, 1955 (Copepoda: Caligidae). In: G. A. BOXSHALL & D. DEFAYE (eds.), Pathogens of wild and farmed fish: sea lice: 16-29. (Ellis Horwood Limited, West Sussex).

— —, 1998. Illustrated encyclopedia of fauna and flora of Korea. Volume **38**. Cirripedia, symbiotic Copepoda, Pycnogonida: 1-1038. (Ministry of Education of Korea, Seoul). [In Korean.]

LIN, C. L. & J.-S. HO, 1993. Life history of *Caligus epidemicus* Hewitt, parasitic on tilapia (*Oreochromis mossambicus*) cultured in brackish water. In: G. A. BOXSHALL & D. DEFAYE (eds.), Pathogens of wild and farmed fish: sea lice: 5-15. (Ellis Horwood Limited, West Sussex).

LIN, C. L., J.-S. HO & S. N. CHEN, 1997. Development of *Caligus multispinosus* Shen, a caligid copepod parasitic on the Black Sea bream (*Acanthopagrus schlegelii*) cultured in Taiwan. J. Nat. Hist., **31**: 1483-1500.

MATSUOKA, S., 1995. Occurrence of viral, parasitic and other non-bacterial diseases in cultured marine fin-fish in Ehime Prefecture from 1961-1993. Suisanzoshoku, **43**: 535-541.

NOĞCHI, T., O. ARAKAWA & T. TAKATANI, 2006. TTX accumulation in pufferfish. Comp. Biochem. Physiol. Part D, **1**: 145-152.

NOĞCHI, T., T. TAKATANI & O. ARAKAWA, 2004. Toxicity of pufferfish cultured in netcages. J. Food Hyg. Soc. Jpn., **45**: 146-149.

OGAWA, K., 1996. Marine parasitology with special reference to Japanese fisheries and mariculture. Veter. Parasitol., **64**: 95-105.

— —, 2006. *Pseudocaligus* infection (Pseudocaligosis). In: K. HATAI & K. OGAWA (eds.), New atlas of fish diseases: 230. (Midori Shobo, Tokyo). [In Japanese.]

OGAWA, K. & K. INOUYE, 1997. Parasites of cultured tiger puffer (*Takifugu rubripes*) and their seasonal occurrences, with descriptions of two new species of *Gyrodactylus*. Fish Pathol., **32**: 7-14.

OGAWA, K. & H. YOKOYAMA, 1998. Parasitic diseases of cultured marine fish in Japan. Fish Pathol., **33**: 303-309.

OHTSUKA, S., I. TAKAMI, B. A. VENMATHI MARAN, K. OGAWA, T. SHIMONO, Y. FUJITA, M. ASAKWA & G. A. BOXSHALL, 2009. Developmental stages and growth of *Pseudocaligus fugu* Yamaguti, 1936 (Copepoda: Siphonostomatoida: Caligidae) host specific to puffer. J. Nat. Hist., **43**: 1779-1804.

OKABE, S., 2003. An ecological study of copepods parasitic on puffer fish (*Takifugu niphobles*) in the Seto Inland Sea, Japan, with note on presence of tetrodotoxin in copepods: 1-62. (Master Dissertation, Hiroshima Univ., Japan). [In Japanese.]

PAPERNA, I., 1975. Parasites and disease of the grey mullet (Mugilidae) with special reference to the seas of the near east. Aquaculture, **5**: 65-80.

PIASECKI, W., 1996. The developmental stages of *Caligus elongatus* von Nordmann, 1832 (Copepoda: Caligidae). Can. J. Zool., **74**: 1459-1478.

PIASECKI, W. & B. M. MACKINNON, 1993. Changes in structure of the frontal filament in sequential developmental stages of *Caligus elongatus* von Nordmann, 1832 (Crustacea, Copepoda, Siphonostomatoida). Can. J. Zool., **71**: 889-895.

— — & — —, 1995. Life cycle of a sea louse, *Caligus elongatus* von Nordmann, 1832 (Copepoda, Siphonostomatoida, Caligidae). Can. J. Zool., **73**: 74-82.

PIKE, A. W., K. MACKENZIE & A. ROWAND, 1993. Ultrastructure of the frontal filament in chalimus larvae of *Caligus elongatus* and *Lepeophtheirus salmonis* from Atlantic salmon, *Salmo salar*. In: G. A. BOXSHALL & D. DEFAYE (eds.), Pathogens of wild and farmed fish: sea lice: 99-113. (Ellis Horwood Limited, West Sussex).

SHIINO, S. M., 1963. On the male of *Pseudocaligus fugu* Yamaguti (Copepoda: Caligoida). Rep. Fac. Fish. Pref. Univ. Mie, **4**(3): 331-334.

— —, 1964. Results of Amami expedition. 6. Parasitic Copepoda. Rep. Fac. Fish. Pref. Univ. Mie, **5**: 243-255.

TAKEGAMI, T., 1986. Difference of host specificity between larva and adult in *Pseudocaligus fugu* Yamaguti (Copepoda, Caligidae): taxonomy and systematics. Zool. Sci., **3**: 1111.

TENSHA, K. & K. MOMOYAMA, 2006. Effects of hydrogen peroxide solution and diluted seawater on detaching the parasitic copepod *Pseudocaligus fugu* from the juvenile tiger puffer *Takifugu rubripes*. Bull. Yamaguchi Pref. Fish. Res. Ctr., **4**: 163-166.

VENMATHI MARAN, B. A., E. IWAMOTO, J. OKUDA, S. MATSUDA, S. TANIYAMA, Y. SHIDA, M. ASAKAWA, S. OHTSUKA, T. NAKAI & G. A. BOXSHALL, 2007. Isolation and characterization of bacteria from the copepod *Pseudocaligus fugu* ectoparasitic on the panther puffer *Takifugu pardalis* with the emphasis on TTX. Toxicon, **50**: 779-790.

YAMAGUTI, S., 1936. Parasitic copepods from fishes of Japan. Part 2. Caligoida, **I**: 1-22. (Published by the author, Kyoto).

First received 15 November 2009.
Final version accepted 11 January 2010.

# BAIT IMPROVEMENT FOR SWIMMING CRAB TRAP FISHERIES

BY


MIGUEL VAZQUEZ ARCHDALE[1]), GUNZO KAWAMURA and KAZUHIKO ANRAKU

Field of Fisheries Engineering, Faculty of Fisheries, Kagoshima University,
Shimoarata 4-50-20, Kagoshima city, Japan 890-0056


## ABSTRACT


Decapod crustaceans are important fisheries resources that are frequently harvested with baited traps. Fishers prefer some bait types to others, but the substances responsible for bait attractiveness are seldom studied. Swimming crabs were used as models for studying methods to improve bait effectiveness. Food searching behavior was examined by exposing crabs to food extracts and to solutions of individual amino acids and saccharides. Results showed that some saccharides elicit stronger responses than amino acids at the same concentrations. Trapping trials targeting crabs confirmed the possibility of increasing the efficiency of fish bait by adding saccharides to it, and traps baited with sugarcane and fish combination had a catch that was double that of traps containing fish bait. Teabags were tried as a novel binding material to enclose fish mince, and this mince bait caught an equivalent amount of crabs to fish bait. This method can be used to recycle fish waste such as skin, heads and viscera. Additional trials tested the possibility of influencing bait attractiveness by mixing other ingredients into the fish mince. Results using traps baited with fish mince enriched with sugar showed that this addition did not affect the number of individuals of the commercial crab species *Portunus pelagicus* in the catch, but it reduced that number for some non-commercial crab species. The addition of substances into the bait may be applied to design species-selective crab baits. Another method tested artificial bait made from fish waste and cornstarch, which does not require refrigeration and is easy and clean to handle. Field trials showed that traps containing artificial bait caught crabs, though less than those containing fish. Considering that the artificial bait consisted only of 40% fish waste, the crab catches were promising and suggest that effective artificial bait might be viable in the future.


## INTRODUCTION

Baited traps are commonly employed in commercial fisheries and aquaculture to harvest crustaceans from the wild and aquaculture ponds. The bait employed ranges from a wide variety of sources, such as dried or salted fish,

---

[1]) Corresponding author; e-mail: miguel@fish.kagoshima-u.ac.jp

animal viscera, cowhides, dried cormorant, frogs and others. Nevertheless, research to determine the active components found in bait is limited. Bait used by commercial fishers consists mostly of frozen fish, but this needs to be handled several times, stored in a freezer, thawed and cut. It is messy and, most of all, wastes valuable fish protein that is fit for human consumption. In this article the authors review several studies they have conducted to improve the effectiveness of conventional fish bait that is placed in the traps that are used to harvest swimming crabs. Four different aspects of improving crab bait are reviewed: determining individual attractive substances present in bait, adding sugars to the fish bait, making fish mince bait and enclosing it in a teabag, and developing an artificial bait from fish waste. First, we started by determining which individual soluble substances are responsible for the attractive nature of the bait, and we found that saccharides as well as amino acids are important. Second, we applied this knowledge to improve bait effectiveness by combining fish with sugarcane; this bait combination increased crab catches. Third, a new bait binding method was developed that facilitates the recycling of fish processing waste by enclosing it in a porous pouch or teabag. Fish waste, consisting of viscera, heads, skin and/or scales, can be easily minced in a meat chopper and enclosed in this pouch. In addition, it is possible to enrich the mince with other substances that enhance its attractiveness to the target animal or reduce it to decrease the capture of non-target species. The possibility of influencing the crab catch composition by blending a substance (sugar) into the mince contained in the teabags was attempted. Sugar was chosen because it is cheap and readily available, and has been found to be attractive to crabs. Fourth, we conducted fishing trials to evaluate the performance of artificial bait made from cornstarch and fish waste and compared the catches of traps baited with artificial bait with those containing fish bait. This artificial bait is dry in appearance, easy and clean to handle, recycles waste products from the fisheries industry, and it does not require refrigeration but can be stored at room temperature for up to half a year.

## DETERMINATION OF ATTRACTIVE INDIVIDUAL SUBSTANCES

When the swimming crabs *Portunus pelagicus* (Linnaeus, 1758) and *Charybdis feriata* (Linnaeus, 1758) are held in tanks, they tend to remain inactive and buried under the sand substrate during the day. Nevertheless, they can be easily induced into activity by adding some fish, clam or other edible items into their tank. Applying a solution of fish muscle homogenate into the tank through a pipette can easily stimulate them to search for food. The behavioral sequence that follows this addition starts with an increase in the crab's rate of antennule flicking, followed by exposure of the mouthparts, and mouthpart movement; after which, emergence from the sand substrate will result. Crawling, prodding the sand substrate with the walking legs, picking up sand particles with the chelipeds and carrying them to the mouth usually terminates this sequence. These last behavioral stages were interpreted as an indication that the substance encountered was attractive. Low molecular weight substances emanate from baits, and they contain amino acids, saccharides and ammonium compounds (Carr, 1988). If these individual substances are obtained in a purified form, diluted in seawater, and introduced through a pipette into the tank containing a crab, they may sometimes elicit the same food searching sequence as the fish muscle homogenate. If this is the case, that substance can be identified as responsible for some of the attractive qualities of the bait.

Results obtained by exposing *P. pelagicus* to solutions of individual amino acids and saccharides showed that the amino acids alanine, arginine, glycine, serine, taurine and betaine were stimulants within the concentration range of $2 \times 10^{-7}$ to $2 \times 10^{-4}$ M, while the saccharides galactose and glucose were more stimulatory than the above amino acids at the same concentrations (Archdale & Nakamura, 1992). Positive responses were also obtained in *C. feriata* to alanine, glycine, serine, galactose and glucose within the same concentration range. These results confirmed that both amino acids and saccharides act as stimulating substances initiating feeding behavior in swimming crabs. Similar results have been obtained in the porcelain crab *Petrolisthes cinctipes* (Randall, 1839) (Hartman & Hartman, 1977), the sand fiddler crab *Uca pugilator* (Bosc, 1802) (Robertson et al., 1981), and the ghost crab *Ocypode quadrata* (Fabricius, 1787) (Trott, 1984), but not in any commercially important crab species.

Examinations of stomach contents of swimming crabs (Williams, 1982; Matsui et al., 1986) showed a prevalence of crustaceans, mollusks and polychaetes among prey species though the percentage composition of the contents was dependent on prey availability. Analysis of amino acids in crab, shrimp, oyster and fish (Carr, 1988) has shown that alanine, arginine, glycine and taurine have the highest concentrations; all these proved to be highly stimulatory to *P. pelagicus*. Squid, which is occasionally used as a bait in the traps for this crab, has a high content of taurine, glycine, alanine and arginine (Mackie, 1973), all of which showed high stimulating effects. As for saccharides, both the blood of fish and the hemolymph of crustaceans are rich in them; hemolymph from crustaceans was shown to be high in glucose and maltose (Meenakshi & Scheer, 1960).

Therefore, it seems that swimming crabs have a very well developed sense of chemoreception that enables them to discriminate their food according to the chemical composition of the substances that emanate from it. The results of this behavioral experiment only dealt with individual substances, but because the attractiveness of saccharides was confirmed, it showed that it might be possible to increase the effectiveness of conventional fish bait by adding sugars to it.

## SUGARCANE AND FISH BAIT COMBINATION

Sugarcane is the raw material from which sugar is extracted. It is easy to grow and found in many tropical and subtropical regions around the world; consequently, it would be a desirable crustacean bait source, if effective. After having determined that the saccharides glucose and galactose induced searching for food in some swimming crab species, the possibility of using sugarcane as new bait was investigated.

Comparative fishing trials are common practice when testing bait effectiveness. Traps are set with different types of bait and placed in the same fishing ground. After retrieving the traps from the sea, the catch number for each species is determined for each different type of bait. Traps baited with sugarcane, fish and a combination of the two were tested in two fishing locations in Japan (Kawamura et al., 1995). The crab catch was very low in sugarcane baited traps when compared to those baited with fish, but the traps baited with the sugarcane and fish combination had almost double the catch of *P. pelagicus* and more than triple that of *Charybdis japonica* (Milne-Edwards, 1861) when compared with the catch of fish baited traps (table I). Additional fishing

### TABLE I

Crab catches obtained using sugarcane, fish and their combination as trap bait

| Fishing ground | Target species | Reference | Bait treatments | | |
|---|---|---|---|---|---|
| | | | Sugarcane | Fish | Combination |
| Kagoshima, Japan | *P. pelagicus* | Kawamura et al., 1995 | 3 | 44 | 71 |
| Nagashima, Japan | *C. japonica* | Kawamura et al., 1995 | 2 | 11 | 36 |
| Panay, Philippines | Swimming crabs | Anraku et al., 2001 | 9 | 27 | 40 |

trials carried out in the Philippines (Anraku et al., 2001) also had similar results, showing that the sugarcane and fish bait combination attracted the largest number of crabs.

These results suggest the possibility of adding sugarcane to fish to improve the effectiveness of the current crab bait. Sugarcane does not require refrigeration, does not spoil as easily as fish, and is easy to handle and prepare when used as bait. The greater crab catch found in traps using this bait combination probably results from a synergistic effect caused by mixing the substances emanating from both bait types, which results in a higher attractiveness. Many countries where swimming crabs are harvested are also producers of sugarcane, so the bait improvement seems very feasible; particularly in Australia, where *P. pelagicus* is commercially exploited.

## DEVELOPMENT OF MINCE IN TEABAG BAIT

Teabags containing shredded frozen fish have been successfully used to bait hooks for the cod longline fishery. Based on this information, an attempt was made to determine if this same bait binding method could be applied to the crab trap fishery. Teabags, or other porous bags, can enclose fish mince made from discarded fish heads, viscera, skin and scales, thus not wasting the valuable meat that is suitable for human consumption. Mince bait releases attractants faster than whole fish, and these attractants distribute themselves along wider areas from which more target animals can be lured. Furthermore, mince bait can be enriched with other substances in the same way vitamins and minerals are blended into aquaculture feeds.

Fishing trials were conducted to test if the teabag is a suitable binding material for fish mince. Mince in teabag bait was manufactured by mincing mackerel through a meat chopper and placing 100 g of mince inside each teabag, which was then frozen for storage until required for the trials. Teabag bait was placed in traps and crab catches were compared with those of non-baited and fish-baited traps, serving as controls. After hauling the traps, it was determined that mince in teabag bait caught an equivalent number of crabs to fish baited traps, while non-baited traps caught only a few (Vazquez Archdale et al., 2008). These results support the conclusion that teabags containing fish mince are just as effective as conventional fish bait.

Following these promising results, an additional set of trials was carried out to determine if adding substances to the bait could influence crab catches. White sugar was blended into the mince bait, packed in teabags and used in the same way as in the previous trials. The crab catches from traps containing mince plus sugar in teabag bait was compared with those of traps containing fish alone. Interestingly, the addition of sugar to the mince had different effects on the various crab species found in the catch. In the case of the commercial crab *P. pelagicus*, catches were similar to those of fish bait, and no notable differences were found compared to the previous mince trials, but the swimming crabs *C. japonica* and *Thalamita prymna* (Herbst, 1803) showed a notable decrease after the addition of sugar that amounted to half the number of crabs in the catch. These results suggest that crabs have different preferences depending on species and keen discriminatory abilities when it comes to food items. In the future, it might be possible to manipulate the substances emanating from bait to create species-specific bait.


## EFFICIENCY OF ARTIFICIAL BAIT

The availability of a certain fish bait is sometimes limited depending on the season of the year. Fresh fish has to be stored in a freezer, and before using it as bait it must be thawed, cut and handled, which is a dirty and tedious work. In the future manufacturing artificial bait industrially that can be stored at room temperature is a necessity, and it should be made into a product that is dry, clean and easy to handle. Furthermore, artificial bait can recycle a large amount of the waste that is generated by the fish processing industry; and consequently help to conserve some of the world's dwindling fisheries resources.

The new artificial bait was manufactured from 30-40% fish waste mixed with cornstarch and processed into pellets. The pellets were enclosed loosely in small meshed bags to allow seawater to soak and flow though the pellets therefore permitting a wider dispersal of their attracting substances.

This artificial bait was tested during fishing trials in the East China Sea on board the training vessel Kagoshima Maru. Trap treatments were assigned 10 traps for each bait type, which consisted of fish, mince in teabag and artificial bait, and 10 traps without bait as control. The amount of bait in each trap was standardized to 100 grams for all treatments. The traps were set following commercial fishing practices, and they were fastened with branch lines to a long bottom line with two anchors at each end. Each anchor was fastened to a line that was tied to a marker buoy. The depth of the sea bottom was approximately 120 m. Traps were soaked for 24 hours and retrieved the following day.

Fig. 1. Trap baited with artificial bait containing captured sand crabs *Ovalipes punctatus*.

Fishing trials were conducted for 3 days and the catch consisted almost entirely of sand crabs *Ovalipes punctatus* (De Haan, 1833). About 50% of the crab catch was captured in the fish baited traps, 30% in the mince in teabag traps and 20% in the artificial bait traps (fig. 1). Non-baited traps hardly caught any crabs. The catch from the artificial bait seems low when compared to that of conventional fish bait, but if we take into account that it only consisted of 40% fish waste while the fish bait was 100% high quality fish suitable for human consumption, the catches were not so poor. Recalculating the catch considering only the amount of fish present in the bait shows that the artificial bait caught almost as many crabs as the fish bait. Furthermore, after hauling the traps, almost all fish and mince in teabag baits were completely consumed by captured organisms and scavengers, while a large proportion of the artificial bait was still present in the traps. If the traps are soaked for longer periods, such as 2-3 days, which is not unusual in crab trap fisheries, the catch might increase due to the longer presence of artificial bait in the traps. On the other hand, fish and mince in teabag baited traps will cease to attract crabs after the bait is consumed and that period will be less than 1 day.

## CONCLUSION

Observations of food searching behavior of swimming crabs towards amino acids and saccharides revealed that both substances play an important role in crab chemoreception. Application of this finding contributed to development of a new baiting method that employs a combination of fish and sugarcane, which doubled the trap catches of swimming crabs. A new baiting method was developed for recycling fish processing waste into bait, which consisted in binding fish mince inside a porous membrane or teabag. Mince in teabag bait proved just as effective as ordinary fish bait; furthermore, it allowed for enrichment with attractants, repellents or other substances by blending them into the bait. Mixing sugar into the bait maintained catches of commercial swimming crabs but reduced those of two non-commercial crab species. This result documents that chemosensory preferences in crabs vary depending on species and this might be used to develop species selective crab baits in the future.

Alternative artificial bait made from fish waste was developed and its crab catching performance was compared to fish bait. The catches of the artificial bait were less than half of those by fish, but when adjusted for crab catch per amount of fish present in the bait, the results were almost the same. In the future, artificial baits might be used as suitable substitutes for conventional fish bait.

## REFERENCES

ANRAKU, K., M. VAZQUEZ ARCHDALE, B. MENDEZ CORTES & R. A. ESPINOSA, 2001. Crab trap fisheries: capture process and an attempt on bait improvement. UPV J. Nat. Sci., **6**: 121-129.

ARCHDALE, M. V. & K. NAKAMURA, 1992. Responses to the swimming crab *Portunus pelagicus* to amino acids and mono- and disaccharides. Nippon Suisan Gakkaishi, **58**: 165.

CARR, W. E. S., 1988. The molecular nature of chemical stimuli in the aquatic environment. In: J. ATEMA, R. R. FAY, A. N. POPPER & W. N. TAVOLGA (eds.), Sensory biology of aquatic animals: 3-20. (Springer-Verlag, New York).

HARTMAN, H. & M. HARTMAN, 1977. The stimulation of filter feeding in the porcelain crab, *Petrolisthes cinctipes* Randall by amino acids and sugars. Comp. Biochem. Physiol., **56**: 19-22.

KAWAMURA, G., T. MATSUOKA, T. TAJIRI, M. NISHIDA & M. HAYASHI, 1995. Effectiveness of a sugarcane-fish combination as bait in trapping swimming crabs. Fish. Res., **22**: 155-160.

MACKIE, A. M., 1973. The chemical basis of food detection in the lobster *Homarus gammarus*. Mar. Biol., **21**: 103-108.

MATSUI, S., Y. HAGIWARA, H. TOU & H. TSUKAHARA, 1986. Study on the feeding habits of the Japanese blue crab *Portunus trituberculatus* (Miers). Sci. Bull. Fac. Agr. Kyushu Univ., **40**: 175-181.

MEENAKSHI, V. R. & B. T. SCHEER, 1960. Metabolism of glucose in the crabs *Cancer magister* and *Hemigrapsus nudus*. Comp. Biochem. Physiol., **3**: 30-41.

ROBERTSON, J. R., J. A. FUDGE & G. K. VERMEER, 1981. Chemical and live feeding stimulants of the sand fiddler crab, *Uca pugilator* (Bosc). J. Exp. Mar. Biol. Ecol., **53**: 47-64.

TROTT, T. J. & J. R. ROBERTSON, 1984. Chemical stimulants of cheliped flexion behaviour by the Western Atlantic ghost crab, *Ocypode quadrata* (Fabricius). J. Exp. Mar. Biol. Ecol., **78**: 237-252.

VAZQUEZ ARCHDALE, M., C. P. AÑASCO & Y. TAHARA, 2008. Catches of swimming crabs using fish mince in "teabags" compared to conventional fish baits in collapsible pots. Fish. Res., **91**: 291-298.

WILLIAMS, M. J., 1982. Natural food and feeding in the commercial sand crab *Portunus pelagicus* Linnaeus, 1766 (Crustacea: Decapoda: Portunidae) in Moreton Bay, Queensland. J. Exp. Mar. Biol. Ecol., **59**: 165-176.

First received 13 November 2009.
Final version accepted 22 December 2009.

# A PRELIMINARY INVESTIGATION INTO THE POTENTIAL VALUE OF GASTRIC MILLS FOR AGEING CRUSTACEANS


BY

J. C. LELAND[1,3]), J. COUGHRAN[2]) and D. J. BUCHER[1])

[1]) Marine Ecology Research Centre, School of Environmental Science & Management, Southern Cross University, Lismore Campus, New South Wales, Australia, 2480
[2]) Environmental Futures Centre, Griffith School of Environment, Gold Coast Campus, Griffith University, Queensland, Australia, 4222



## ABSTRACT

A preliminary study was conducted to assess the usefulness of calcified ossicles from the gastric mills of decapod crustaceans for ageing, based on the application of a traditional fisheries biology method for the analysis of concentric growth marks in scales, bones and otoliths. Transverse cross sections of prepyloric, zygocardiac and pyloric ossicles of gastric mills from *Cherax quadricarinatus*, *Euastacus valentulus*, *Ranina ranina*, *Scylla serrata* and *Thenus orientalis* showed the presence of alternating translucent and opaque growth rings. Growth ring analysis indicated a growth record in two series, each likely representing different periodicities. Comparisons of growth ring counts showed close agreement between all gastric ossicles examined. Progressive sectioning of zygocardiac ossicles from *C. quadricarinatus* and *S. serrata* showed that growth ring counts varied along the length of these structures and demonstrated the need for species-specific assessment of gastric mill components. In conclusion, further empirical research is warranted to validate the periodicity of the gastric ossicle growth record to further elucidate its potential for ageing crustaceans.


## INTRODUCTION

One of the most commonly cited problems regarding age determination in crustaceans is the absence of any calcified structures containing growth rings (Farmer, 1973; Sheehy, 1990; France et al., 1991; Belchier et al., 1998; Hartnoll, 2001; Kodama et al., 2006). This is because in order to grow these animals must moult (Mauchline, 1976; Holdich, 2002; Kuballa & Elizur,

---

[3]) Corresponding author; e-mail: jesse.leland@scu.edu.au

 New frontiers in crustacean biology: 57-68

2007), a process during which they shed their entire exoskeleton (Street, 1966; Holdich & Lowery, 1988; Holdich, 2002). However, even the ephemeral gastroliths of crayfish record time at a daily resolution (Scudamore, 1947), indicating that any calcified structure might be expected to contain a record of growth. Furthermore, Bazin (1970) described a calcified 'ovoid body' in Nephropid lobsters that is retained for the lifetime of the animal, and contains concentric layers of unknown periodicity (Hartnoll, 2001). However, there appears to have been no further research into the potential use of calcified structures for ageing crustaceans since these earlier studies. In short, because most calcified structures are lost at each moult their potential use for age estimation has apparently been overlooked.

Any calcified structure subject to biotic processes and abiotic variables could be expected to show a record of development and growth at some resolution (Thompson et al., 1980; Ewing et al., 2003; King, 2007). Fisheries biologists routinely examine calcified structures (e.g., otoliths) that contain growth records of different resolutions (Pannella, 1971; Campana & Neilson, 1985; Gray et al., 1997; Campana, 2001). Age determination of fish has been comparatively straightforward as they retain calcified structures for the duration of their lifetime and growth rings found in these structures are frequently annular (Ewing et al., 2003; Smith & Deguara, 2003; Stewart & Hughes, 2007). However, estimated ages derived from such studies must be validated, preferably by two or more methods (Beamish & McFarlane, 1983; Campana, 2001). In contrast, most crustacean ageing studies have lacked such stringent age validation.

Presently, there are two commonly applied methods of age estimation for crustaceans, size frequency analysis (France et al., 1991) and lipofuscin assays (Sheehy et al., 1994; Belchier et al., 1998; Kodama et al., 2006). Although both methods have advanced crustacean aging research they are subject to various limitations (Farmer, 1973; France et al., 1991; Hartnoll, 2001) and could be strengthened with further validation (Campana, 2001; Hartnoll, 2001). Therefore, if there are any calcified structures in crustaceans that contain growth rings, they could potentially complement or improve current ageing techniques. Crustaceans have numerous internal (Martin et al., 1998) and external calcified structures, however a review of the literature indicates no attempt to apply cross-sectional analysis to these structures to ascertain if they contain growth rings.

The foregut of decapod crustaceans can be divided into three regions, the esophagus, anterior chamber and posterior chamber (Chisaka & Kozawa,

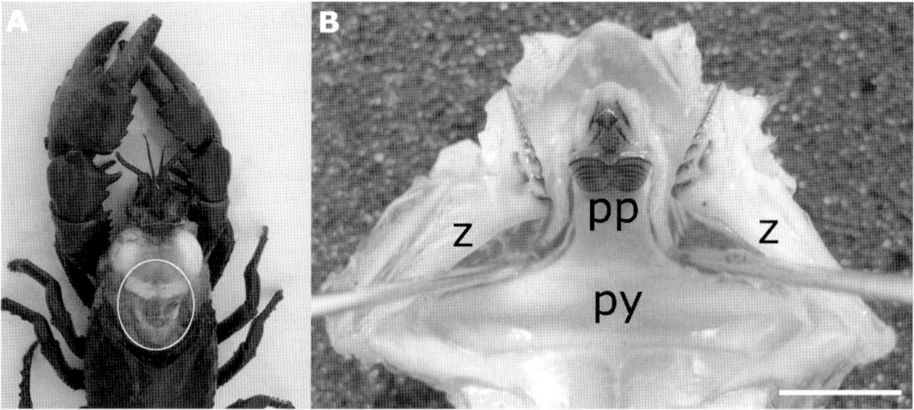


Fig. 1. Decapod crustacean (*Euastacus valentulus*) showing the: A, in-situ position of the gastric mill (indicated by white oval); B, extracted gastric mill. (pp) prepyloric ossicle; (z) left and right zygocardiac ossicles; (py) pyloric ossicle (scale bar equals 100 mm).

2003; McGaw, 2006). Housed within the anterior chamber are calcified gastric mills that masticate food items to adequate dimensions for passage through the digestive system (Holdich & Lowery, 1988; McGaw, 2006). The main functional components of the gastric mill are the opposing zygocardic ossicles, central prepyloric ossicle, and the pyloric ossicle (Chisaka & Kozawa, 2003) (fig. 1).

The question addressed by this preliminary study is: do the zygocardic, prepyloric and pyloric ossicles of the gastric mill of decapod crustaceans contain any record of growth? Answering this question is a rudimentary step to broaden our current understanding of calcified structures in reference to ageing crustaceans.

## MATERIALS AND METHODS

### Study specimens

Five species of decapod crustaceans were selected for cross section analysis: (1) Redclaw crayfish *Cherax quadricarinatus* (von Martens, 1868), (2) Powerful crayfish *Euastacus valentulus* Riek, 1951, (3) Spanner crab *Ranina ranina* (Linnaeus, 1758), (4) Mud crab *Scylla serrata* (Forskål, 1775) and (5) Moreton Bay bug *Thenus orientalis* (Lund, 1793). Two representatives of each species were sourced with the exception that only one *S. serrata* individual was available for study. Freshwater specimens *C. quadricarinatus* and *E. valentulus* were acquired from Southern Cross University and marine specimens *R. ranina*, *S. serrata* and *T. orientalis* were obtained from a commercial source. All specimens were euthanased by placement in a freezer at −10°C.

## Morphometric measurements

Measurements of wet weight (WWT), body length (BL, except for *S. serrata*), carapace length (CL), orbital carapace length (OCL) and carapace width (CW) were recorded for all specimens (table I). Each specimen was towel dried and WWT was determined using a top loading pan balance to the nearest 0.01 g. All other morphometric measures were taken dorsally with electronic vernier callipers to the nearest 0.1 mm. Measures of BL, CL, OCL and CW were respectively defined as the distance measured from the: anterior-most tip of the rostrum or spine to the longest point of the telson; anterior-most tip of the rostrum or spine to the posterior margin of the carapace; posterior edge of the eye socket to the posterior margin of the carapace; and the widest point of the carapace. Of the morphometric measures taken, BL was chosen as a standard measure to compare with primary growth ring counts.

## Gastric mill extraction

The carapace of each specimen was removed and the exposed cardiac stomach extracted intact. Gastric mills were dissected out under stereoscope magnification and the prepyloric, zygocardiac and pyloric ossicles were disconnected. Organic materials were removed from all gastric ossicles and stored

TABLE I

Biometric values of wet weight (WWT), body length (BL), carapace length (CL), orbital carapace length (OCL) and carapace width (CW) of all study specimens (nd, no data)

| Specimen | WWT (g) | BL (mm) | CL (mm) | OCL (mm) | CW (mm) |
|---|---|---|---|---|---|
| *Cherax quadricarinatus* | 41.5 | 132 | 63 | 43 | 18 |
| *Cherax quadricarinatus* | 42.7 | 132 | 62 | 42 | 17 |
| *Euastacus valentulus* | 21.6 | 85 | 40 | 34 | 18 |
| *Euastacus valentulus* | 314.1 | 186 | 92 | 80 | 53 |
| *Ranina ranina* | 434.8 | 165 | 106 | 98 | 96 |
| *Ranina ranina* | 980.1 | 199 | 135 | 121 | 125 |
| *Scylla serrata* | 723.6 | nd | 105 | 97 | 157 |
| *Thenus orientalis* | 49.1 | 157 | 66 | 45 | 96 |
| *Thenus orientalis* | 139.2 | 182 | 80 | 50 | 113 |

in 70 percent ethanol. Digital images were taken with Quick Capture (2001-2003© QImaging v.2.68.2) software operating on an Apple Macintosh™ computer, and adjusted for maximum clarity with Adobe Photoshop™ 7.0.

## Cross section analysis

Ossicles of the gastric mill were embedded in clear casting polyester resin and one centrally positioned transverse section (200 $\mu$m thickness) of each ossicle was cut on a low speed Buehler™ lapidary saw. One prepyloric and zygocardiac ossicle was sectioned for each species and selected pyloric ossicles were opportunistically sectioned. One zygocardiac ossicle from *C. quadricarinatus* and *S. serrata* was progressively sectioned in 400 $\mu$m increments from the posterior margin towards the distal edge of the ossicle. Cross sections were mounted on slides for viewing under compound microscope magnification ($\times$40-$\times$100) and the maximum number of primary growth rings in each section was recorded. Growth rings observed in cross sections of reduced quality where precise counts were not possible, were simply denoted as 'present'.

## RESULTS

### Cross section analysis

Growth rings were present in transverse cross sections of the prepyloric, zygocardiac and pyloric ossicles of *C. quadricarinatus*, *E. valentulus*, *R. ranina* and *S. serrata*, and in the zygocardiac ossicles of *T. orientalis*. The maximum number of growth rings present in the gastric ossicles sectioned for both juvenile *C. quadricarinatus* specimens showed close agreement with a maximum variance of ±2 growth rings between specimens of equal TL. The juvenile specimen of *E. valentulus* showed close agreement (±1 growth ring) between prepyloric and zygocardiac ossicles sectioned whilst the mature specimen showed higher maximum growth ring counts that were more variable (±3 growth rings) (table II).

*Cherax quadricarinatus* and *E. valentulus* were the only species for which good quality sections were produced in all ossicles sectioned (fig. 2). The best quality sections from *R. ranina*, *S. serrata* and *T. orientalis* were derived from the zygocardiac ossicles (fig. 3). Calcified structures such as fish otoliths are generally concentric in structure (Stewart & Hughes, 2007; Jonsdottir et al., 2006; Panfili et al., 2002). However, the prepyloric, zygocardiac and pyloric ossicles of gastric mills have a more complex morphology (see Morgan, 1997)

TABLE II

Body lengths (BL) of all study specimens and the maximum number, presence (p), or absence (a) of primary growth rings observed in sectioned prepyloric (PP), zygocardiac (ZC) and pyloric (PY) ossicles (nd, no data; ons, ossicle not sectioned)

| Specimen | BL (mm) | PP | ZC | PY |
|---|---|---|---|---|
| *Cherax quadricarinatus* | 132 | 10 | 11 | 10 |
| *Cherax quadricarinatus* | 132 | 12 | 12 | 10 |
| *Euastacus valentulus* | 85 | 8 | 7 | p |
| *Euastacus valentulus* | 186 | 14 | 11 | 12 |
| *Ranina ranina* | 165 | p | p | ons |
| *Ranina ranina* | 199 | p | 23 | ons |
| *Scylla serrata* | nd | p | 14 | p |
| *Thenus orientalis* | 157 | a | 5 | ons |
| *Thenus orientalis* | 182 | a | 6 | ons |

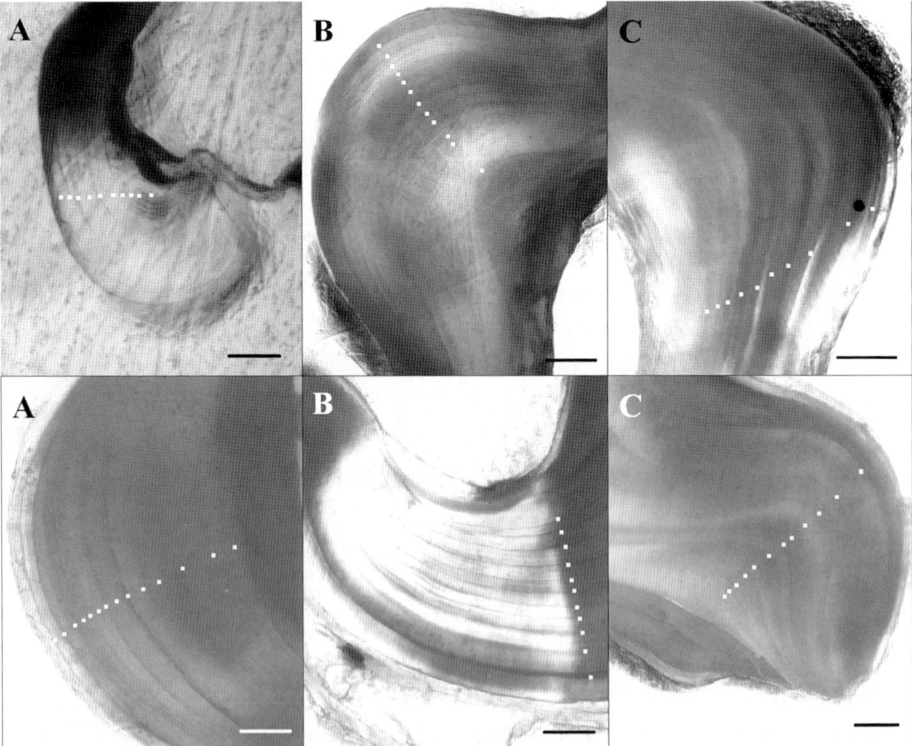


Fig. 2. Transverse cross sections of *Cherax quadricarinatus* (top) and *Euastacus valentulus* (bottom) gastric ossicles showing growth rings observed in the: A, prepyloric ossicle; B, zygocardiac ossicle; C, pyloric ossicle (white squares approximately demarcate primary growth rings; scale bars equal 150 $\mu$m).

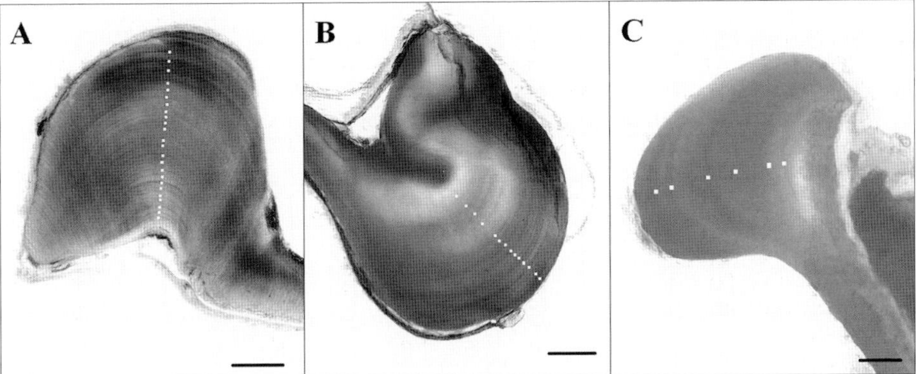

Fig. 3. Transverse cross sections showing growth rings observed in the zygocardiac ossicles of: A, *Ranina ranina*; B, *Scylla serrata*; C, *Thenus orientalis* (white squares approximately demarcate primary growth rings; scale bars equal 100 $\mu$m).

that caused the formation of air pockets in the resin during mounting, and therefore variability in cross-sectional quality. This limited the number of within-species comparisons of growth ring counts.

## Progressive cross sectioning

Incremental cross sectioning of zygocardiac ossicles from *C. quadricarinatus* and *S. serrata* showed that the number of growth rings in each section varied with increasing distance from the posterior margin of the structure. The *C. quadricarinatus* specimen showed an initial increase in the number of growth rings, and then decreased anteriorly, with a maximum count positioned 1.2 mm from the posterior edge of the zygocardiac ossicle. *Scylla serrata* showed greater variability in the number of growth rings along the length of the ossicle, but a clearly defined maximum count was recorded 3.6 mm from the posterior edge (fig. 4).

## DISCUSSION

Transverse cross sections of the gastric ossicles from all species examined revealed the presence of alternating translucent and opaque growth rings, similar in appearance to those that are routinely examined in fish otoliths (Ewing et al., 2003; Smith & Deguara, 2003; Stewart & Hughes, 2007). Furthermore, a secondary series of growth rings were present between the more prominent primary series, indicating two resolutions within the growth

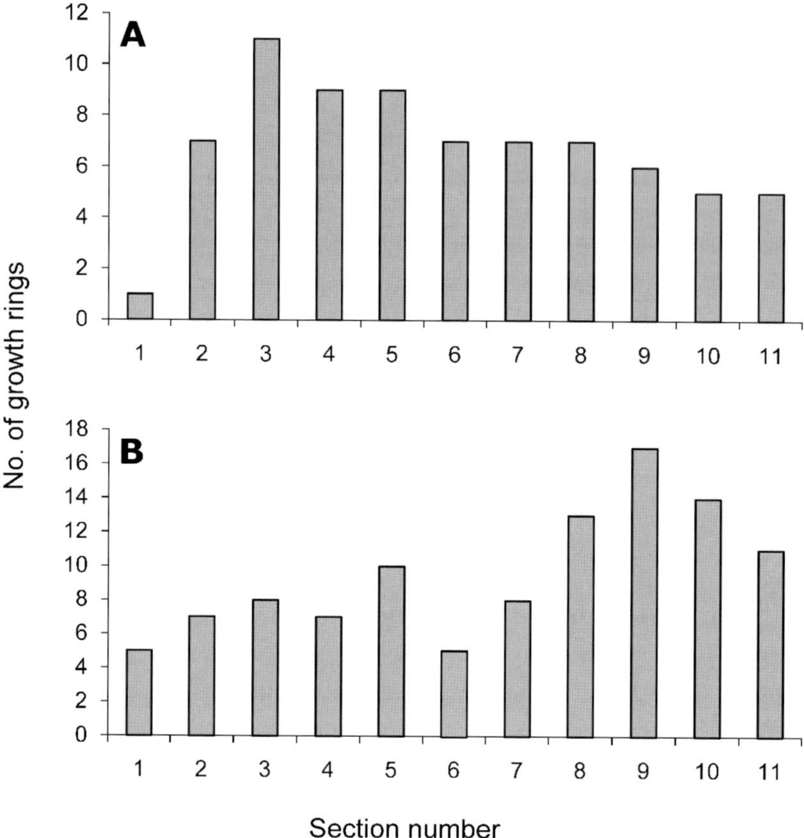



Fig. 4. Growth ring counts derived from eleven posteriorly progressive transverse cross sections (400 $\mu$m increments) of zygocardiac ossicles from: A, *Cherax quadricarinatus*; B, *Scylla serrata*.

record observed. This pattern is commonly reported for calcified structures in other animals (Nolan & Clarke, 1993; Admassu & Casselman, 2000; Goodwin et al., 2001), the primary series represents growth on a broader scale and the secondary series is indicative of fine-scale microstructural growth (Neilson & Geen, 1982; Campana, 1984; Lee & Byun, 1996). To our knowledge this is the first time that transverse cross-sectional analysis has revealed the presence of growth rings of any nature in crustaceans.

The specimens of *C. quadricarinatus* were of equal BL and showed close agreement in growth ring counts for all ossicles. Conversely, *E. valentulus* was represented by specimens of different BL. The juvenile was approximately half the BL of the adult and had approximately half the number of growth rings. This similarity between BL and growth ring counts was also apparent for *T.*

*orientalis* individuals of different BL. These preliminary findings, although limited by small sample size, may indicate a relationship between BL and the number of growth rings in the gastric ossicles of crustaceans, and certainly warrants further empirical research.

Comparison of growth ring counts derived from sectioned prepyloric, zygocardiac and pyloric ossicles of all specimens showed agreement within the acceptable constraints of rigorous ageing studies (Stewart & Hughes, 2007; Horn, 2002), and also demonstrated that ossicles of the gastric mill show differences in the maximum number of growth rings present. This was highlighted by incremental sectioning of the zygocardiac ossicles of *C. quadricarinatus* and *S. serrata*, which showed that growth ring counts vary along the length of the ossicle. Progressive sectioning of zygocardiac ossicles also indicated that growth centres of gastric ossicles are species-specific. Consequently, the ossicle with the highest number of growth rings, and the location of the growth centre within that ossicle, must both be determined as an obligatory step prior to applying cross-sectional analysis to crustaceans. Failure to meet these requirements would certainly affect the precision of growth ring counts (Beamish & McFarlane, 1983; Campana, 2001). Indeed, slight discrepancies in cross-sectional placement may account for some of the within-species variability of growth ring counts in the present study.

This study has demonstrated that there is a clear record of growth contained within the gastric ossicles of crustaceans. However, determining the periodicity of the primary series of growth rings is essential to elucidate what the gastric ossicle growth record represents. It may simply be representative of episodic, post-moult growth of the ossicle itself, alternatively it may be a record of moult history or a cumulative record of the animal's growth over time. This will be a key area of future research by the authors, in order to establish the potential usefulness of the gastric ossicle growth record in resolving the long existing problem of determining the time elapsed during inter-moult periods. Furthermore, the gastric ossicle growth record may be useful as a reference measure in currently applied ageing techniques including size frequency analysis and lipofuscin assays.


## CONCLUSION

This study has identified two incremental records of growth contained within the prepyloric, zygocardiac and pyloric ossicles of crustaceans. This gastric ossicle growth record could be of considerable value in furthering crustacean ageing research given two prerequisites are met: (1) rigorous assessment of all gastric mill components to determine the structure that shows the greatest number of growth rings, and the location of the growth centre within that structure, and (2) validation of the periodicity of primary growth ring formation. Elucidation of the gastric ossicle growth record may present the opportunity to reduce the current uncertainties in modelling non-uniform growth in crustaceans, and provide greater compatibility across currently applied methods of age determination.

## ACKNOWLEDGEMENTS

This study was undertaken at Southern Cross University (Lismore, NSW) and the authors would like to thank the following laboratory staff for technical assistance: C. Taylor, L. Taylor, P. Bligh-Jones and M. Dawes. We also thank J. M. Furse for providing useful comments on the original manuscript.

## REFERENCES

ADMASSU, D. & J. M. CASSELMAN, 2000. Otolith age determination for adult tilapia, *Oreochromis niloticus* L. from Lake Awassa (Ethiopian Rift Valley) by interpreting biannuli and differentiating biannual recruitment. Hydrobiologia, **418**: 15-24.

BAZIN, F., 1970. Étude comparée de l'organe deutocérébral des macroures reptantia et des anomoures (Crustacés Décapodes). Arch. Zool. Éxp. Gen., **111**: 245-264.

BEAMISH, R. J. & G. A. MCFARLANE, 1983. The forgotten requirement for age validation in fisheries biology. Trans. Amer. Fish. Soc., **112**(6): 735-743.

BELCHIER, M., L. EDSMAN, M. R. J. SHEEHY & P. M. J. SHELTON, 1998. Estimating age and growth in long-lived temperate freshwater crayfish using lipofuscin. Freshwater Biol., **39**: 439-446.

CAMPANA, S. E., 1984. Microstructural growth patterns in the otoliths of larval and juvenile starry flounder, *Platichthys stellatus*. Can. J. Zool., **62**: 1507-1512.

— —, 2001. Accuracy, precision and quality control in age determination, including a review of the use and abuse of age validation methods. J. Fish Biol., **59**: 197-242.

CAMPANA, S. E. & J. D. NEILSON, 1985. Microstructure of fish otoliths. Can. J. Fish Aquat. Sci., **42**: 1014-1032.

CHISAKA, H. & Y. KOZAWA, 2003. Fine structure and mineralisation of the gastric mill in the crayfish *Procambarus clarkii* during the inter-moult stage. J. Crust. Biol., **23**(2): 371-379.

EWING, G. P., D. C. WELSFORD, A. R. JORDAN & C. BUXTON, 2003. Validation of age and growth estimates using thin otolith sections from the purple wrasse, *Notolabrus fucicola*. Mar. Fresh. Res., **54**: 985-993.

FARMER, A. S., 1973. Age and growth in *Nephrops norvegicus* (Decapoda: Nephropidae). Mar. Biol., **23**: 315-325.

FRANCE, R., J. HOLMES & A. LUNCH, 1991. Use of size-frequency data to estimate the age composition of crayfish populations. Can. J. Fish. Aquat. Sci., **48**: 2324-2332.

GOODWIN, D. H., K. W. FLESSA, B. R. SCHONE & D. L. DETTMAN, 2001. Cross-calibration of daily growth increments, stable isotope variation, and temperature in the Gulf of California bivalve mollusc *Chione cortezi*: implications for paleoenvironmental analysis. Palaios, **16**: 387-398.

GRAY, A. P., R. SEED & C. A. RICHARDSON, 1997. Reproduction and growth of *Mytilus edulis chilensis* from the Falkland Islands. Sci. Mar., **61**(2): 39-48.

HARTNOLL, R. G., 2001. Growth in Crustacea — twenty years on. Hydrobiologia, **449**: 111-122.

HOLDICH, D. M. (ed.), 2002. Biology of freshwater crayfish: 1-702. (Blackwell Science, Oxford).

HOLDICH, D. M. & R. S. LOWERY (eds.), 1988. Freshwater crayfish: biology management and exploitation: 1-498. (The University Press, Cambridge).

HORN, P. L., 2002. Age estimation of barracouta (*Thyrsites atun*) off southern New Zealand. Mar. Fresh. Res., **53**: 1169-1178.

JONSDOTTIR, I. G., S. E. CAMPANA & G. MARTEINSDOTTIR, 2006. Otolith shape and temporal stability of spawning groups of Icelandic cod (*Gadus morhua* L.). ICES J. Mar. Sci., **63**: 1501-1512.

KING, M., 2007. Fisheries biology, assessment and management (2$^{nd}$ ed.): i-xiv, 1-382. (Blackwell Publishing Ltd, Oxford).

KODAMA, K., H. SHIRAISHI, M. MORITA & T. HORIGUCHI, 2006. Verification of lipofuscin-based crustacean ageing: seasonality of lipofuscin accumulation in the stomatopod *Oratosquilla oratoria* in relation to water temperature. Mar. Biol., **150**: 131-140.

KUBALLA, A. & A. ELIZUR, 2007. Novel molecular approach to study moulting in crustaceans. Bull. Fish. Res. Ag., **20**: 53-57.

LEE, T. W. & J. S. BYUN, 1996. Microstructural growth in otoliths of conger eel (*Conger myriaster*) leptocephali during the metamorphic stage. Mar. Biol., **125**: 259-268.

MARTIN, J. W., P. JOURHARZADEH & P. H. FITTERER, 1998. Description and comparison of major foregut ossicles in hydrothermal vent crabs. Mar. Biol., **131**: 259-267.

MAUCHLINE, J., 1976. The Hiatt growth diagram for Crustacea. Mar. Biol., **35**: 79-84.

MCGAW, I. J., 2006. Feeding and digestion in low salinity in an osmoconforming crab, *Cancer gracilis*, II: gastric evacuation and motility. J. Exp. Biol., **209**: 3777-3785.

MORGAN, G. J., 1997. Freshwater crayfish of the genus *Euastacus* Clark (Decapoda: Parastacidae) from New South Wales, with a key to all species of the genus. Rec. Aust. Mus., (Supplement) **23**: 1-110.

NEILSON, J. D. & G. H. GEEN, 1982. Otoliths of chinook salmon (*Oncorhynchus tshawytscha*): daily growth increments and factors influencing their production. Can. J. Fish. Aquat. Sci., **39**: 1340-1347.

NOLAN, C. P. & A. CLARKE, 1993. Growth in the bivalve *Yoldia eightsi* at Signy Island, Antarctica, determined from internal shell increments and calcium-45 incorporation. Mar. Biol., **117**: 243-250.

PANFILI, J., H. DE PONTUAL, H. TROADEC & P. J. WRIGHT (eds.), 2002. Manual of fish sclerochronology: 1-463. (IFEMER Publishing, Brest).

PANNELLA, G., 1971. Fish otoliths: daily growth layers and periodical patterns. Science, **173**: 1124-1127.

SCUDAMORE, H. H., 1947. The influence of the sinus gland upon molting and associated changes in the crayfish. Physiol. Zool., **20**(2): 187-208.

SHEEHY, M. R., 1990. Widespread occurrence of fluorescent morphological lipofuscin in the crustacean brain. J. Crust. Biol., **10**(4): 613-622.

SHEEHY, M. R. J., J. G. GREENWOOD & D. R. FIELDER, 1994. More accurate chronological age determination of crustaceans from field situations using the physiological age marker, lipofuscin. Mar. Biol., **121**: 237-245.

SMITH, K. A. & K. DEGUARA, 2003. Formation and annual periodicity of opaque zones in sagittal otoliths of *Mugil cephalus* (Pisces: Mugilidae). Mar. Fresh. Res., **54**: 57-67.

STEWART, J. & J. M. HUGHES, 2007. Age validation and growth of three commercially important hemiramphids in south-eastern Australia. J. Fish Biol., **70**: 65-82.

STREET, P., 1966. The crab and its relatives (2nd ed.): 1-167. (Latimer Trend and Co. Ltd, Plymouth).

THOMPSON, I., D. S. JONES & D. DREIBELBIS, 1980. Annual internal growth banding and life history of the ocean quahog *Artica islandica* (Mollusca: Bivalvia). Mar. Biol., **57**: 25-34.

First received 15 November 2009.
Final version accepted 19 December 2009.

＃ THE EFFECT OF MALE SIZE AND SPERMATOPHORE CHARACTERISTICS ON REPRODUCTION IN THE CARIBBEAN SPINY LOBSTER, *PANULIRUS ARGUS*


BY

MARK J. BUTLER IV[1,3]), JAMIE S. HEISIG-MITCHELL[1]),
ALISON B. MACDIARMID[2]) and R. JAMES SWANSON[1])

[1]) Department of Biological Sciences Old Dominion University, Norfolk,
Virginia 23435, U.S.A.
[2]) National Institute for Water & Atmospheric Research, Wellington, New Zealand


## ABSTRACT


The average size of spiny lobsters (Decapoda; Palinuridae) has decreased dramatically worldwide as a result of the over-fishing of large individuals. Average male size is usually diminished more than that of females because of sexual dimorphism and this can impact reproductive success through sperm limitation. Using laboratory mating experiments and field comparisons of fished and unfished populations, we studied differences in spermatophore characteristics that may influence reproductive success in the Caribbean spiny lobster, *Panulirus argus*, in the Florida Keys, Florida (U.S.A.). We found that large males produce larger spermatophores with more sperm, resulting in operational sperm:egg ratios (range: 21-37:1) that were 40% lower in fished areas. Our experiments show that female mating receptivity is suspended upon receipt of a spermatophore and that this behavior is controlled by a combination of chemical and physical stimuli provided by the spermatophore. The distribution of sperm within spermatophores indicates that the fertilization of multiple clutches from one spermatophore is unlikely, as confirmed by laboratory observations. These results highlight the importance of spermatophore characteristics on fertilization success in spiny lobsters and suggest that reduced male size in the wild may limit reproductive success.


## INTRODUCTION

Among the most universal effects of over-fishing is the depletion of large individuals within exploited populations, and spiny lobsters (Decapoda; Palinuridae) are no exception. Most fisheries for decapod crustaceans include a

---


[3]) Corresponding author; e-mail: mbutler@odu.edu


   New frontiers in crustacean biology: 69-84

minimum size limit to protect non-reproductive individuals, but this focuses fishing effort on larger lobsters. Male lobsters grow larger than females, so when male size is reduced by over-fishing there are potential consequences on the reproductive success of the population. In several crab fisheries, for example, only large males attain the legal size limit and thus only they are extracted from the population (Sainte-Marie, 1993; Paul, 1984; Smith & Jamieson, 1991; Kendall & Wolcott, 1999; Sato et al., 2005, 2006, 2007; Sato & Goshima, 2006). This in turn leads to the increased participation of small males in reproduction (Smith & Jamieson, 1991; Sainte Marie, 1993; Kendall & Wolcott, 1999; Kendall et al., 2002). Similar situations exist for spiny lobster.

In Florida and New Zealand, male spiny lobsters (Caribbean spiny lobster *Panulirus argus* and Red Rock Lobster, *Jasus edwardsii*; respectively) are much larger than females in unfished populations (e.g., males are typically five times the mass of females), but in fished areas males and females are of equivalent size (MacDiarmid, 1989; Bertelsen & Matthews, 2001). When large males are unavailable, female fecundity can plummet (Sainte-Marie, 1993; Kendall et al., 2002; Sato et al., 2005, 2006, 2007; Sato & Goshima, 2006). In fact, when females of some spiny lobster species go unmated they release unfertilized eggs (e.g., *P. argus*), whereas other species (e.g., *J. edwardsii*) forgo egg deposition (MacDiarmid & Butler, 1999). The consequences of these actions range from the loss of a single clutch during one reproductive season in which multiple clutches will be produced (*P. argus*), to a severe reduction in future reproductive success resulting from damage to reproductive organs (*J. edwardsii*).

Lost mating opportunities are not the only reproductive problem faced by female lobsters in fished regions. Decapods may also experience sperm limitation if ejaculate size scales with male body size, mating history, expected female output, or future mating opportunities (reviewed by MacDiarmid & Sainte-Marine, 2006). Indeed, in the spiny lobsters *P. argus* and *J. edwardsii*, male size and spermatophore size are correlated, and spermatophore size explains nearly half of the variance in clutch size (i.e., number of eggs fertilized per clutch) (MacDiarmid & Butler, 1999). Appropriate sperm:egg ratios (S:E) are also crucial for the maintenance of high rates of fertilization, but S:E ratios vary greatly among organisms with external fertilization. In sea lamprey (*Petromyzon marinus*), for example, a S:E ratio of 50 000 provides maximal fertilization rates (Ciereszko et al., 2000), whereas a S:E ratio of 15 000:1 is optimal in African catfish (*Clarias gariepinus*) (Rurangwa et al., 1998). Studies of fertilization in the Crown-of-Thorns starfish, *Acanthaster planci* reveal that fertilization rates can be high over a wide range of S:E ratios but drop precipitously at ratios less than 50. Estimates of S:E in decapod crustaceans range from 70:1 (*C. opilio*; Sainte-Marie & Lovrich, 1994), to "several":1 (*Libinia emarginata*; Hinsch, 1971), to approximately 2.5:1 (*Chionoecetes bairdi*; Paul, 1984). Inappropriately low S:E ratios can reduce the probability of fertilization success to such an extent that the females may delay egg deposition, as occurs in snow crabs when the S:E ratio drops below 7:1 (Sainte-Marie & Lovrich, 1994).

Sufficient access to suitable mates is clearly of prime importance for females and thus has strong selective consequences for female mate choice. In contrast, behaviors or mechanisms that ensure sole or majority paternity of offspring is important for males. Many brachyuran crabs and homarid (clawed) lobster males employ proximate mate guarding to reduce the likelihood of sperm competition and enhance their fertilization success (*C. sapidus*: Jivoff, 1997; *C. opilio*: Sainte-Marie et al., 1997; *H. americanus*: Atema & Voigt, 1995). An indirect means of post-copulatory mate guarding employed by many arthropods is the production of spermatophores that emit chemical signals or provide a physical cue that inhibits copulation (Roth, 1962; Sugawara, 1979). Whether similar mechanisms exist in Palinurid lobsters is unknown.

Given prior evidence that male size influences the reproductive success of spiny lobsters (MacDiarmid & Butler, 1999), we experimentally determined the impact of male size on sperm attributes in the Caribbean spiny lobster. We then examined the distribution of sperm cells within intact and used spermatophores to evaluate whether a single spermatophore can be used to fertilize multiple clutches, and tested whether the deposition of a spermatophore inhibits further mating through chemical or mechanical cues. Finally, we determined the operational S:E ratio of lobster populations in fished and unfished areas to assess the potential for fishing-induced sperm limitation.


## MATERIALS AND METHODS

### Male size and spermatophore characteristics

To determine the impact of male size on sperm and spermatophore attributes, we conducted mating experiments in the laboratory controlling for male and female size. Divers collected male and female lobsters from an unfished marine reserve (Dry Tortugas National Park, Florida, U.S.A.) and from various fished areas near the Florida Keys, Florida (U.S.A.) during February and March of 1999-2001 prior to the reproductive season. The lobsters were transported in aerated live-wells to a laboratory at the Florida Fish and Wildlife Conservation Commission field laboratory in Marathon, FL where all experiments were conducted.

Laboratory mating experiments were designed to simulate the size structure of the reproductively active lobsters in both regions, and thus field mating size structure. Thus, small females (71-90 mm CL) were mated with small males (92-100 mm CL) and large females (95-140 mm CL) were mated with large males (120-155 mm CL). Males and females were kept in separate experimental tanks until approximately 1 week after capture when a single male and five females of the appropriate size class were placed in round experimental tanks (1.75 m diameter; 1500 liter) receiving aerated, filtered seawater from a flow-through system. Seawater temperatures (23-32°C) and photoperiod (12-13.5 hours of daylight) were at ambient conditions and lobsters were fed frozen squid and shrimp ad libitum. The lobsters mated in these tanks, and females were examined daily for the presence of a spermatophore — a paired, external sperm packet deposited by the male on the female's sterum. We used flat forceps to remove intact fresh spermatophores after they had hardened for 24 hours. The spermatophores were weighed to the nearest hundredth of a gram and stored at 5°C in labeled vials with $\sim$10 ml of sterile seawater (filtered to 0.2 $\mu$m) until the number of sperm cells contained within could be derived using the sperm counting procedure described below.

Spermatophores were sliced laterally into thin sections (approximately 0.5 mm) using a scalpel. The sections were placed in a known volume (10-15 ml) of sterile seawater and mechanically shaken for 3 minutes. Aliquots of 9.8 $\mu$l were removed from the shaken sample and placed into the wells of a hemacytometer. The number of sperm cells was then enumerated within each of four subsamples viewed in the hemacytometer. A preliminary evaluation of this sperm liberation technique, where a subset of eight spermatophores were subjected to sequential washings till less than 1% of cells remained in them, indicated that an average of 73% of the sperm within a spermatophore were removed using our technique. However, sperm counts were not adjusted in the formal study because all analyses were comparative. We then determined the weight (in g; measured with a top-loading balance) of the spermatophores, as well as the total sperm number and density of sperm per gram of spermatophore using only data collected from the first spermatophores deposited by each male (n = 28). The relationship between male carapace length and these three spermatophore variables (i.e., weight, sperm number, sperm concentration) was examined in three separate linear regression analyses; all regression assumptions were met, therefore no data transformations were necessary.

## Sperm cell distribution

We examined the distribution of sperm cells within sections of spermatophores removed from females captured in the wild. In March of 1998, divers obtained spermatophore-bearing females from the Florida Keys and the Dry Tortugas National Park. The perimeters of the spermatophores on each female were traced onto acetate sheets and the intact spermatophores removed. The spermatophores were then bisected longitudinally and each resulting half was further divided into four to five sections laterally (fig. 3 inset).

The spermatophore sections were weighed and sperm density and total sperm number were then obtained from each individual section (n = 5) in a similar manner to that described above for whole spermatophores. We compared spermatophore weight, sperm density, and total number of sperm between left and right halves of the spermatophore using a 1-factor randomized block MANOVA. All MANOVA assumptions were met with the raw data, so no transformations were necessary.

## Inhibition of female receptivity

We conducted a laboratory study to examine the influence of spermatophore presence on female receptivity to further mating. Spermatophore-bearing females were obtained by divers from the Florida Keys in May of 1999 and randomly distributed to each of four treatment groups: control, mechanical cue alone, chemical cue alone and a combination of chemical cue and mechanical cue.

For the control treatment, we removed the spermatophores from the females as described above and then reattached them in the same location on each female using a cyanoacrylate adhesive. The mechanical cue treatment was designed to provide a physical representation (stimulus) of a spermatophore without the chemical signals a spermatophore may provide. The mechanical cue was created by replacing spermatophores on females with artificial spermatophores made of silicone that were shaped, hardened in air, and cured overnight in seawater before being glued to the sternum of females. The chemical treatment provided the chemical stimulus of a spermatophore without the physical presence of a spermatophore on the female's sternum. To represent this, we removed the spermatophore from each female's sternum and reattached it with adhesive on the side of the same female's carapace. Presumably, a chemical signal emitted by the spermatophore could be detected by the female via chemosensory aesthetascs on her pereiopods and antennules. To determine if the inhibition of female receptivity was controlled by a combination of chemical and mechanical signals, we devised a treatment that allowed a chemical signal to penetrate the sternum while preventing the detection of chemical stimuli to the female's pereiopods if she stroked the spermatophore. This treatment also provided a mechanical input simulating the presence of a spermatophore on the sternum. The combination treatment "experimental spermatophore" was created by first removing the spermatophore from the female (as described above), cutting away it's dorsal (outer) portion, and then completely covering it's outer surface with an artificial spermatophore made of silicone. Thus creating a bilayered spermatophore with an inner "natural" layer and an outer artifical layer. This was then reapplied back to the sternum of the original female.

Following the application of the spermatophore treatments to the experimental females, one female from each treatment was placed in an experimental mating tank (as previously described) with a single male (i.e., 4 females/male). The lobsters were checked daily to determine the presence of new spermatophores or damage to the manipulated spermatophore. If the lobsters had removed or severely damaged the spermatophore, this was noted and the treatment was reapplied accordingly. Animals were fed frozen squid and shrimp *ad libitum*. Differences in the frequency of mating by females, as determined by the presence of a newly-deposited spermatophore, in the four treatments were to be analyzed using a $2 \times 4$ contingency table analysis (n = 24; 6 per treatment).

## Sperm:Egg ratios

We determined the operational S:E in fished and unfished wild populations of *P. argus* by enumerating the number of sperm in spermatophores found on females obtained by divers from the Florida Keys (n = 62) and the Dry Tortugas National Park (n = 57) during the spring of 2000. The spermatophores were removed from females and the sperm cells were counted as described above. Using an equation developed by Bertelsen and Matthews (2001) for lobsters from these regions, the expected egg output of each female was calculated based upon female carapace length. Based on the preliminary study of the sperm liberation technique outlined above, the S:E ratios were adjusted to compensate for the discrepancy between the number of cells enumerated and the actual number present in the spermatophore. We used a two-sample t-test on log-transformed data to examine the differences in the mean S:E ratios between fished and unfished populations.


## RESULTS

### Male size and spermatophore characteristics

Male size had a significant effect on the weight of the spermatophore and number of sperm cells contained therein (fig. 1). Spermatophore weight increased with increasing male carapace length, and male size explained approximately 63% of the variation in spermatophore weight ($r^2 = 0.6330$; $F = 44.84$; $P < 0.0005$; df $= 1$) (fig. 2a). We also found a significant association ($r^2 = 0.2378$; $F = 8.110$; $P = 0.0085$; df $= 1$) between the number of sperm cells and male size with the total number of sperm cells increasing with increasing male size (fig. 2b). There was, however, no relationship between sperm cell density and male size (fig. 3c; $r^2 = 0.0299$; $F = 0.790$; $P = 0.382$; df $= 1$), perhaps a result of the greater rate of increase of spermatophore weight with male size than the rate associated with sperm number.

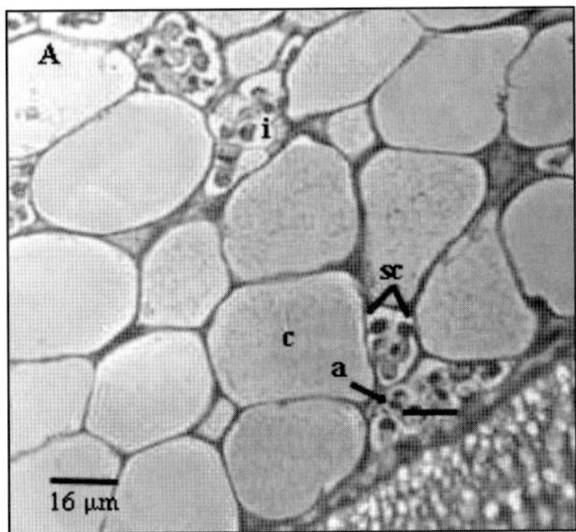

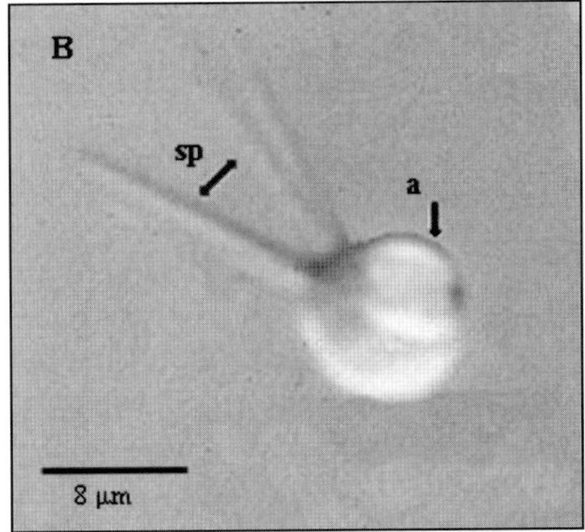


Fig. 1. Micrographs of spermatophore matrix with embedded sperm cells and an individual sperm cell. A, cross-section of primary spermatophore layer demonstrating association of sperm cell clusters with the acellular cavities in *P. argus*: a = acrosome, c = acellular cavity, i = interstices of cavities, sc = sperm cluster. 200×, hematoxylin and eosin staining; B, Nomarski-phase interference micrograph of *P. argus* spermatozoa: a = acrosome, sp = nuclear spikes. 1000×.

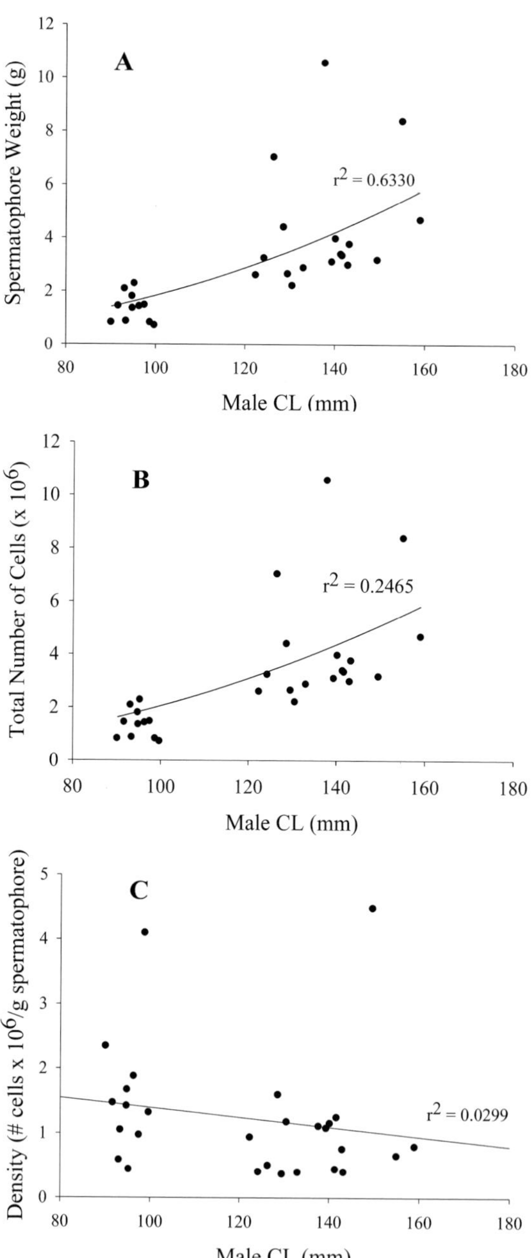

Fig. 2. The effect of male size on three spermatophore attributes. A, spermatophore weight; B, number of sperm cells; C, sperm cell density. Large males (>120 mm CL) were mated with large females (>95 mm CL) and small males (<100 mm CL) were mated with small females (<95 mm CL) (total n = 28) in these laboratory experiments and data on spermatophore attributes were measured for the first mating.

## Sperm cell distribution

An analysis of sperm cell distribution with spermatophores revealed that the sperm cells are concentrated in the mid-posterior region of the spermatophore (fig. 3). There are noticeably fewer sperm cells in the anterior region, the region that often remains after spermatophore utilization at oviposition. Left and right spermatophore halves were not significantly different in weight, number of sperm cells, or sperm density despite the two halves of the spermatophore developing separately in the paired testes (table I; $F = 0.068$; $P = 0.974$; $df = 3$).

## Inhibition of female receptivity

The only females to remate in the experiment where spermatophores were manipulated to test for post-copulatory mate inhibitory cues were those in the mechanical cue and chemical cue treatments (table II); half of the females in these treatments remated. The females assigned to the control treatment (intact spermatophore) and the chemical/mechanical cue combination treatment did not remate.

## Sperm:Egg ratios

The mean S:E ratio in the fished population of lobsters in the Florida Keys was $21.5 \pm 10.6 : 1$ (mean $\pm$ 95% C.I.; $n = 62$). In the unfished lobster population in the Dry Tortugas Marine Reserve, the average S:E ratio was 40% higher ($37.4 \pm 18.3 : 1$; mean $\pm$ 95% C.I.; $n = 57$) (fig. 4), but this difference was not significant due to the large variability ($t = 0.075$, $df = 100$, $P = 0.941$). The lowest S:E ratios were similar in both populations, with several females in each population having an S:E ratio less than 5:1. However, only 3% of the females in the fished population had a S:E ratio greater than 50:1, in contrast to 23% of the females in the unfished population. The mean S:E ratio from females from fished populations that we collected with only partially eroded spermatophores was substantially lower than that seen in intact spermatophores from the same population of animals, averaging $3.8 \pm 3.2 : 1$ (mean $\pm$ 95% C.I., $n = 8$).

## DISCUSSION

The impetus for this study was a concern that a reduction in the size of male spiny lobsters in populations subject to fishing could adversely affect

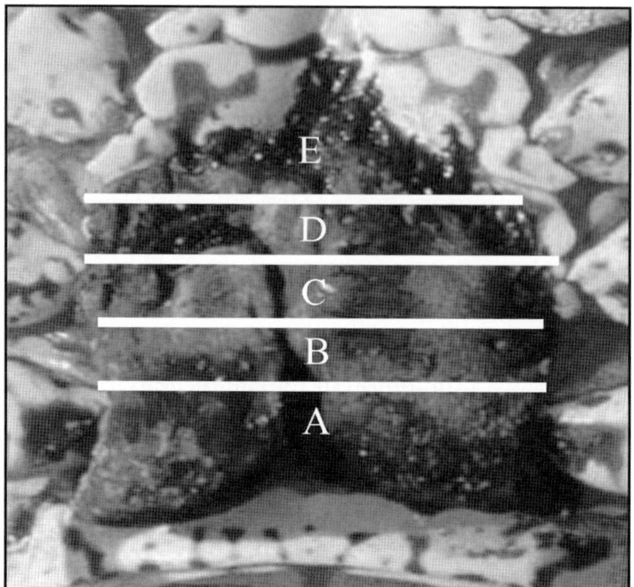


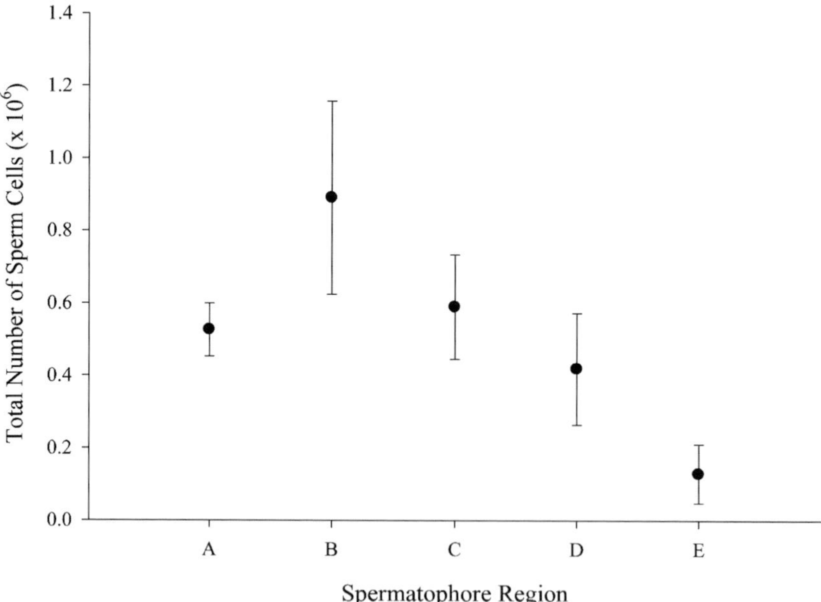

Fig. 3. Sperm cell abundance and distribution within the spermatophore of *P. argus*. Top: photo of a spermatophore attached to the ventral surface of the carapace of a female lobster showing the relative orientation of the regions sampled (A–E). Each region was roughly equal in area. Bottom: the number of sperm cells per region (mean $\pm$ 1 standard error of the mean).

TABLE I

Results of a MANOVA examining the difference between the left and right spermatophore halves for each of three spermatophore attributes: spermatophore weight, number of sperm, and sperm density. Pillai's Trace, Wilks' Lambda, Hotelling's Trace and Roy's Largest Root were calculated in the MANOVA. As all test statistics provided identical results, only those results associated with the Wilks' Lambda are displayed

|  | Hypothesis df | Error df | F Value | Wilks' Lambda | P |
|---|---|---|---|---|---|
| Spermatophore Half | 3.000 | 4.000 | 0.068 | 0.951 | 0.974 |

TABLE II

Results of experiment examining the role of spermatophores in controlling female mating. Whether female *P. argus* remated or not in four spermatophore manipulation treatments is shown

| Experimental Outcome | Spermatophore Manipulation Treatment | | | |
|---|---|---|---|---|
|  | Control | Chemical Cue | Chemical+ Mechanical Cue | Mechanical Cue |
| Female Remated | 0 | 3 | 0 | 3 |
| Female Did Not Remate | 6 | 3 | 6 | 3 |

fertilization success, hence the realized fecundity of exploited populations. Our results indicate that small males indeed deliver smaller quantities of sperm and smaller quantities of spermatophore matrix, that consequently may lead to lower S:E ratios in fished populations. Females are unable to counter a loss of large males by using a single spermatophore more than once, and we found no evidence that females mate more than once per clutch.

Our finding that male size significantly impacts spermatophore weight and the number of sperm cells transferred to female *P. argus* in Florida is consistent with findings by MacDiarmid & Butler (1999), who observed a positive relationship between spermatophore area and male size in both *P. argus* and the southern temperate rock lobster *J. edwardsii*. They also noted that spermatophore area varied more for large males, who apportioned spermatophores of different sizes to females of different sizes: large spermatophores to large females and vice versa. However, they did not measure spermatophore weight or enumerate sperm. Nor did Mauger (2001) when demonstrating that male size in *J. edwardsii* was positively correlated with vas defrens weight and with ejaculate recharge rate. The results for spiny lobsters contrast with findings in studies of the crabs *C. opilio*, *C. bairdi*, and *C. sapidus*, which show that

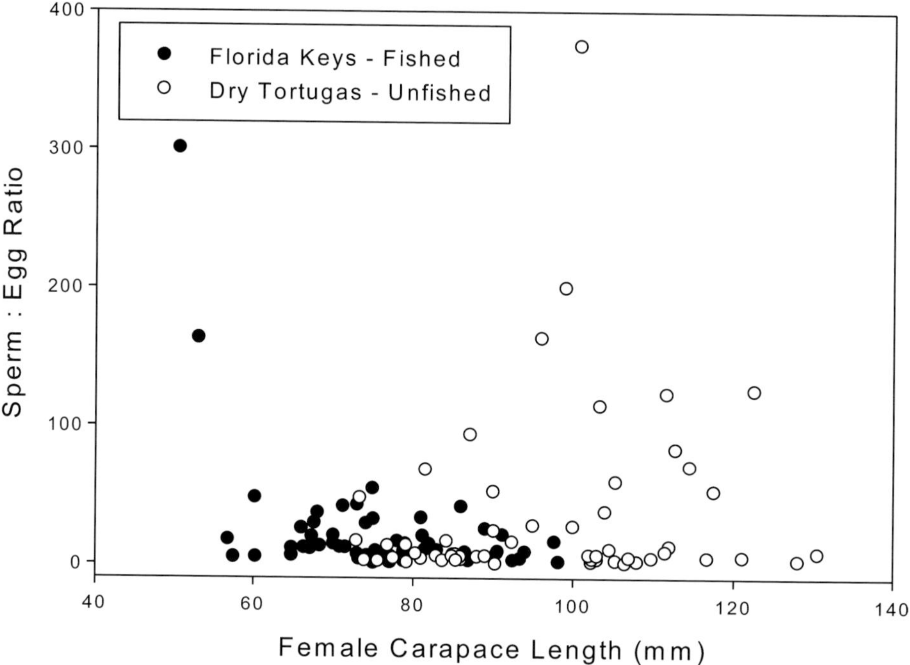


Fig. 4. Operational sperm:egg ratios plotted against female size (carapace length; mm) for *P. argus* collected from fished (closed circles; Florida Keys; n = 62) and unfished populations (open circles; Dry Tortugas; n = 57).

the number of sperm cells received by females is independent of either male and/or female size (Sainte-Marie & Lovrich, 1994; Kendall et al., 2002).

Although female *P. argus* mate many times in their lifetime, often more than once a season if they are large, our results suggest that only one spermatophore can be used to fertilize a single clutch of eggs. There are accounts of female *P. argus* possessing multiple spermatophores layered one atop another (e.g., Mota-Alves & Paiva, 1976), but our observations indicate that these instances simply represent the deposition of a second spermatophore over a previously used but incompletely removed one. When females fertilize their eggs, they use their legs to scratch open and often completely remove the posterior portion of the spermatophore to release the stored sperm, leaving an "eroded spermatophore" with the anterior-most region of the spermatophore intact. Eroded spermatophores are often found alone or beneath newly deposited spermatophores on females in the field, but it is unlikely that an eroded spermatophore could be used to successfully fertilize a second clutch. Sperm cells are evenly distributed among left and right halves of the spermatophore (similar to observations in both crabs and other lobsters; Sainte-Marie &

Lovrich, 1994; Mauger, 2001) and the majority (~80%) are located in the middle to posterior region of the spermatophore, leaving relatively few cells in the anterior region — the only portion that remains in eroded spermatophores.

Unlike the situation in brachyuran crabs and clawed lobsters where females mate multiple times and store sperm before fertilization (Sainte-Marie & Lovrich, 1994; Aiken & Waddy, 1980), our experiments show that the presence of an intact spermatophore inhibits further mating by females. That experiment also suggests that inhibition requires sternal contact from a chemical in the spermatophore.

The S:E ratios that we observed in *P. argus* (generally <40:1) are lower than those seen in most other species and >50% lower in heavily exploited populations subject to fishing compared to an unexploited population. In broadcast spawners, one would expect higher S:E ratios to combat the effect of dilution because fertilization declines with decreasing sperm concentration in the sea (e.g., Levitan et al., 1991; Benzie & Dixon, 1994; Tvedt et al., 2001). For example, a study on the fertilization kinetics in sea urchins revealed optimal fertilization rates at a S:E ratio of 72 000:1 (Levitan et al., 1991), whereas some starfish can tolerate greater dilutions of sperm, with S:E ratios as low as 50:1 (Benzie & Dixon, 1994). Studies of fertilization success in broadcast-spawning marine fish indicate that maximal fertilization occurs within a S:E ratio range of $9 \times 10^5$ to $5 \times 10^4$ (Tvedt et al., 2001; Ciereszko et al., 2000).

Because mating by *P. argus* involves a direct transfer of gametes, one assumes that the S:E may be lower than that of broadcast spawners. Yet, in humans, 200-300 million spermatozoa are deposited per ejaculate to fertilize a single egg, although only 300-500 of these survive to reach the site of fertilization (Sadler, 1990). In vivo studies of laboratory rodents also indicate that the S:E ratio at the site of fertilization is often only 1:1 (Gomendio et al., 1998). Gomendio and colleagues (1998) speculate that the high initial S:E ratios in these organisms are designed to combat low survival rates encountered in hostile reproductive tracts. During fertilization, decapod sperm encounter neither extreme dilution nor an environmentally hostile internal reproductive tract. Thus, it is not surprising that decapod S:E ratios are consistently low, with ratios varying from about 70:1 or lower in snow crabs (Sainte-Marie & Lovrich, 1994, 1999) to <5:1 in spider and tanner crabs (Hinsch, 1971; Paul, 1984). The high S:E ratio estimates obtained in a few in vitro fertilization studies of crabs and lobsters are probably an artifact, because the vast majority of the fertilized eggs observed in those studies were polyspermic and not incubated long enough to obtain cleavage, thus yielded unreliable assessments of successful fertilization (Gomendio et al., 1998; Yanagimachi, 1994; Talbot et al., 1991).

## CONCLUSIONS

There is growing evidence that over-fishing of decapod crustaceans can lead to reduced reproductive success, not only because of smaller female size but also due to sperm-limitation, which may ensue when small males come to predominate in the population (MacDiarmid & Saint-Marie, 2006). In not all situations, however, has intense exploitation resulted in sperm-limitation of fecundity, apparently because of differences among species in spermatophore characteristics and mating dynamics that in some instances may mitigate the effects of fishing on male size. Detailed investigations of the reproductive attributes and mating behavior of individuals in fished and un-fished areas are necessary if we are to discover why some decapods are more susceptible to fishery-induced sperm limitation of fecundity than others. What is clear, however, is that the role of male size in ensuring fertilization success, once viewed as demographically irrelevant, is indeed important to Palinurid population viability.

## ACKNOWLEGEMENTS

We are appreciative of the laboratory and field assistance provided by: D. Behringer, S. Donahue, J. Goldstein, E. Ricelet and D. Robertson. John Hunt of the Florida Fish and Wildlife Conservation Commission graciously afforded us access to their laboratory in Marathon, FL for much of this work. This research was supported by a grant to M. Butler from The National Science Foundation (No. INT-9418306).

## REFERENCES

AIKEN, D. E. & S. L. WADDY, 1980. Reproductive biology. In: J. S. COBB & B. F. PHILLIPS (eds.), The biology and management of lobsters, 1: 215-276. (Academic Press, New York).
ATEMA, J. & R. VOIGT, 1995. Behavior and sensory biology. In: J. R. FACTOR (ed.), Biology of the lobster, *Homarus americana*: 313-348. (Academic Press, New York).
BENZIE, J. A. H. & P. DIXON, 1994. The effects of sperm concentration, sperm: egg ratio, and gamete age on fertilization success in crown-of-thorn starfish (*Acanthaster planci*) in the laboratory. Biol. Bull., **186**: 139-152.

BERTELSEN, R. D. & T. R. MATTHEWS, 2001. Fecundity dynamics of female spiny lobster (*Panulirus argus*) in a south Florida fishery and Dry Tortugas National Park lobster sanctuary. Mar. Freshw. Res., **52**: 1559-1565.

CIERESZKO, A., J. GLOGOWSKI & K. DABROWSKI, 2000. Fertilization in landlocked sea lamprey: storage of gametes, optimal sperm: egg ratio, and methods of assessing fertilization success. J. Fish. Biol., **56**: 495-505.

GOMENDIO, M., A. H. HARCOURT & E. R. S. ROLDAN, 1998. Sperm competition in mammals. In: T. R. BIRKHEAD & A. P. MØLLER (eds.), Sperm competition and sexual selection: 667-756. (Academic Press, San Diego).

GOUDEAU, H. & M. GOUDEAU, 1986. Electrical and morphological responses of the lobster egg to fertilization. Develop. Biol., **114**: 325-335.

— — & — —, 1989. A long-lasting electrically mediated block, due to the egg membrane hyperpolarization at fertilization, ensures physiological monospermy in eggs of the crab *Maia squinado*. Develop. Biol., **133**: 348-360.

HINSCH, G. W., 1971. Penetration of the oocyte envelope by spermatozoa in the spider crab. J. Ultrastruc. Res., **35**: 86-97.

JIVOFF, P., 1997. The relative roles of predation and sperm competition on the duration of the post-copulatory association between the sexes in the blue crab, *Callinectes sapidus*. Behav. Ecol. Sociobiol., **40**: 175-185.

KENDALL, M. S. & T. G. WOLCOTT, 1999. The influence of male mating history on male–male competition and female choice in mating associations in the blue crab, *Callinectes sapidus* (Rathburn). J. Exp. Mar. Biol. Ecol., **239**: 23-32.

KENDALL, M. S., D. L. WOLCOTT, T. G. WOLCOTT & A. H. HINES, 2002. Influence of male size and mating history on sperm content of ejaculates of the blue crab, *Callinectes sapidus*. Mar. Ecol. Prog. Ser., **230**: 235-240.

LANGLOIS, T. J., M. J. ANDERSON & R. C. BABCOCK, 2005. Reef-associated predators influence adjacent soft-sediment communities. Ecology, **86**: 1508-1519.

LEVITAN, D. R., M. A. SEWELL & F. CHIA, 1991. Kinetics of fertilization in the sea urchin *Strongylocentrotus franciscanus*: interaction of gamete dilution, age and contact time. Biol. Bull., **181**: 371-378.

LIPCIUS, R. N., 1985. Size-dependent reproduction and molting in spiny lobsters and other long-lived decapods. In: A. M. WENNER (ed.), Factors in adult growth. Crustacean Issues, **3**: 129-148. (A. A. Balkema, Boston).

MACDIARMID, A. B., 1989. Size at onset of maturity and size-dependent reproductive output of female and male spiny lobsters *Jasus edwardsii* (Hutton) (Decapoda: Palinuridae) in northern New Zealand. J. Exp. Mar. Biol. Ecol., **127**: 229-243.

MACDIARMID, A. B. & M. J. BUTLER IV, 1999. Sperm economy and limitation in spiny lobsters. Behav. Ecol. Sociobiol., **146**: 14-24.

MACDIARMID A. B. & B. SAINTE-MARIE, 2006. Reproduction. In: B. PHILLIPS (ed.), Lobsters: biology, management, aquaculture and fisheries: 45-77. (Blackwell Publishing, Oxford).

MAUGER, J. W., 2001. Sperm depletion and regeneration in the spiny lobster *Jasus edwardsii*. (Unpublished M.Sc. Thesis, University of Auckland).

MOTA-ALVES, M. I. & M. P. PAIVA, 1976. Frequencia de acasalamentos em lagostas do genero *Panulirus* White (Decapoda, Palinuridae). Arquiv. Ciênc. Mar., **16**: 61- 63.

PAUL, A. J., 1984. Mating frequency and viability of stored sperm in the tanner crab *Chionoecetes bairdi* (Decapoda, Majidae). J. Crust. Biol., **4**: 375-381.

ROBLES, C. D., R. SHERWOOD-STEVENS & M. ALVARADO, 1995. Responses of a key intertidal predator to varying recruitment of its prey. Ecology, **76**: 565-579.

ROTH, L. M., 1962. Hypersexual activity induced in females of the cockroach *Nauphoeta cinerea*. Science, **138**: 1267-1269.

RURANGWA, E., I. ROELANTS, G. HUYSKENS, M. EBRAHIMI, D. E. KIME & F. OLLEVIER, 1998. The minimum effective spermatozoa: egg ratio for artificial insemination and the effects of mercury on sperm motility and fertilization ability in *Clarias gariepinus*. J. Fish Biol., **53**: 402-413.

SADLER, T. W., 1990. Langman's medical embryology (6th ed.): i-xii, 1-411. (Williams and Wilkins Baltimore).

SAINTE-MARIE, B., 1993. Reproductive cycle and fecundity of primiparous and multiparous female snow crab, *Chionoecetes opilio*, in the Northwest Gulf of Sainte Lawrence. Can. J. Fish. Aquat. Sci., **50**: 2147-2156.

SAINTE-MARIE, B. & G. A. LOVRICH, 1994. Delivery and storage of sperm at first mating of female *Chionoecetes opilio* (Brachyura: majidae) in relation to size and morphometric maturity of male parent. J. Crust. Biol., **14**: 508-521.

SAINTE-MARIE, B., J.-M. SÉVIGNY & Y. GAUTHIER, 1997. Laboratory behavior of adolescent and adult males of the snow crab (*Chionoecetes opilio*) (Brachyura: Majidae) mated noncompetitively and competitively with primiparous females. Can. J. Fish. Aquat. Sci. **54**: 239-248.

SAINTE-MARIE, G. & B. SAINTE-MARIE, 1999. Reproductive products in the adult snow crab (*Chionoecetes opilio*). II. Multiple types of sperm cells and of spermatophores in the spermathecae of mated females. Can. J. Zool., **77**: 451-462.

SATO, T., M. ASHIDATE, T. JINBO & S. GOSHIMA, 2006. Variation of sperm allocation with male size and recovery rate of sperm numbers in spiny king crab *Paralithodes brevipes*. Mar. Ecol. Prog. Ser., **312**: 189-199.

— —, — —, — — & — —, 2007. Does male-only fishing influence reproductive success of female spiny king crab, *Paralithodes brevipes*? Can. J. Fish. Aquat. Sci., **64**: 735-742.

SATO, T., M. ASHIDATE, S. WADA & S. GOSHIMA, 2005. Effects of male mating frequency and male size on ejaculate size and reproductive success of female spiny king crab, *Paralithodes brevipes*. Mar. Ecol. Prog. Ser., **296**: 251-262.

SATO, T. & S. GOSHIMA, 2006. Impacts of male-only fishing and sperm limitation in manipulated populations of an unfished crab, *Hapalogaster dentata*. Mar. Ecol. Prog. Ser., **313**: 193-204.

SMITH, B. D. & G. S. JAMIESON, 1991. Possible consequences of intensive fishing for males on mating opportunities of Dungeness crabs. Trans. Amer. Fish. Soc., **120**: 650-653.

SUGAWARA, P., 1979. Stretch reception in the bursa copulatrix of the butterfly *Pieris rapae crucivora*, and its role in behavior. J. Comp. Physio., **130**: 191-199.

TALBOT, P., W. POOLSANGUAN, B. POOLSANGUAN & H. AL HAJJ, 1991. In vitro fertilization of lobster oocytes. J. Exp. Zool., **258**: 104-112.

TVEDT, H. B., T. J. BENFEY, D. J. MARTIN-ROBICHAUD & M. POWER, 2001. The relationship between sperm density, spermatocrit, sperm motility and fertilization success in Atlantic Halibut *Hippoglossus hippoglossus*. Aquaculture, **194**: 191-200.

YANAGIMACHI, R., 1994. Mammalian fertilization. In: E. KNOBIL & J. D. NEILL (eds.), The physiology of reproduction, **1**: 189-318. (Raven Press, New York).


First received 2 November 2009.
Final version accepted 20 December 2009.

# FISHERIES MANAGEMENT OF THE SNOW CRAB, *CHIONOECETES OPILIO*, OFF KYOTO PREFECTURE IN THE WESTERN SEA OF JAPAN, WITH EMPHASIS ON ITS RESOURCE RECOVERY

BY


ATSUSHI YAMASAKI[1])

Kyoto Prefectural Agriculture, Forestry & Fisheries Technology Research Center,
Fisheries Technology Research Department


## ABSTRACT


Since its inception, the snow crab *Chionoecetes opilio* in the western Sea of Japan have been overfished by the Danish seine fleet. The landings have decreased markedly since 1970, in spite of various restrictions such as a shorter fishing season, an increase in the minimum legal size of landed crabs, and a reduction of the maximum landed catch per trip. The failure of the stock to recover has been complicated by the inability of the Danish seine fleet to avoid capture of undersized crabs during both the crab and the flounder fishing seasons. To reduce the fishing pressure, six preserved areas totaling $67.8 \text{ km}^2$, equivalent to about 4.4% of the total crab fishing grounds, were established off Kyoto Prefecture between 1983 and 2007. Artificial reefs made of concrete blocks were constructed in each of the six areas. Furthermore, to avoid bycatch of the crabs during the closed fishing season, fishing at depths of 220-350 m has been prohibited from September to October since 1979, and at depths of 230-350 m from April to May since 1994. An improved seine net, with a separator panel, has proven effective for retaining the target flounder species and yet allows most crabs to escape. All fishermen of Kyoto Prefecture have used this improved seine net for flounder fishing since 2003 and there has been a marked recovery of crab catches in subsequent years.


## INTRODUCTION

Snow crab, *Chionoecetes opilio* O. Fabricius, 1788, are distributed in depths between 200 and 450 m in the western Sea of Japan (Kuwahara et al., 1995) where it is the most important resource for the Danish seine fleet. Landings have decreased substantially since 1970 despite various management restrictions, such as: a shorter fishing season; an increase of the minimum

---


[1]) e-mail: a-yamasaki20@pref.kyoto.lg.jp





New frontiers in crustacean biology: 85-94

legal size; and reduction of the maximum catch per trip. Also, the bycatch of snow crab in other fisheries has contributed to the overfishing of this species. A number of management strategies were implemented by fishermen in the western Sea of Japan in the 1980's; and the resource subsequently began to recover (Yamasaki, 2002). This report provides selected examples of fishery management measures that were applied to restore the snow crab resource in the waters off Kyoto Prefecture.

## REGULATIONS OF SNOW CRAB FISHERY

Danish seines are the principal gear used to fish snow crab in the western Sea of Japan. The fishery is regulated at two levels. The first level has been part of the jurisdiction of the Ministry of Agriculture, Forestry and Fisheries since 1955, and specify the timing of the fishery, the categories of crab that can be landed, and the total allowable catch (TAC) since 1997. The second level of regulation consists of voluntary measures put in place by the fishermen. The fishermen are members of the Special Committee for the Snow Crab Fishery in the Sea of Japan, which has been in existence since 1964. At this level of regulation, the previous measures are restated and the permissible trip limits are determined (table I). In addition, the permissible landings of soft-shelled males and females are determined and they vary according to the number of days at sea on a specific trip.

TABLE I

Regulations for snow crab fishery by Danish seine fleet in the western Sea of Japan

| Management item | | | Contents | |
|---|---|---|---|---|
| Fishing season | Male | Hard-shelled | : 6 Nov.–20 Mar. | |
| | | Soft-shelled | : 21 Dec.–20 Mar. | |
| | Female | | : 6 Nov.–10 Jan. | |
| Prohibited categories of crab from fishing | Male | | Less than 90 mm CW | |
| | Female | | Immature and adult with orange-colored eggs | |
| Maximum catch per trip (number of individual) | Male | Soft-shelled | <24 hrs./trip | : <1000 |
| | | | 24-48 hrs./trip | : <2000 |
| | | | >48 hrs./trip | : <3000 |
| | Female | | <24 hrs./trip | : <6000 |
| | | | 24-48 hrs./trip | : <10 000 |
| | | | >48 hrs./trip | : <20 000 |

## FISHERY MANAGEMENT IN THE WATERS OFF KYOTO PREFECTURE

A number of studies were conducted, and methods were examined and proposed to fishermen, to determine how to effectively manage the snow crab resource. Meetings were then held with fishermen to encourage the adoption of each new management methods. The strategies voluntarily adopted were: to avoid overfishing; the establishment of zero-catch sanctuaries in 1983; and the landing catch of soft-shelled males was prohibited in 2008. Generally, only hard-shelled, terminally molted male crabs participate in reproduction, therefore the last management greatly reduces the landing of crabs that have not yet mated. In addition, the unit price of soft-shelled males in fish markets is less than 1/10 of hard-shelled males. In eastern Canada, Canadian Fisheries Resource Conservation Council (Anonymous, 2005) proposed that the harvesting soft-shelled males be prohibited because this practice is a waste of the resource and represents a significant threat to the conservation of snow crab stocks.

To avoid bycatch of snow crabs in the closed season, areas where all fishing is prohibited were established in 1979, and an improved Danish seine net (Miyajima et al., 2007) has been in use since 2003. Since then, similar measures have been adopted by other prefectures in the western Sea of Japan.

## ESTABLISHMENT OF A PRESERVED AREA, "SANCTUARY"

To aid in rehabilitation of the snow crab resource, six preserved areas were established between 1983 and 2007 (fig. 1). Together, they represent 67.8 km$^2$ or 4.4% of the total crab fishing grounds. The preserved areas are located in depths between 235 and 300 m in the waters off Kyoto Prefecture. Between 57 and 579 concrete blocks were placed on the sea floor in each preserved area to obstruct fishing with Danish seines. The concrete blocks were cube-shaped, 3.25 m to a side, and weighted 13.0 tons. Because the sea floor is covered with soft sediment (silt and clay) (Hayashi & Kiyono, 1984), each concrete block was carefully lowered to the bottom. Concrete blocks were spaced about 250 m apart in the outer and about 300 m apart in the inner area. The concrete blocks placed in 1983 remained in good condition based on observations taken in 1989 by the deep-sea research submersible (Yamasaki, 1990). Many concrete blocks in the preserved area could be clearly identified, and the height of each block could be estimated by echosounder as about 3 m.

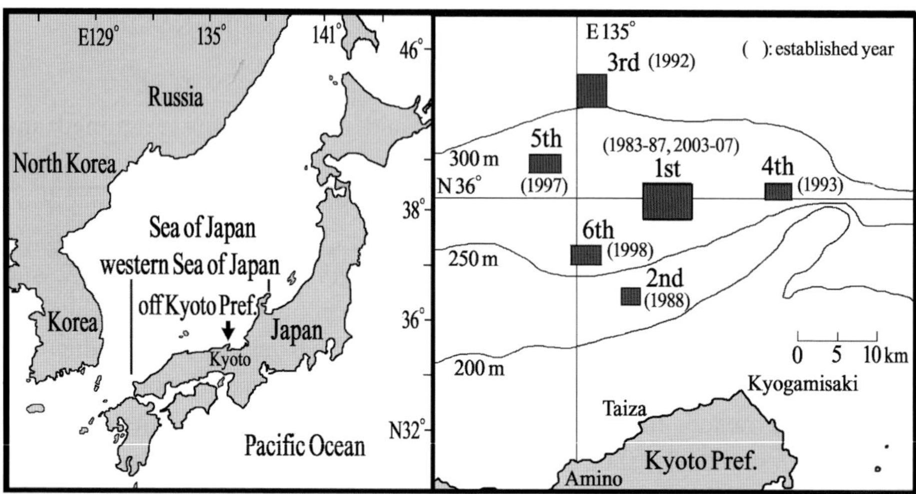



Fig. 1. Locations of preserved areas for snow crabs off Kyoto Prefecture in the western Sea of Japan.

In order to examine the population density of males and females in the first preserved area and in the surrounding area between 260 and 280 m depth, samples were collected between August and October from 1984 to 2003 during the closed crab season, using conical traps (Sinoda et al., 1987). The traps used were 82 cm in diameter on the upper surface, 130 cm on the lower surface, 43 cm high, and had a 42 cm diameter opening at the top. Twelve or twenty-four traps were attached at 50 m intervals to a ground line during the years 1984 to 1988. Sixteen or fifty traps were attached at 100 m intervals to a ground line during the years 1991 to 2003. The catch per unit effort (CPUE per 10 traps) of males >90 mm carapace width (CW) and adult females were calculated and compared between the preserved area and the surrounding area (fig. 2). The CPUE of males ranged between 8 and 128 crabs (mean = 55.3) in the preserved area and 4 and 46 crabs (mean = 16.4) in the surrounding area. The CPUE of females ranged between 5 and 515 crabs (mean = 134.1) in the preserved area and 9 and 112 crabs (mean = 48.0) in the surrounding area. The CPUE of males and females was significantly higher in the preserved area than in the surrounding area.

Mortality rates were estimated from recaptures of tagged crabs released both inside and outside the preserved area. The tagged crabs were recaptured by the commercial fleet on fishing grounds located outside the preserved area. Yamasaki (2002) reported that the survival rate of crabs released in the preserved area was 0.26 year$^{-1}$ for males and 0.43 year$^{-1}$ for females.

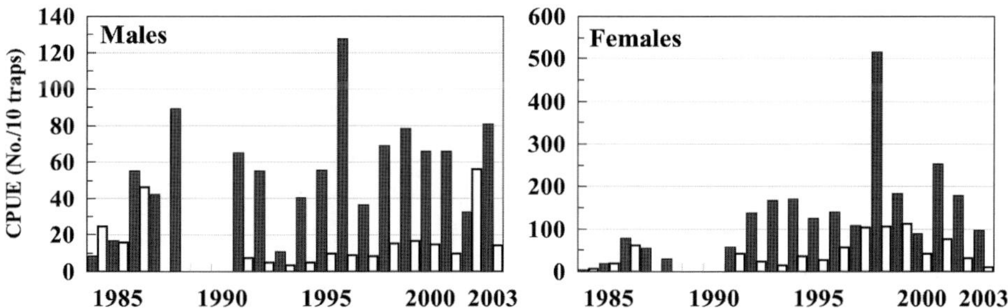

Fig. 2. Fluctuations in CPUE (per 10 traps) for males and females. Solid and open columns indicate inside and outside the first preserved area, respectively.

The survival rate for crabs released outside the preserved area was lower at 0.20 year$^{-1}$ for males and 0.31 year$^{-1}$ for females.

The egg hatching and subsequent mating of multiparous females occurs in winter, coincident with the fishing season (Ito, 1967; Kon, 1980). Based on crab-trap catches, functionally mature males and multiparous female form mating aggregations in the first preserved area (Yamasaki et al., 1994) and this was the reason that preserved area was selected. However, it is important to afford some projection to all lifecycle stages, particularly undersized crabs. It is well known that the areas inhabited vary with the growth stage (Kon, 1980). Therefore, the other five preserved areas were established in places where juvenile and young crabs, soft-shelled crabs are mainly distributed. Evidence to date suggests this exclusion of Danish seiners from about 4.4% of the fishing grounds has contributed to the replenishing of the snow crab stock (Yamasaki, 2002).

The establishment of a snow crab preserved areas in the waters off Kyoto Prefecture was a test case for this management technique. Consequently, several such areas have since been established elsewhere in the western Sea of Japan.

## PROHIBITED FISHING AREA DURING THE CLOSED SEASON OF THE SNOW CRAB

The Danish seine fleet targets species other than snow crab. There are spring, autumn, and winter fishing seasons, and from June to August no fishing is permitted for any species (fig. 3). In most years, fishing for demersal species (e.g., as the flathead flounder *Hippoglossoides dubius* Schmidt, 1904 and northern shrimp *Pandalus eous* Makarov, 1935) occurs in spring and autumn

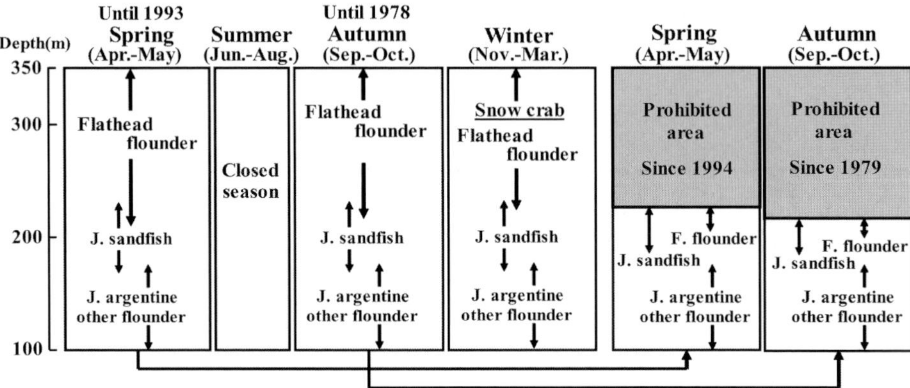


Fig. 3. The seasonal shift in fishing activity by depth and target fish for the Danish seine fleet and prohibited fishing areas in spring and autumn seasons off Kyoto Prefecture.

at depths deeper than 200 m. Consequently, a large number of crabs are swept into the net and discarded along with other inedible animals. Therefore, the bycatch and survival rates of these crabs were assessed for spring and autumn seasons of 1989 and 1990 in the waters off Kyoto Prefecture. On average, 76 to 1673 crabs were discarded per haul during spring compared with 125 to 591 crabs per haul during autumn (Yamasaki, 1994). There was a particularly large number of discarded crabs at depths between 230 and 260 m. The male crabs were between 40 and 150 mm CW, and from 40 to 90 mm CW for females (Yamasaki, 1994). The survival rate of crabs, that were carefully discarded was 0.87-0.99 in the spring season but dropped to 0-0.15 in the autumn (Yamasaki, 1994). It is reasonable to assume survival rates were lower for crabs discarded during commercial operations. The poor autumn survival is most likely a function of the difference between surface and bottom temperatures and the damage sustained by soft shell crabs due to crushing.

Because it is essential to reduce the bycatch of snow crabs during the closed season, fishing at depths of 220 to 350 m have been closed during the autumn since 1979 and between 230 and 350 m depths during the spring since 1994 (fig. 3). The number of discarded crabs decreased as a direct effect of these prohibitions. Tagged female crabs were released on these fishing grounds before and after the establishment of a prohibited fishing area in the spring. Natural mortality coefficients of females before and after the establishment of the prohibition were estimated at 0.716 and 0.241-0.439 year$^{-1}$, respectively (Yamasaki et al., 2001).

## DEVELOPMENT AND USE OF THE IMPROVED SEINE NET

An improved Danish seine net, which reduced the catch of non-target species, was developed to address the issue of immature crabs being caught and discarded. To create the improved net, the traditional Danish seine was modified by adding a separator panel. Specifically, the improved seine net has double floor-nets. The upper floor-net consists of a slope net, a separator panel, and a partition. The lower floor-net has outer (fig. 4). Opening mesh sizes of the slope net, separator panel net, and partition net are approximately 50, 600, and 37 mm, respectively. The length of the outlet is approximately 2.3 m. Generally, fish such as flounder swim along the belly of the separator panel and enter the codend. Species with limited to no swimming ability (e.g., snow crab, sea stars, and small prawns) tend to pass though the mesh of the separator panel net onto the lower panel and then are swept along to the outlet and are able to escape.

To assess the selectivity of the improved seine net, test fishing with a cover net over the seine was carried out in the waters off Kyoto Prefecture in 2003-2004. On average, 86% (range: 74-98%) of the snow crabs passed through the net while 23% (range: 12-33%) of the flounder were lost (Miyajima et al., 2007). Thus the improved seine net met the goal of reducing the bycatch with an acceptable reduction in catches of the target species. All Danish seine fishermen in Kyoto Prefecture have been using the improved seine net for flathead flounder fishery since 2003.

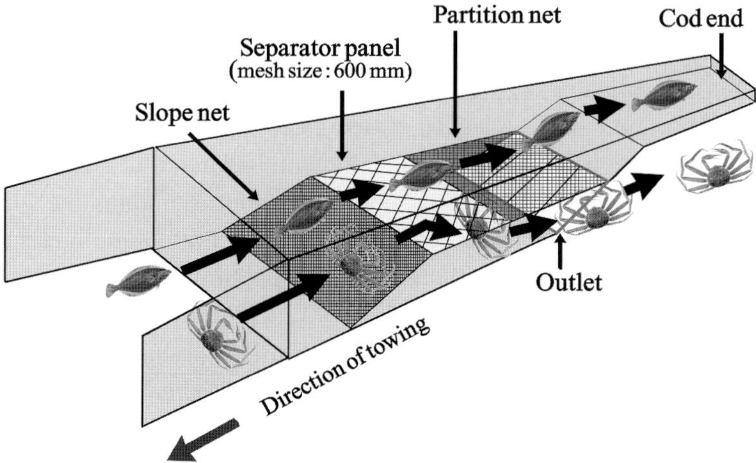


Fig. 4. An illustration of the improved seine net used in the flounder fishery.

Secondary benefits to reducing the bycatch of crab in the flounder fishery included: a reduction by half of time spent sorting the catch; less damage to the flounder from contact with the hard carapace of the crab; and an increase in per unit value of the flounder due to the reduced number and extent of injuries to the fish.

## POSSIBLE EFFECT OF THE MANAGEMENT STRATEGIES

A similar pattern of overfishing has clearly occurred elsewhere such as in the snow and tanner crabs fisheries in the eastern Bering Sea. Typically, there is an initial burst of landings, followed by an abrupt decline, and a moderate recovery with landings stabilizing at approximately 25% less than the initial peak (Anonymous, 2008). These stocks were declared as overfished in 1999 and severe harvesting restrictions were applied reducing the harvest rate to 15-20% of the population biomass (Anonymous, 2008). In eastern Canada, based on the sudden decline observed in the late 1980's, a series of strict management measures were necessary, and enforced, to reverse the effects of overfishing on the snow crab stocks (Anonymous, 2005). There is no exception for the western Sea of Japan snow crab fisheries and progressive effort for recovery of the stocks has been deployed in the last decade.

The effects for the various management efforts can be seen in the annual catch of snow crabs from fishing grounds off Kyoto Prefecture (fig. 5). First, there was the establishment of prohibited fishing areas during the autumn in 1979, the creation of the first preserved area in 1983 followed by a progressive establishment of the other five preserved areas, and then the spring-time closed fishing areas established in 1994. The trend of decreasing catches in the 1980's came to an end and the catch was maintained at a low level. However, the catch has since been increasing gradually with annual fluctuations. Recent catches are approximately 130 tons per year, about 2.5 times the minimum catch in 1980. The recovery of catch sizes off Kyoto Prefecture in recent years is regarded as the synergistic effects of the establishment of preserved areas during the crab fishery and additional closed areas during the off-season for the crab. However, more detailed analysis is still necessary to elucidate the direct benefit of each new management strategies to the snow crab population by dissociating the effects of new management from natural fluctuations in the population size. In the meantime, the current effort for rebuilding overfished snow crab stocks merits continuing.

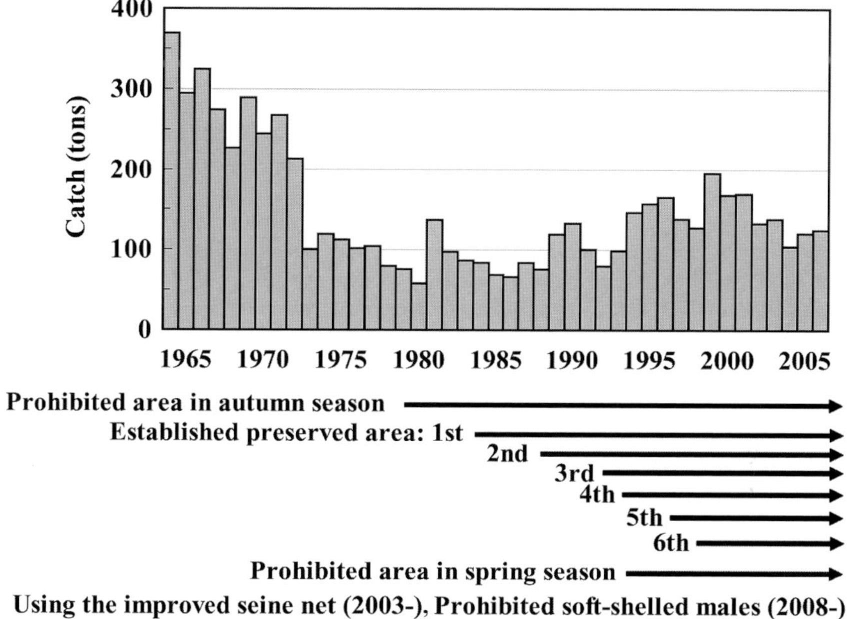



Fig. 5. Annual catch of snow crabs and various voluntary management strategies employed by fishermen off Kyoto Prefecture.

## ACKNOWLEDGMENTS

The author wishes to thank Drs. M. Moriyasu, M. Hanson and C. Ferron for their helpful discussion and critical reading of the manuscript. The author also thanks the researchers of the Fisheries and Marine Research Division, Fisheries Technology Department, Kyoto Prefectural Agriculture, Forestry and Fisheries Technology Center, and the captain and crew of the R/V Heian-maru, of this center, for their assistance in the field work. This paper is contribution No. 164 from this center.

## REFERENCES


ANONYMOUS, 2005. Strategic conservation framework for Atlantic snow crab. (Fs 158-1/2005E, Fisheries Resource Conservation Council, Ottawa Ontario, Canada).
— —, 2008. Stock assessment and fishery evaluation report for the king and tanner crab fisheries of the Bering Sea and Aleutian Islands Region. (North Pacific Fishery Management Council, Anchorage, AK, U.S.A.).
HAYASHI, I. & S. KIYONO, 1984. Macrobenthos in and offshore of Wakasa Bay in the Japan Sea. Mem. Coll. Agri. Kyoto Univ., **123**: 1-26.
ITO, K., 1967. Ecological studies on the edible crab, *Chionoecetes opilio* in the Japan Sea-1. Bull. Japan Sea Reg. Fish. Lab., **17**: 67-84.

KON, T., 1980. Studies on the life history of the zuwai crab, *Chionoecetes opilio* (O. Fabricius). Spec. Publ. Sado Mar. Biol. Sta. Niigata Univ., **2**: 1-64. [In Japanese.]

KUWAHARA, A., M. SINODA, A. YAMASAKI & S. ENDO, 1995. Management of the snow crab resource in the western Sea of Japan. Fish. Res. Lib., **44**: 2.

MIYAJIMA, T., A. IWAO, N. YAGISHITA & A. YAMASAKI, 2007. Bycatch exclusion of snow crab using separator panel in seine net for flounder fishery off Kyoto Prefecture. Nippon Suisan Gakkaishi, **73**: 8-17.

SINODA, M., T. IKUTA & A. YAMASAKI, 1987. On changing the size selectivity of fishing gear for *Chionoecetes opilio* in the Japan Sea. Nippon Suisan Gakkaishi, **53**: 1173-1179.

YAMASAKI, A., 1990. Ecological observations of the snow crab, *Chionoecetes opilio*, in the preserved area in the sea off Kyoto Prefecture. JAMSTECTR Deepsea Res., **6**: 335-340.

— —, 1994. Studies on stock management of snow crab *Chionoecetes opilio* based on biology. Bull. Kyoto Inst. Oceanic Fish. Sci. Spec. Rep., **4**: 1-53.

— —, 2002. Establishment of preserved area for snow crab *Chionoecetes opilio* and consequent recovery of the crab resources. Fish. Sci., (Supplement) **68**: 1699-1702.

YAMASAKI, A., S. OHKI & E. TANAKA, 2001. Estimation of mortality coefficient of adult female snow crab *Chionoecetes opilio* from tag recoveries in the waters off Kyoto Prefecture. Nippon Suisan Gakkaishi, **67**: 244-251.

First received 27 October 2009.
Final version accepted 7 December 2009.

# REVIEW OF THE CURRENT STATUS OF THE SNOW CRAB *CHIONOECETES OPILIO* (O. FABRICIUS, 1788) FISHERIES AND BIOLOGICAL KNOWLEDGE IN EASTERN CANADA

BY

MIKIO MORIYASU[1])

Department of Fisheries and Oceans, Oceans and Science Branch, Aquatic Resources Division, Snow Crab Section, Gulf Fisheries Centre, 343 Université Avenue, Moncton, New Brunswick, E1C 5K4 Canada

## ABSTRACT

In Atlantic Canada, the snow crab, commercial fishery began in the mid-1960's in the southern Gulf of St. Lawrence and rapidly expanded to cover fishing grounds in the northern Gulf of St. Lawrence and Atlantic shore from southern Nova Scotia to Labrador. These fishing areas are currently divided into 61 management zones with total landings at approximately 93 000 tons in 2008. This is a male only trap fishery with a minimum size limit of 95 mm carapace width. The historical fluctuation of landings and/or biomass estimates in certain fisheries in eastern Canada showed a cyclic-like pattern with a 4-6 year ascending period followed by a 4-6 year descending period. Eastern Canada's snow crab fisheries are deemed to be fully exploited.

The biological knowledge of the species has rapidly progressed since the finding of a terminal molt in male snow crab in 1986. Since then, numerous studies have been conducted on a wide range of subjects. In this review, the development of snow crab fishery and the life history of snow crab in eastern Canada are described and the perspective of the eastern Canadian snow crab stocks is discussed.

## INTRODUCTION

The snow crab, *Chionoecetes opilio* (O. Fabricius, 1788) (Brachyura: Majoidea), is a subarctic species commonly found in the northwestern Atlantic (from Greenland to the Gulf of Maine), the northern Pacific, the Bering Sea, the Arctic Ocean and the Sea of Japan. In the Atlantic Ocean, the snow crab is the only *Chionoecetes* species present, while in the Sea of Japan and the Pacific

---

[1]) e-mail: mikio.moriyasu@dfo-mpo.gc.ca



New frontiers in crustacean biology: 95-107"

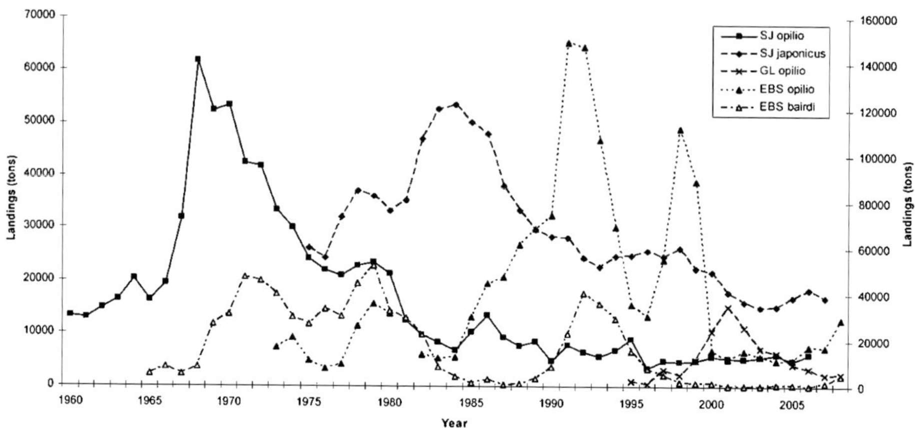

Fig. 1. Historical landings of *Chionoecetes* fisheries in the eastern Bering Sea (EBS *opilio* & EBS *bairdi*), in the Sea of Japan (SJ *opilio* & SJ *japonicus*), in the western Greenland (GL *opilio*). Right ordinate for EBS *opilio* and *bairdi*, left ordinate for SJ *opilio* & *japonicus*, & GL *opilio*.

Ocean four other species and one subspecies of *Chionoecetes* can be found. *C. opilio elongatus* Rathbun, 1924 and *C. japonicus* Rathbun, 1932 are present in the Sea of Japan. The distribution of *C. bairdi* Rathbun, 1924 extends from the Bering Sea to California and that of *C. angulatus* Rathbun, 1924 from the Bering Sea to Oregon. *C. tanneri* Rathbun, 1893 is found along the coasts of British Columbia and Washington and Oregon states. The historical landing records of these species have shown a typical pattern of boom and bust (fig. 1). The snow crab fisheries in the eastern Canada were not an exception. The fishing crisis in the Canadian fishery during the late 1980's brought demands for actions from all industries and the Department of Fisheries and Oceans (DFO) in developing new management approaches to rebuild the fishery. The history of these fisheries and advancement of biological knowledge on snow crab in eastern Canada is reviewed.

## HISTORY OF ATLANTIC SNOW CRAB FISHERIES

Currently eastern Canada's snow crab fisheries are managed by four administrative regions, northern Gulf of St. Lawrence (nGSL) with eight management areas and a co-management area between Newfoundland and Labrador (NL), southern Gulf of St. Lawrence (sGSL) with four management areas, eastern Nova Scotia (eNS) with four management areas, and NL with 44 management areas (fig. 2).

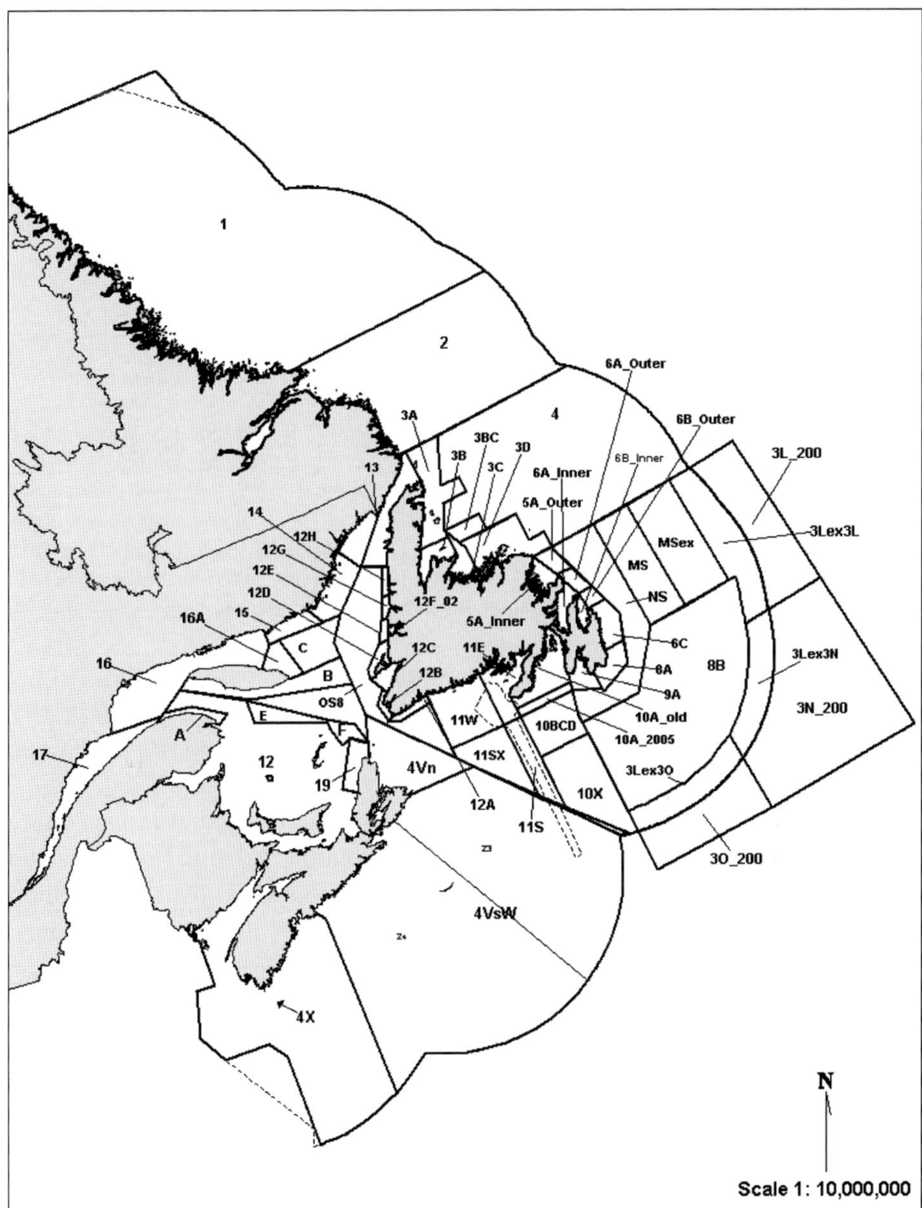


Fig. 2. Snow crab, *Chionoecetes opilio* (O. Fabricius), fishery management zones in eastern Canada.

Snow crab landings in Canada (fig. 3) were first reported from the Gaspé area in the Gulf of St. Lawrence between 1960 and 1966. These landings were mainly by-catches from groundfish trawlers (Hare & Dunn, 1993). In 1965,

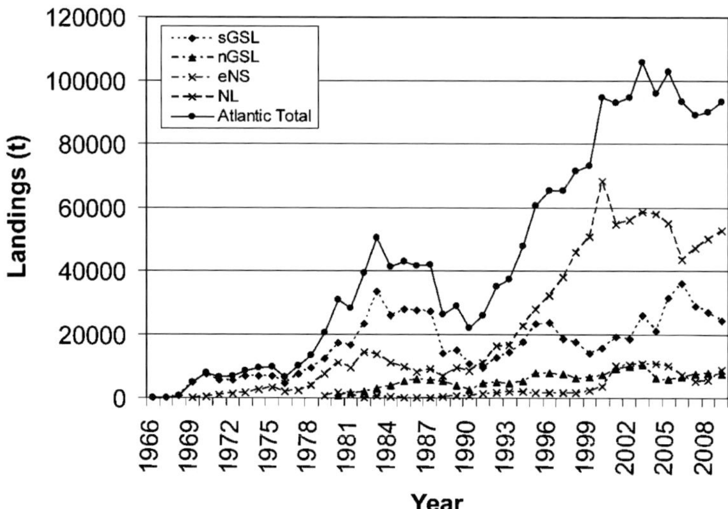


Fig. 3. Historical landings of *Chionoecetes opilio* (O. Fabricius) in eastern Canada (nGSL: northern Gulf of St. Lawrence, sGSL: southern Gulf of St. Lawrence, eNS: eastern Nova Scotia, NL: Newfoundland and Labrador).

a Danish seine fishery for snow crab was started off western Cape Breton Island, which landed about 6.8 t (Hare & Dunn, 1993). The snow crab fishery in NL commenced in 1968 in Trinity Bay. New fisheries off western Cape Breton (sGSL) and off eNS began in the late 1970's.

Through the 1970s, the fishery in eastern Canada continued to expand along the north shore of the nGSL and spread throughout the Atlantic Provinces and Quebec. By 1974, all fisheries indicators already pointed to over exploitation (Hare & Dunn, 1993). Through the 1970's, intensive discussions continued with the goal of setting a sound regulatory framework for the sGSL snow crab fishery (Elner, 1982; Hare & Dunn, 1993; Anonymous, 2005), which eventually resulted in the basis of the current fishing regulations: i.e., prohibition of landing females, minimum size limit at 95 mm carapace width (CW), minimum mesh size of 130 mm, the setting of TACs (Total Allowable Catch), and a control rule for reducing the catch of soft shell crab. In 1979, the total landings in eastern Canada reached 30 800 t (fig. 3) with 358 vessels, i.e., 28 in nGSL (Dufour, 1995), 160 in sGSL, 119 in eNS and 51 in NL (Elner, 1982).

The snow crab fishery has grown steadily to become the economic mainstay of a number of regions on Canada's Atlantic coast. In 1982, landings in eastern Canada totaled 48 300 t (fig. 3) with a landed value exceeding $40 million. The number of licenses (including exploratory and temporary permits) has continuously increased through the 1980's to reach 1178 in 1989 (136, 246,

102, and 694 for the nGSL, sGSL, eNS and NL, respectively). The catch began to decrease in all fisheries in the late 1980s to 22 400 t in 1989, especially in the sGSL, when an abrupt decrease in landings from 27 000 t to 14 000 t was observed in 1989. This sharp decline occurred before the implementation of TACs in most crab fishing areas.

Since then, snow crab landings in eastern Canada have increased to 95 000 t in 1999 (fig. 3). Notable increases in landings occurred off eNS and off NL. In both fisheries, the landings increased 6-8 fold and reached 3700 t and 69 100 t (the highest historical landing record), respectively. In 1999, the number of licenses (including temporary and exploratory permits) reached 4288 (136, 325, 164, and 3663 for the nGSL, sGSL, eNS and NL, respectively).

Through the 2000's, three of four fisheries in eastern Canada have experienced their highest annual landings (fig. 3). In the nGSL, the landings reached a historical high in 2002 at 10 000 t with an average landing through the 2000's of 7800 t. In the sGSL, the landings reached a historical high in 2005 at 36 000 t but have been continuously decreasing since, with average annual landings through the 2000's at 26 000 t. The higher landings observed through the 2000's in the sGSL were due to higher harvesting rates applied to the biomass estimates compared to the previous decade when lower harvesting rates were applied to much higher biomass estimates (Hébert et al., 2009). In NL, an average landing through the 2000's was at 53 000 t. In eNS, a snow crab bottom trawl survey introduced in 1997 showed unexpected amounts of available resource and the fishery quota was increased accordingly commencing in 2000, which resulted in a significant landing increase (highest historical landings at 10 900 t in 2002). The average landings through the 2000's were at 8800 t. In both the NL and eNS fisheries, expansion of bottom covered by favorably cold water, decrease in predator (groundfish) abundance, and timely larval flow from other locations may have contributed to the significant increase in crab availability. In 2008, the number of licenses (including exploratory permits) reached 4368 (199, 482, 202, and 3485 for the nGSL, sGSL, eNS and NL, respectively).

Additional restrictions have been recently added, e.g., partial closure of the fishery in case of high incidence of newly molted crabs, and restraints on fishing seasons. In certain cases, weekly harvest limits or vessel trip limits were also introduced.

The biomass estimates in the sGSL showed a cyclic-like fluctuation pattern with 4-6 years of strong recruitment to the population followed by 4-6 years of lean recruitment period. This recruitment pattern is apparently common to many (if not all) snow crab populations in the North Atlantic (Sainte-Marie et al., 2008) with some exceptions in NL (E. Dawe, pers. comm.). Eastern Canada's snow crab fisheries are deemed to be fully exploited.


## LIFE CYCLE OF SNOW CRAB IN ATLANTIC CANADA

The original description of life cycle of snow crab in the northwestern Atlantic (Anonymous, 1990) was updated based on the current biological knowledge of the species, especially since the publication of the existence of a terminal molt in male snow crab by Conan & Comeau (1986).


### Distribution and habitat

In the northwest Atlantic, snow crabs live on muddy bottoms at depths between 20 and 1400 m in NL (Dawe & Colburn, 2002), but mostly from 70 to 280 m and are distributed from Greenland to the Gulf of Maine (Squires, 1990). Snow crab is a stenothermic species (Foyle et al., 1989) and has been observed in water ranging from $-1.4°C$ (Lovrich et al., 1995) to $9.7°C$ (Powles, 1968), but is most commonly found at bottom temperatures of $<3°C$ (Tremblay, 1997). Bottom temperature has been suggested to play an important role in its life history events (Sainte-Marie et al., 2008). There seems to be a thermal limit for snow crab around $7°C$ and at temperatures greater than $12°C$ adult snow crab stop feeding (Foyle et al., 1989).


### Growth

Male and female snow crabs are characterized by a conspicuous sexual size/age dimorphism (Sainte-Marie et al., 2008). For the males, there are three stages: i.e., immature, when reproductive organs do not function; adolescent, when reproductive organs do function and a slight change in chela-CW relationship occurs (Comeau & Conan, 1992) but claws are not clearly enlarged; and adult, when claws are clearly differentiated after a terminal molt (Conan & Comeau, 1986). The molt frequency depends on water temperature. Instars II-VI protract intermolt periods spanning 8-10 months and they molt approximately once a year after. It takes at least 8.7 to 11 years to reach commercial size at 95 mm CW (Sainte-Marie et al., 1995). Throughout its adolescent phase, a crab may skip a regular molting season (called 'skip molter' (Hébert et al., 2002)). Male crabs may go through terminal molt at a wide range of sizes (Conan & Comeau, 1986) varying from 40 (Sainte-Marie

& Hazel, 1992) to 150 mm CW (Bailey & Elner, 1989). The life expectancy of terminally molted males can reach up to 7.7 years and the reproductive value may be highest at 2-5.5 years both after terminal molt (Fonseca et al., 2008).

In females, the immature stage is characterized by a narrow abdomen and no detectable ovaries; the prepubescent has ovaries that begin to develop and, adult when their abdomen gets broader and they have the ability to reproduce. Females become sexually mature at their last (terminal) molt, mostly at CW between 30 to 95 mm. Average size at maturity may be temperature dependent, and within cohort, larger females may mature earlier than smaller females (Alunno-Bruscia & Sainte-Marie, 1998).

## Reproduction

There are two types of mating (Moriyasu & Conan, 1988), i.e., mating of post-molted pubescent female (nulliparous) which become primiparous after extruding the first clutch of fertilized eggs; and mating of multiparous female (already spawned at least once). The mating season occurs from January to mid-March after terminal molt for the former, and in late-May to early June following egg hatching for the latter. In certain areas, a spring migration of multiparous mating pairs toward shallow waters is reported (Sainte-Marie et al., 2008). A male guards a female that is in preparation for molting (pubescent female) or egg hatching (multiparous female) (Sainte-Marie et al., 2008). The competition for females may be fierce and males and females often become injured in the process. Females may be quite selective when choosing a mating partner (Sainte-Marie et al., 2008). It is evident that a soft-shell adolescent male cannot mate, but whether a hard-shell adolescent male (either skip molted or molted in much earlier date than normal molting season) can mate appears to depend on female reproductive stage (Elner & Beninger, 1992, 1995). The complexity and plasticity of the polygynandrous mating system of snow crab is detailed by Sainte-Marie et al. (2008).

A female can produce from 20 000 to 150 000 eggs depending on her size (Anonymous, 1990). Primiparous females are up to about 32% less fecund than multiparous females at constant CW (Sainte-Marie, 1993). The duration of embryonic development is 2 years for multiparous female and 2.3 years for primiparous female. However, the bottom water temperature influences the developmental process and a continuously warmer temperature ($\geqslant 1.8°C$) may shorten the development duration to 1 year (Moriyasu & Lanteigne, 1998). After the larvae are released, the female may mate again or use the stored sperm to fertilize future clutches of eggs (Conan & Comeau, 1986; Sainte-Marie & Carrière, 1995). Larval hatching occurs in April-June phased with the sedimentation peak of phytoplankton particles originating from the surface spring bloom with a lag of 2-3 weeks (Starr et al., 1994). After spring hatching, snow crabs go through a larval phase for 12-20 weeks then settle to the bottom (M. Biron, pers. comm.).

## Natural mortality

Density dependent intra-cohort cannibalism by instar V against instar I was reported by Sainte-Marie & Lafrance (2002). Cannibalism rates were significantly lower by larger males (50-120 mm CW) than by males of 8-50 mm CW (Lovrich & Sainte-Marie, 1997) and females are more frequently cannibalistic than males (Squires & Dawe, 2003). This size-dependent cannibalism may serve as an intrinsic density-dependent mechanism for adjustment of recruitment periodicity (Sainte-Marie et al., 1996). Dutil et al. (1997) observed a peak cannibalism activity by large males (>110 mm CW) on prey sizes of 30-37 mm CW.

Bitter crab disease (BCD) caused by a hemo-parasitic dinoflagellate of the genus *Hematodinium* commonly infects snow crab off NL. This fatal disease occurs predominantly in recently molted crabs of both sexes and intermediate-sized crab (Dawe, 2002). Until now there is no confirmed record of infected crabs other than off Newfoundland.

## PERSPECTIVE

In all *Chionoecetes* fishery in the world, there is a similarity in the fluctuation pattern of its historical landings. In general the landings gradually and continuously increase to the historical peak due to industrial development and increase in fishing technology, then followed by an abrupt decrease in landings due to a lack of timely sound management strategy. Unexpected decline of landings results in setting more precautionary harvesting approaches and consequently the landings start to increase and peak again but not as high as the first peak level. Then the subsequent oscillations of landings reach its peaks only at the 10-25% level compared to the first peak and remain low for longer period. In some fisheries more serious strategies are applied to rebuild the depressed stock, e.g., Eastern Bering Sea (EBS) *C. bairdi* stock (Anonymous,

2008), and *C. opilio* stock off Kyoto Prefecture in the Sea of Japan (A. Yamasaki, pers. comm.). However, the recovery of these stocks to a reasonable level has not yet been seen.

The snow crab stocks in eastern Canada are not exempt of this overfishing threat. Although more rigorous stock management regimes were introduced to the eastern Canadian snow crab stocks since the stock decline in the late 1980's, fishing effort increase has never been ceased due to an increase in the availability of snow crab stock over continental shelf off NL as well as the discovery of high abundance of crab off eNS without comprehensive harvesting rules such as EBS snow and Tanner crab harvesting rules (Anonymous, 2008). In fact in the sGSL snow crab stock for which an accurate biomass estimates are available for the last two decades, there seems to be an imminent threat of stock depression due to the recent aggressive harvesting practice. This stock has already experienced a competitive harvest practice which brought this stock down to a sudden fall of landings in the later 1980's followed by a precautionary stock rebuilding approach by applying a stock exploitation rate of 20-40% (corrected for natural mortality) throughout the 1990's. However, the recent aggressive harvesting with the exploitation rate varying between 45 and 70% seems to be detrimental to the stock that already had suffered from overfishing through the 1980's. For this stock, alarming tendencies in some stock parameters have been observed, e.g., a continuous decrease in total mature female abundance and population fecundity since the late 1980's, 5-folds less abundance of new recruitment to the population which will be a major component of the commercial stock towards 2015 compared to the level of the previous recruitment pulses observed in 1993-1997 (Hébert et al., 2006).

Recent advancement on snow crab biology revealed their complex mating system and shed some light onto the possible process of male only harvesting towards over-fishing by alternation of population reproductive potential. Sainte-Marie et al. (2008) concluded that a large-male only fishery may modify the intensity of male competition, shift the balance of fitness from large to small adult males, resulting in sperm limitation, and reduce female opportunity for male choice, even causing female mortality during the mating process. What have been observed in the sGSL stock since the 1980's may insinuate such consequences of aggressive harvesting of the stock, although no clear relationship between large male harvesting practice and decline of female abundance and its reproductive potential has been demonstrated for the stock in question.

There are many challenges for establishment of proper management of the eastern Canadian snow crab stocks. Outcomes of given stock management regime can only be evaluated after 10-20 years due to the complexity of stock fluctuation pattern and a long duration of recruitment process. In such a case, economical and social demands often prevail over stock conservation needs. An establishment of comprehensive harvesting rules such as one deployed for EBS crab species by scrutinizing different harvesting regimes historically deployed and its results is also urged for the eastern Canadian snow crab stocks.

## ACKNOWLEDGMENTS

I thank Dr. A. Burmeister (Greenland Institute of Natural Resources), M. Biron, Dr. J. Choi, D. Collister, E. Dawe, M. Hébert, J. Lambert, T. Perry, Dr. B. Sainte-Marie, L. Sonsini, D. White, B. Zisserson (DFO, Canada), Dr. A. Yamasaki (Kyoto Prefectural Agriculture, Forestry & Fisheries Technology Research Center, Japan), & Dr. I. Yosho (Japan Sea Regional Fisheries Research Laboratory) for their valuable information, M. Biron, E. Dawe, M. Hébert, & Dr. B. Sainte-Marie for their review of the earlier version of the manuscript, and T. Dykens (DFO) for her help in preparation of the manuscript.

## REFERENCES

ANONYMOUS, 1990. Underwater world, the Atlantic snow crab. (DFO/4339, UW/6, Communication Directorate, Department of Fisheries and Oceans, Ottawa, Ontario).
— —, 2005. Strategic conservation framework for Atlantic snow crab. (FRCC.05.R1, Fisheries Resource Conservation Council, Ottawa, Ontario, Canada).
— —, 2008. Stock Assessment and fishery evaluation report for the King and Tanner crab fisheries of the Bering Sea and Aleutian Islands Regions. (Compiled by The Plan Team for the King and Tanner Crab Fisheries of the Bering Sea and Aleutian Islands, North Pacific Fishery Management Council, Anchorage, AK, U.S.A.).
ALUNNO-BRUSCIA, M. & B. SAINTE-MARIE, 1998. Abdomen allometry, ovary development, and growth of female snow crab, *Chionoecetes opilio* (Brachyura, Majidae), in the northwestern Gulf of St. Lawrence. Can. J. Fish. Aquat. Sci., **55**: 459-477.
BAILEY, R. F. J. & R. W. ELNER, 1989. Northwest Atlantic snow crab fisheries: lessons in research and management. In: J. F. CADDY (ed.), Marine invertebrate fisheries: their assessment and management: 261-280. (John Wiley and Sons, New York).
CHABOT, D., B. SAINTE-MARIE, K. BRIAND & J. M. HANSON, 2008. Atlantic cod and snow crab predator-prey relationship in the Gulf of St. Lawrence, Canada. Mar. Ecol. Prog. Ser., **363**: 227-240.
COMEAU, M. & G. Y. CONAN, 1992. Morphometry and gonad maturity of male snow crab, *Chionoecetes opilio*. Can. J. Fish. Aquat. Sci., **49**: 2460-2458.

COMEAU, M., M. STARR, G. Y. CONAN, G. ROBICHAUD & J.-C. THERRIAULT, 1999. Fecundity and duration of egg incubation for multiparous female snow crabs (*Chionoecetes opilio*) in the fjord of Bonne Bay, Newfoundland. Can. J. Fish. Aquat. Sci., **56**(6): 1088-1095.

CONAN, G. Y. & M. COMEAU, 1986. Functional maturity and terminal molt of snow crab, *Chionoecetes opilio*. Can. J. Fish. Aquat. Sci., **49**: 2460-2468.

CONAN, G. Y., M. STARR, M. COMEAU, J.-C. THERRIAULT, F. X. MAYNOU, I. HERNÀNDEZ & G. ROBICHAUD, 1996. Life history strategies, recruitment fluctuations, and management of the Bonne Bay fjord Atlantic snow crab (*Chionoecetes opilio*). In: B. BAXTER (ed.), Proceedings of the International Symposium on the Biology, Management and Economics of Crab from High Latitude Habitats. Lowell Wakefield Fish. Sym. Ser., Alaska Sea Grant College Program Rep., **96-02**: 59-97.

DAWE, G. E., 2002. Trends in prevalence of bitter crab disease caused by *Hematodinium* sp. in snow crab (*Chionoecetes opilio*) throughout the Newfoundland and Labrador continental shelf. In: A. J. PAUL, E. G. DAWE, R. ELNER, G. S. JAMIESON, G. H. CRUSE, R. S. OTTO, B. SAINTE-MARIE, T. C. SHIRLEY & D. WOODBY (eds.), Proceedings of the International Symposium on the Biology, Management and Economics of Crab from High Latitude Habitats: 385-400. (AK-SG-02-01, Lowell University of Alaska Sea Grant, Fairbanks).

DAWE, G. E. & E. B. COLBOURNE, 2002. Distribution and demography of snow crab (*Chionoecetes opilio*) males on the Newfoundland and Labrador Shelf. In: A. J. PAUL, E. G. DAWE, R. ELNER, G. S. JAMIESON, G. H. CRUSE, R. S. OTTO, B. SAINTE-MARIE, T. C. SHIRLEY & D. WOODBY (eds.), Proceedings of the International Symposium on the Biology, Management and Economics of Crab from High Latitude Habitats: 577-594. (AK-SG-02-01, Lowell University of Alaska Sea Grant, Fairbanks).

DUFOUR, R., 1995. Le crabe des neiges de l'estuaire et du nord du Golfe Saint-Laurent: État des populations en 1994. (DFO Atl. Fish. Res. Doc. 95/96).

DUTIL, D., J. MUNRO & M. PÉLOQUIN, 1997. Laboratory study of the influence of prey size on vulnerability to cannibalism in snow crab (*Chionoecetes opilio* O. Fabricius, 1780). J. Exp. Mar. Biol. Ecol., **212**: 81-94.

ELNER, R. W., 1982. Overview of the snow crab *Chionoecetes opilio* fishery in Atlantic Canada. In: B. MELTEFF (ed.), Proceedings of the International Symposium on the Genus *Chionoecetes*. Alaska Sea Grant College Program, Univ. Alaska, Fairbanks, Alaska, Alaska Sea Grant Rep., **82-10**: 4-19.

ELNER, R. W. & P. B. BENINGER, 1992. The reproductive biology of snow crab, *Chionoecetes opilio*: a synthesis of recent contributions. Am. Zool., **32**: 524-533.

— — & — —, 1995. Multiple reproductive strategies in snow crab, *Chionoecetes opilio*: physiological pathways and behavioral plasticity. J. Exp. Mar. Biol. Ecol., **193**: 93-112.

FONSECA, D. B., B. SAINTE-MARIE & F. HAZEL, 2008. Longevity and change in shell condition of adult male snow crab *Chionoecetes opilio* inferred from dactyl wear and mark-recapture data. Trans. Amer. Fish. Soc., **137**: 1029-1043.

FOYLE, T. P., R. K. O'DOR & R. W. ELNER, 1989. Energetically defining the thermal limits of the snow crab. J. Exp. Biol., **145**: 371-393.

HARE, G. M. & D. L. DUNN, 1993. A retrospective analysis of the Gulf of St. Lawrence snow crab (*Chionoecetes opilio*) fishery 1965-1990. In: L. S. PARSONS & W. H. LEAR (eds.), Perspective on Canadian marine fisheries management. Can. Bull. Fish. Aquat. Sci., **226**: 177-192.

HÉBERT, M., K. BENHALIMA, G. MIRON & M. MORIYASU, 2002. Moulting and growth of male snow crab, *Chionoecetes opilio* (O. Fabricius, 1788) (Decapoda, Majidae), in the southern Gulf of St. Lawrence. Crustaceana, **75**(5): 671-702.

HÉBERT, M., E. WADE, M. BIRON, P. DeGRâCE, R. SONIER & M. MORIYASU, 2009. The 2008 assessment of snow crab, *Chionoecetes opilio*, stocks in the southern Gulf of St. Lawrence (Areas 12, 19, E and F). (Can. Sci. Adv. Secr., Res. Doc. 2009/53).

HÉBERT, M., E. WADE, T. SURETTE, P. DeGRâCE, R. RUEST & M. MORIYASU, 2006. The 2005 assessment of snow crab, *Chionoecetes opilio*, stocks in the southern Gulf of St. Lawrence (Areas 12, 19, E and F). (Can. Sci. Adv. Secr., Res. Doc. 2006/29).

LOVRICH, G. A. & B. SAINTE-MARIE, 1997. Cannibalism in the snow crab, *Chionoecetes opilio* (O. Fabricius) (Brachyura: Majidae), and its potential importance to recruitment. J. Exp. Mar. Biol. Ecol., **211**: 225-245.

LOVRICH, G. A., B. SAINTE-MARIE & B. D. SMITH, 1995. Depth distribution and seasonal movements of *Chionoecetes opilio* (Brachyura: Majidae) in Baie Sainte-Marguerite, Gulf of St. Lawrence. Can. J. Zool., **73**: 1712-1726.

MORIYASU, M. & G. Y. CONAN, 1988. Aquarium observations on mating behavior of snow crab, *Chionoecetes opilio*. (Int. Counc. Explor. Sea, CM 1988/K:9).

MORIYASU, M. & C. LANTEIGNE, 1998. Embryo development and reproductive cycle in the snow crab, *Chionoecetes opilio* (Crustacea: Majidae), in the southern Gulf of St. Lawrence, Canada. Can. J. Zoology, **76**(11): 2040-2048.

POWLES, H., 1968. Distribution and biology of the spider crab, *Chionoecetes opilio*, in the Magdalen Shallows, Gulf of St. Lawrence. Fish. Res. Brd. Canada, Manuscr. Rep. Ser., **997**: 1-106.

SAINTE-MARIE, B., 1993. Reproductive cycle and fecundity of primiparous and multiparous female snow crab, *Chionoecetes opilio*, in the northwest Gulf of St. Lawrence. Can. J. Fish. Aquat. Sci., **50**: 2147-2156.

SAINTE-MARIE, B. & C. CARRIÈRE, 1995. Fertilization of the second clutch of eggs of snow crab, *Chionoecetes opilio*, from females mated once or twice after their molt to maturity. Fish. Bull., U.S., **93**: 759-764.

SAINTE-MARIE, B., T. GOSSELIN, J.-M. SÉVIGNY & N. URBANI, 2008. The snow crab mating system: opportunity for natural and unnatural selection in a changing environment. Bull. Mar. Sci., **83**(1): 131-161.

SAINTE-MARIE, B. & F. HAZEL, 1992. Moulting and mating of snow crab, *Chionoecetes opilio* (O. Fabricius), in shallow waters of the northwestern Gulf of St. Lawrence. Can. J. Fish. Aquat. Sci., **49**: 1282-1293.

SAINTE-MARIE, B. & M. LAFRANCE, 2002. Growth and survival of recently settled snow crab *Chionoecetes opilio* in relation to intra- and intercohort competition and cannibalism: a laboratory study. Mar. Ecol. Prog. Ser., **244**: 191-203.

SAINTE-MARIE, B., S. RAYMOND & J.-C. BRÊTHES, 1995. Growth and maturation of the benthic stages of male snow crab, *Chionoecetes opilio* (Brachyura: Majidae). Can. J. Fish. Aquat. Sci., **52**: 903-924.

SAINTE-MARIE, B., J.-M. SÉVIGNY, B. D. SMITH & G. A. LOVRICH, 1996. Recruitment variability in snow crab *Chionoecetes opilio*: pattern, possible causes, and implications for fishery management. In: B. BAXTER (ed.), Proceedings of the International Symposium on the Biology, Management and Economics of Crab from High Latitude Habitats. Lowell Wakefield Fish. Sym. Ser., Alaska Sea Grant College Program Rep., **96-02**: 451-478.

SQUIRES, H. J., 1990. Decapod Crustacea of the Atlantic coast of Canada. Can. Bull. Fish. Aquat. Sci., **221**: i-viii, 1-532. (Department of Fisheries and Oceans, Ottawa).

SQUIRES, H. J. & E. G. DAWE, 2003. Stomach contents of snow crab (*Chionoecetes opilio*, Decapoda, Brachyura) from the northeast Newfoundland Shelf. J. Northw. Atl. Fish. Sci., **32**: 27-38.

STARR, M., J.-C. THERRIAULT, G. Y. CONAN, M. COMEAU & G. ROBICHAUD, 1994. Larval release in a sub-euphotic zone invertebrate triggered by sinking phytoplankton particles. J. Plankton Res., **16**: 1137-1147.

TREMBLAY, M. J., 1997. Snow crab (*Chionoecetes opilio*) distribution limits and abundance trends on the Scotian Shelf. J. Northwest Atl. Fish. Sci., **128**: 7-22.

First received 14 November 2009.
Final version accepted 29 January 2010.

# THE CURRENT STATUS OF THE FISHERIES FOR *CHIONOECETES* SPP. (DECAPODA, OREGONIIDAE) IN ALASKAN WATERS

BY


G. BISHOP[1,5]), J. ZHENG[2]), L. M. SLATER[3]), K. SPALINGER[3]) and R. GUSTAFSON[4])

[1]) Southeast Regional Office, Commercial Fisheries Division, Alaska Department of Fish and Game, P.O. Box 110024, Juneau, AK 99811-0024, U.S.A.

[2]) Commercial Fisheries Division, Alaska Department of Fish and Game, P.O. Box 115526, Juneau, AK 99811-5526, U.S.A.

[3]) Westward Regional Office, Commercial Fisheries Division, Alaska Department of Fish and Game, 211 Mission Road, Kodiak, AK 99615, U.S.A.

[4]) Commercial Fisheries Division, Alaska Department of Fish and Game, 3298 Douglas Place, Homer, AK 99603, U.S.A.


## ABSTRACT


An overview of commercial fisheries for *Chionoecetes* crabs in Alaskan waters is provided, including stock assessment, harvest strategies, and trends in harvest and population size. Although a sum of peak harvests for individual *Chionoecetes* fisheries total more than 200 000 metric tons (t), the harvest from these fisheries during the 2008/09 season was less than 30 000 t. Most Alaskan *Chionoecetes* fisheries have stock assessment surveys that provide high quality data, although the data for some stocks is available for only short time periods. There is a need for improvements to habitat mapping and estimation of stock assessment model parameters, as well as a review of the effectiveness of proxy thresholds.


## INTRODUCTION

Alaskan *Chionoecetes* fisheries are prosecuted in 11 major management areas (fig. 1). Tanner crab (*Chionoecetes bairdi* (Rathbun, 1924)) are harvested from the eastern Bering Sea (EBS) to Southeast Alaska, snow crab (*Chionoecetes opilio* (Fabricius, 1788)) from the EBS, and grooved (*Chionoecetes tanneri* (Rathbun, 1893)) and triangle (*Chionoecetes angulatus* (Rathbun, 1893)) Tanner crab from the EBS and Aleutian Islands to Southeast Alaska.

---


[5]) Corresponding author; e-mail: gretchen.bishop@alaska.gov

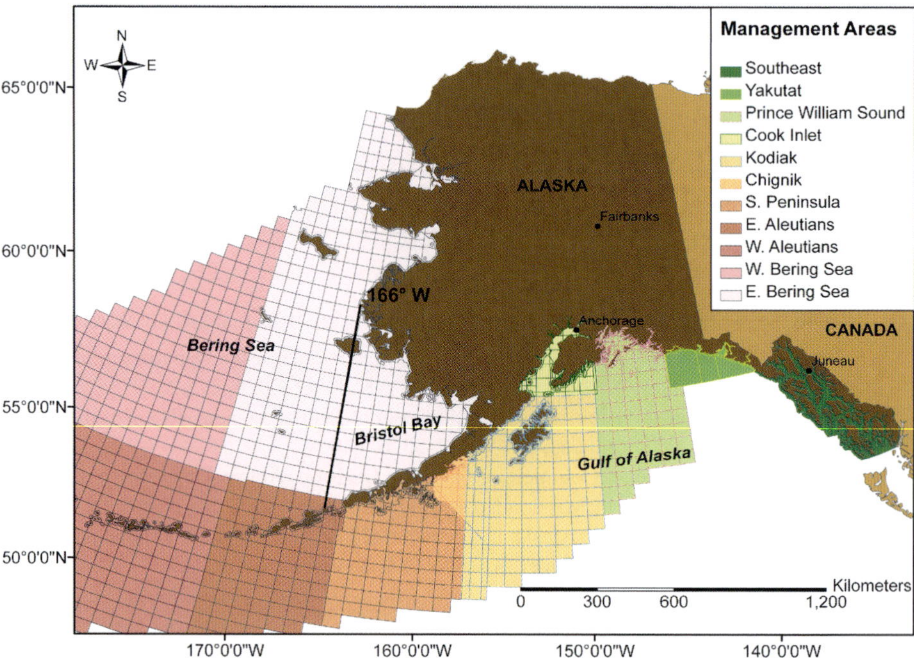


Fig. 1. ADF&G management areas for Tanner (*Chionoecetes bairdi* (Rathbun, 1924)) and snow crab (*Chionoecetes opilio* (Fabricius, 1788)) fisheries in Alaska.

The State of Alaska has jurisdiction over all fisheries inside territorial waters up to three miles offshore. These fisheries are managed by the Alaska Department of Fish and Game (ADF&G), according to regulations promulgated by the Alaska Board of Fisheries. Federal water jurisdiction extends from three to 200 miles offshore. In general, fisheries in Federal waters are managed by NOAA Fisheries according to regulations put in place by the North Pacific Fisheries Management Council (NPFMC). However, the State has sole management responsibility for *Chionoecetes* stocks, except for Tanner and snow crab in the EBS, where joint management occurs.

The base harvest strategy for *Chionoecetes* fisheries includes a closure during the male molt period, a male-only fishery, and a minimum size limit. The size limit for Tanner crab is 140 mm carapace width (CW) including spines, except in Prince William Sound, where it is 135 mm CW; for snow crab it is 78 mm CW, although industry preference is at least 102 mm CW.

The Alaskan Tanner crab fishery was pioneered by the Japanese and Russian fleets in the Bering Sea in 1965. Harvest methods were rapidly exported to the rest of the state, with domestic fisheries in the Kodiak area beginning in 1967, Cook Inlet and Prince William Sound in 1968, Southeast Alaska in

1971, Yakutat in 1973, and the EBS in 1974. The domestic fishery for snow crab began in 1977 (Otto, 1989) and for grooved and triangle Tanner crabs, in the early 1980s, incidental to the Aleutian Islands golden king crab (*Lithodes aequispinus* (Benedict, 1894)) fishery.

Although peak harvests of 57 669, 149 073, and 630 metric tons (t) were achieved for Tanner crab, snow crab, and combined grooved and triangle Tanner crab fisheries, in 1977/78, 1990/91 and 1995/96 respectively, their combined harvest in Alaskan waters totaled only 27 992 t in 2008/09. Nonetheless, *Chionoecetes* fisheries remain economically important to the State, with the highest annual exvessel value of any Alaskan crab genus — $83.54 million USD for 2008/09 alone.

We review the stock assessment, harvest strategies, and harvest and population trends for *Chionoecetes* fisheries in Alaskan waters.

# FISHERIES

## Eastern Bering Sea Tanner and snow crabs

### Stock assessment

Stock assessment data for eastern Bering Sea Tanner and snow crabs come from the summer trawl survey, commercial harvest reports, and bycatch observations. Surveys have been conducted annually by NOAA Fisheries using a standard Eastern otter trawl with an 83-ft headrope and a 112-ft footrope since 1968. A systematic grid of 376 37-km$^2$ stations are sampled (Chilton et al., 2009). Abundance is estimated using an area-swept approach (Alverson & Pereyra, 1969). Recruit class-specific harvest and bycatch data are obtained for stock assessments. Area-swept estimates of abundance are used to set total allowable catch (TAC) for Tanner crab while a stock assessment model is applied to multiple years of survey and fishery data to estimate abundance for snow crab (Turnock & Rugolo, 2008).

### Harvest strategy

A two-part harvest strategy is in place for Tanner and snow crabs in the EBS (table I). First, the Federal government has the annual responsibility, under National Standard 1 of the Magnuson-Stevens Fishery Conservation and Management Act, to determine whether stocks are overfished, whether overfishing (Total Fishing Mortality > Overfishing Level (OFL)) occurred

TABLE I

Current federal overfishing level (OFL) determination process in use by NOAA Fisheries and State harvest strategies for Tanner crab (*Chionoecetes bairdi* (Rathbun, 1924)) and snow crab (*Chionoecetes opilio* (Fabricius, 1788)) in the eastern Bering Sea. Abbreviations are MM (mature males), MMB (mature male biomass), TMB (total mature biomass), MFB (mature female biomass), MMM (molting mature males), LM (legal males), and ELM (exploitable legal males)

| | | | Snow crab (*Chionoecetes opilio* (Fabricius)) | Tanner crab (*Chionoecetes bairdi* (Rathbun)) |
|---|---|---|---|---|
| Legal size (mm CW) | | | 78 | 140 |
| 50% Functional mature size (mm CW) | | | 75 | 113 |
| Federal | Assumed M | | 0.23 for mature male and immature crab, and 0.29 for mature females | |
| | Assumed discard mortality | Groundfish trawl | 80% | |
| | | Crab pot | 50% | 50% |
| | 2008/09 Season stock status thresholds | Stock status level c | $B < \frac{1}{4}B_{35\%}$: 36 001 t MM on 2/15 | $B < \frac{1}{4}B_{MSY}^{prox}$: 21 519 t MM on 2/15 |
| | | Stock status level b | $B > \frac{1}{4}B_{35\%}$ & $B < B_{35\%}$ | $B > \frac{1}{4}B_{MSY}^{prox}$ & $B < B_{MSY}^{prox}$ |
| | | Stock status level a | $B > B_{35\%}$: 144 106 t MM on 2/15 | $B > B_{MSY}^{prox}$: 86 074 t MM on 2/15 |
| | | Total OFL | Total OFL $= (1 - e(-F_{OFL}))*$ $(MMB_{survey} * e(-M/2))$: 35 063 t MM on 10/15 | Total OFL $= (1 - e(-F_{OFL}))*$ $(MMB_{survey} * e(-M/2))$: 7040 t MM on 10/15 |
| | 2008/09 Season stock level | | b | b |
| | $F_{OFL}$ | Stock status level c | $F = 0$ | $F = 0$ |
| | | Stock status level b | $F_{OFL} = F_{35\%}*$ $(((B/B_{35\%}) - 0.1)/(1 - 0.1)))$: 0.55 for 2008/09 | $F_{OFL} = M*$ $(((B/B_{MSY}^{prox}) - 0.1)/(1 - 0.1))$: 0.12 for 2008/09 |
| | | Stock status level a | $F_{OFL} = F_{35\%}$ | $F_{OFL} = M$ |
| | | $F_{rebuild}$ | $0.75 * F_{OFL}$ | $0.75 * F_{OFL}$ |

TABLE I
(Continued)

| | | | Snow crab (*Chionoecetes opilio* (Fabricius)) | Tanner crab (*Chionoecetes bairdi* (Rathbun)) |
|---|---|---|---|---|
| State | Assumed M | | 0.3 | Estimated annually in model |
| | Assumed discard mortality | Crab pot | 25% | 20% |
| | Thresholds | Closure | <104 508 t TMB | <9525 t MFB |
| | | Tier 1 | 104 508-418 000 t TMB | 9525-20 412 t MFB |
| | | Tier 2 | >418 000 t TMB | ⩾20 412 t MFB |
| | Exploitation rate | Tier 1 | $E_{MMB} = 0.1 + (TMB - 104.5)*$ (0.125/313.5) | $E_{MMM} = 10\%$ or 5% if below threshold last year |
| | | Tier 2 | $E_{MMB} = 22.5\%$ | $E_{MMM} = 20\%$ or 10% if below threshold last year |
| | | Maximum | 58% ELM | 50% ELM or 25% ELM if below threshold last year |
| | | Minimum TAC | 7560 t | n.a. |
| | Other provisions | | ELM = new shell LM ⩾ 102 mm CW plus a percentage corresponding with fishery selectivity (usually 25%) of old shell LM ⩾ 102 mm CW | MMM = 100% new shell males > 112 mm CW plus 15% old shell males > 112 mm CW; ELM = new shell LM plus 32% old-shell LM |

during the previous season, and to set OFLs for the upcoming season. To do this, the mature male biomass (MMB) at maximum sustainable yield ($B_{MSY}$), a minimum stock size threshold (MSST) of half $B_{MSY}$, and the fishing mortality at $B_{MSY}$ are established. Currently, the MMB at $F_{35\%}$ (the fishing mortality rate at which the equilibrium spawning biomass per recruit is reduced to 35% of its value in the equivalent unfished stock) is used as a proxy of $B_{MSY}$, and $F_{35\%}$ as a proxy of $F_{MSY}$ for snow crab (table I); these values are recalculated annually. Because a length-based model is not available for the whole EBS Tanner crab stock, natural mortality was used as a proxy of $F_{MSY}$, and the mean MMB for 1969-1980 was used as a proxy of $B_{MSY}$, for Tanner crab; these are static values (table I).

If a stock is overfished (MMB < ½$B_{MSY}$), or the OFL is exceeded, the NPFMC must develop an amendment to the Fisheries Management Plan (FMP) within two years to provide for rebuilding. The rebuilding plan must have at least a 50% chance of rebuilding the stock within a specified maximum time. If the plan fails, then the fishing mortality rate must not exceed the minimum of the rebuilding F or 75% of $F_{OFL}$ until a revised plan is implemented. The process of establishing OFLs and rebuilding plans may change over time.

The second part of the harvest strategy is the State's process of setting TAC based on a population threshold, variable mature male exploitation rates, and a cap on legal male exploitation rates. For EBS Tanner crab, the State harvest strategy is complex. It includes a threshold of 9525 t of mature female (>79 mm CW) biomass (MFB) at the time of the survey (table I). Separate TACs east and west of 166°W longitude are computed as the product of exploitation rate and molting mature male (MMM) abundance. The exploitation rate is a function of MFB at the time of the survey; however, TACs are capped at a percentage of exploitable legal male (ELM) abundance. If MFB was below threshold in the preceding year, then exploitation rates and the percentage cap on ELM are reduced by one-half (table I). Finally, TACs by weight are computed as the product of TACs in numbers and mean legal weight.

The State harvest strategy for EBS snow crab is simpler. A threshold of 104 500 t of total mature biomass (TMB) is in effect. Exploitation rates are also a function of TMB (table I). The TAC is the product of the exploitation rate ($E_{MMB}$) and the MMB. A maximum of 58% of ELM abundance may be harvested and a minimum TAC of 7560 t is required for the fishery to open (table I). For both EBS Tanner and snow crabs, the sum of the State TAC and the fishery discard losses must be less than the respective Federal Total OFL.

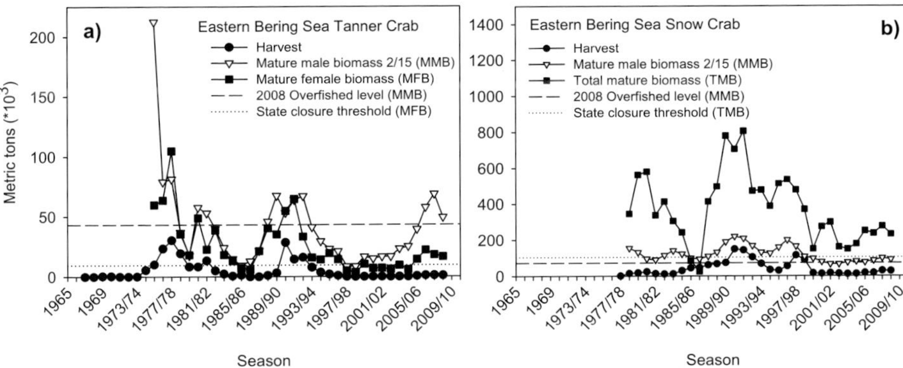

Fig. 2. Interannual trends in harvest and population estimates, and Federal and State thresholds for: a, Tanner crab (*Chionoecetes bairdi* (Rathbun, 1924)); b, snow crab (*Chionoecetes opilio* (Fabricius, 1788)), in the eastern Bering Sea.

## Harvest and population trends

Eastern Bering Sea Tanner crab harvest and population size has fluctuated greatly (fig. 2); MMB peaked at 212 354 t in 1975/76 and harvest at 30 232 t in 1977/78. Biomass declined to 9466 t by 1985/86 and no fishing was allowed in 1985/86 or 1986/87. The fishery reopened in 1987/88 but closed from 1996/97-2003/04, and the stock was declared overfished in 1999. Recently, abundance has increased slowly, and the fishery reopened in 2005, reaching rebuilt status in 2007. However, due to changes in terminal molt size (Zheng, 2008) most male crab currently do not attain legal size, and the exploitable abundance remains lower than that of the 1970s.

Annual harvest of EBS snow crab has varied greatly, following population trends. MMB and harvest peaked at respectively 218 631 t and 149 073 t in 1990/91 (fig. 2). The stock was declared overfished in 1999 and MMB reached its lowest level of 65 317 t in 2002/03 (fig. 2). Abundance has increased recently, but MMB has remained below $B_{MSY}$ (Turnock & Rugolo, 2008) and the stock failed to rebuild by 2009/10.

## Kodiak Tanner crab

### Stock assessment

Tanner crab in Kodiak, Chignik, South Peninsula, and Eastern Aleutian Districts are assessed annually through trawl surveys. A standardized 400-mesh eastern otter trawl is towed for 1.852 km at approximately 380 stations

(Spalinger & Cavin, 2004). There is no survey for the Western Aleutians District.

Population size is estimated as the product of trawl survey density and habitat area, using area-swept methods (Alverson & Pereyra, 1969). Habitat area has been determined both as the area deeper than 20 fathoms and by analyzing bottom type (Kally Spalinger, personal communication).

Harvest strategy

The harvest strategy for Kodiak Tanner crab is well developed. MSSTs are defined as half the mean mature male (MM) abundance from 1973-1998 for Kodiak, 1974-1998 for Chignik and South Peninsula (Urban & Vining, 1999), and 1990-2000 for Eastern Aleutians. A stock below threshold reopens only when the guideline harvest level (GHL) is at least twice the MSST, and then only half that GHL may be harvested (Urban et al., 1999) (table II). Exploitation rates are 10% of MMM when populations are above MSST, but below the mean MM abundance, and 20% when above the mean MM abundance. Harvest is capped at 30% of legal males. GHLs are the product of the exploitation rate, the number of MMM, and the mean legal weight. Minimum GHLs are in effect for some areas. Western Aleutians open only by emergency order and have no defined harvest strategy.

Harvest and population trends

For the Kodiak area, trends in Tanner crab harvest and population size vary by District. Western Aleutians harvest has generally been incidental to the directed red king crab fishery, and peaked at 380 t in 1981/82; there has been no harvest since 1995/96 (fig. 3a). Eastern Aleutians harvest peaked at 1134 t in 1977/78 but declined to 23 t by 1990/91 (fig. 3b) and the fishery was closed for 1994/95-2002/03. Harvest in South Peninsula peaked at nearly 4000 t in 1978/79; the fishery was closed from 1989/90-1999/00 (fig. 3c). The fishery in Chignik peaked at 5000 t in 1975/76, and was closed for 1989/1990-2003/04 (fig. 3d). Kodiak harvest peaked at 15 000 t in 1977/78 and was closed for 1994/95-1999/2000 (fig. 4a).

Cook Inlet and Prince William Sound Tanner crab

Stock assessment

Pot surveys indexed Tanner crab abundance in Cook Inlet (Kimker, 1985) and Prince William Sound (Donaldson, 1988) from 1975-1991, but were

TABLE II

Harvest strategies used by the State for management of Tanner crab (*Chionoecetes bairdi* (Rathbun, 1924)) in Alaska. Additional provisions exist for some areas. Abbreviations are MM (mature males), MMM (molting mature males), and LM (legal males)

| Registra-tion area | District | Section | Legal size (mm CW) | Mature size (mm CW) | Threshold ($*10^6$) | | | Exploitation rate | | | Minimum GHL (t) |
|---|---|---|---|---|---|---|---|---|---|---|---|
| | | | | | Closure | Tier 1 | Tier 2 | Tier 1 | Tier 2 | Maximum | |
| Westward | Western Aleutians | | 140 | | | | | | | | |
| | Eastern Aleutians | Akutan | 140 | 115 | <0.200 MM | 0.2-0.4 MM | >0.4 MM | 10% MMM (100% of new shell male crab and 15% of old or very old shell male crab >114 mm CW) | 20% MMM | 30% LM | 16 |
| | | Unalaska/ Kalekta Bay | 140 | 115 | <0.065 MM | 0.065-0.13 MM | >0.13 MM | | | | 16 |
| | | Makushin/ Skan Bay | 140 | 115 | <0.045 MM | 0.045-0.09 MM | >0.09 MM | | | | 16 |
| | South Peninsula | Eastern | 140 | 115 | <2.0015 MM | 2.0015-4.003 MM | >4.003 MM | | | | 91 |
| | | Western | 140 | 115 | <1.25 MM | 1.25-2.5 MM | >2.500 MM | | | | 91 |
| | Chignik | | 140 | 115 | <0.973 MM | 0.973-1.946 MM | >1.946 MM | | | | 91 |

TABLE II
(Continued)

| Registra-tion area | District | Section | Legal size (mm CW) | Mature size (mm CW) | Threshold ($*10^6$) | | | Exploitation rate | | | Minimum GHL (t) |
|---|---|---|---|---|---|---|---|---|---|---|---|
| | | | | | Closure | Tier 1 | Tier 2 | Tier 1 | Tier 2 | Maximum | |
| | Kodiak | Northeast | 140 | 115 | <1.123 MM | 1.123-2.246 MM | >2.246 MM | | | | Each section, 45; At least 2 sections must open; District must total 181 |
| | | Eastside | 140 | 115 | <1.552 MM | 1.552-3.104 MM | >3.104 MM | | | | |
| | | Southeast | 140 | 115 | <0.733 MM | 0.733-1.466 MM | >1.466 MM | | | | |
| | | Southwest | 140 | 115 | <1.236 MM | 1.236-2.472 MM | >2.472 MM | | | | |
| | | Westside | 140 | 115 | <0.764 MM | 0.764-1.528 MM | >1.528 MM | | | | |
| | | North Mainland | 140 | 115 | <1.469 MM | 1.469-2.938 MM | >2.938 MM | | | | |
| Cook Inlet | Southern | | 140 | n.a. | 0.5 LM | 0.5-1.0 LM | >1.000 LM | 15% LM | 25% LM | | |
| | Kamishak | | 140 | n.a. | 0.7 LM | 0.7-1.4 LM | >1.400 LM | 15% LM | 25% LM | | |
| Prince William Sound | | | 135 | | | | | | | | |
| Yakutat | | | 140 | | | | | | | | |
| Southeast Alaska | | | 140 | 108 | <0.001043 t MM | 0.001043-0.002495 t MM | ≥0.002495 t MM | | | | |

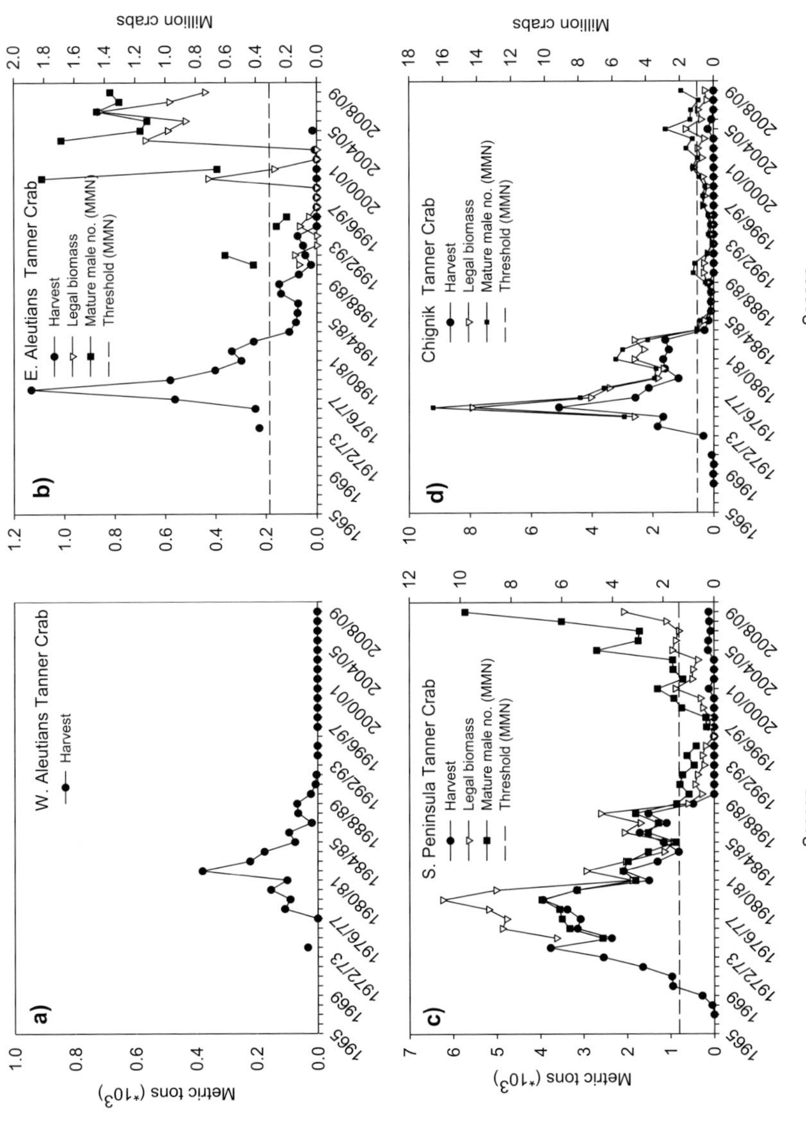



Fig. 3. Interannual trends in harvest and population estimates, and State thresholds for Tanner crab (*Chionoecetes bairdi* (Rathbun, 1924)) in: a, Western Aleutians; b, Eastern Aleutians; c, South Peninsula; d, Chignik.

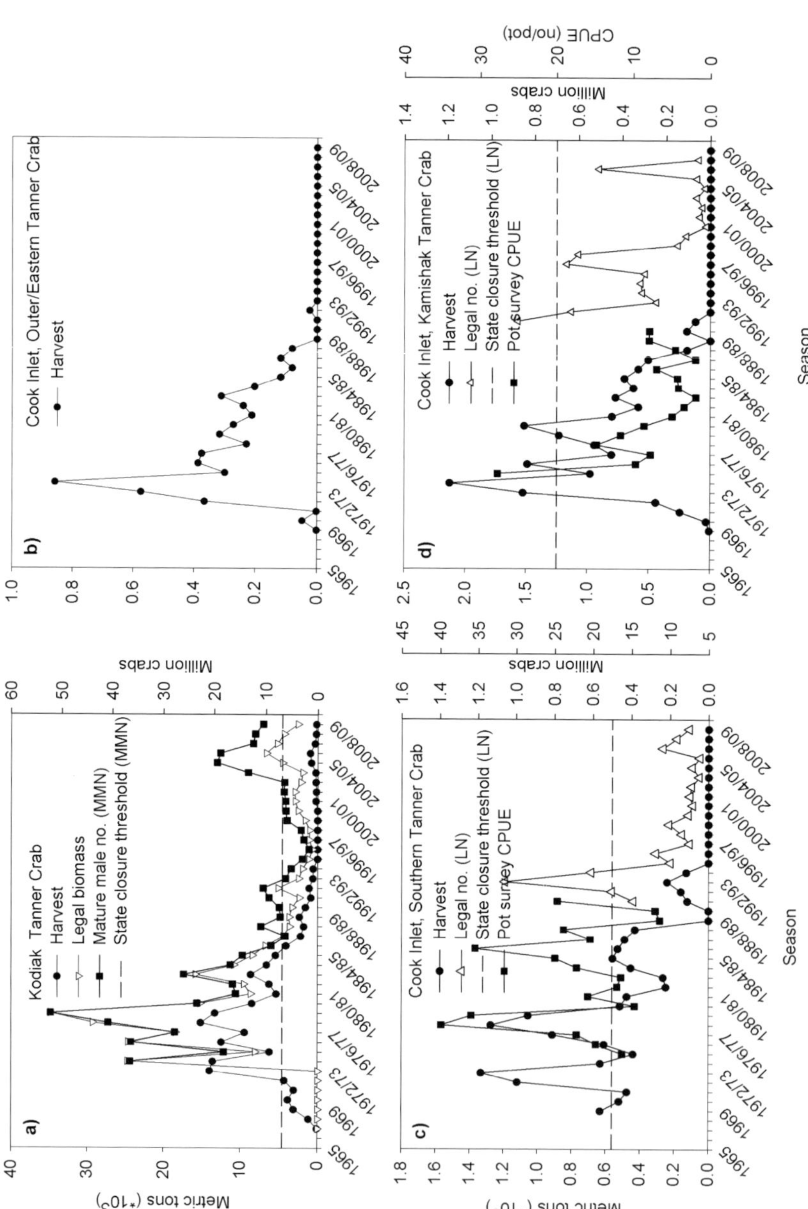



Fig. 4. Interannual trends in harvest, population estimates, and indices; and State thresholds for Tanner crab (*Chionoecetes bairdi* (Rathbun, 1924)) in: a, Kodiak; b, Cook Inlet, Outer/Eastern; c, Cook Inlet, Southern; d, Cook Inlet, Kamishak. No survey was conducted in Kamishak Bay for 2008/09.

replaced by trawl surveys in 1991 (Bechtol, 1999, 2000). A 400-mesh eastern otter trawl is used, and Southern, Kamishak and Barren Islands in Cook Inlet are currently surveyed annually while Northern and Hinchinbrook in Prince William Sound were surveyed annually for 1991-1995 and biennially thereafter. In years when both areas are surveyed, 60-90 1.852-km tows are conducted (Bechtol, 1999, 2000). Legal abundance is estimated using area-swept methods (Alverson & Pereyra, 1969).

Harvest strategy

The Cook Inlet Tanner crab harvest strategy addresses commercial and non-commercial fisheries, with thresholds, tiered exploitation rates, and a cap on the non-commercial exploitation rate when the commercial fishery is closed (table II). The commercial threshold is half the mean legal population estimate from 1990-2001 (Bechtol et al., 2002). The commercial fishery does not open if attaining the GHL will reduce the population below threshold, or if the 12-hour harvest capacity exceeds the GHL. The non-commercial fishery in each of Kachemak Bay and Kamishak is closed if the abundance of legal males is below any one of three uniquely calculated thresholds (table II).

Prince William Sound Tanner crab stock levels have been depressed since the mid-1980's, and no harvest strategy has been developed (table II).

Harvest and population trends

Cook Inlet and Prince William Sound commercial harvests peaked at respectively 3614 t in 1973/74 (fig. 4) and 6318 t in 1972/73 (fig. 5). Harvest subsequently declined, and these management areas were closed, by ADF&G in 1988/89, and by the BOF in 1998/99. Legal population sizes remain below the commercial threshold, but above the non-commercial threshold for Southern and combined Kamishak and Barren Islands. Prince William Sound commercial fisheries likewise remain closed, while the subsistence fisheries have re-opened in most areas.

Southeast Alaska and Yakutat Tanner crab

Stock assessment

Tanner crab stocks in Southeast Alaska are assessed during two annual surveys in which approximately 860 pots are set; the sampling design employs randomly determined pot locations within strata. The summer survey covers

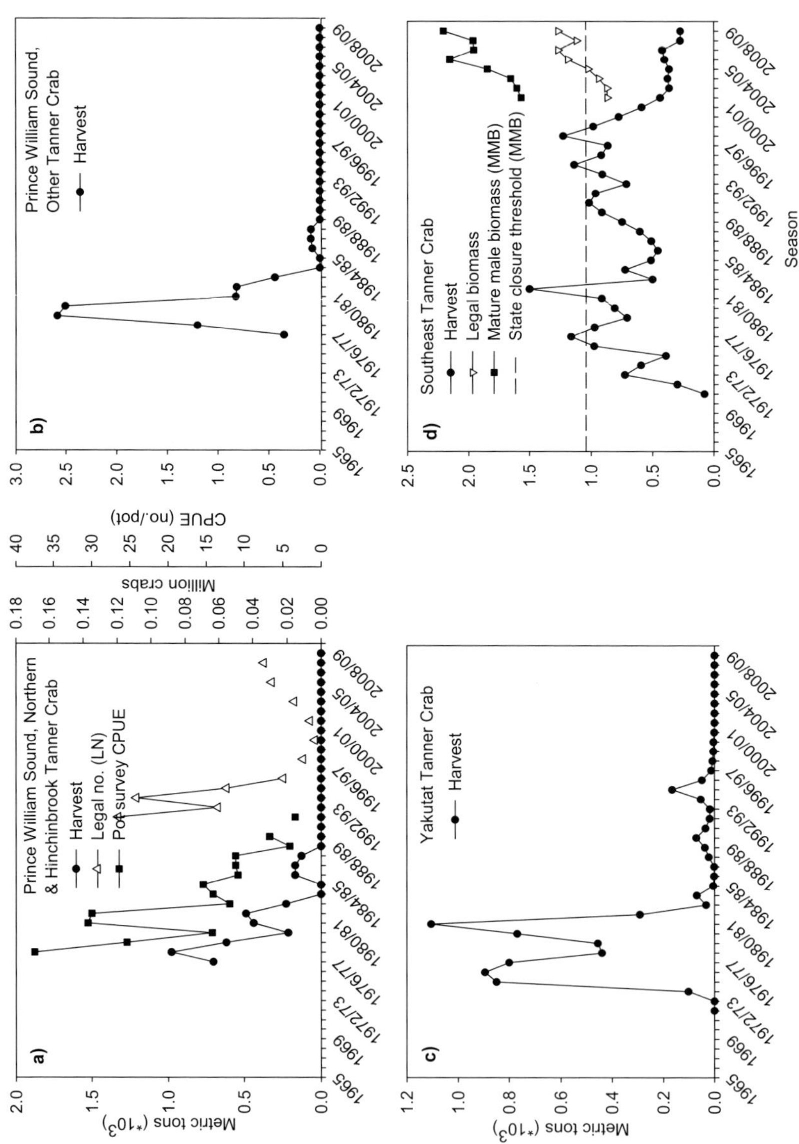

Fig. 5. Interannual trends in harvest, population estimates and indices, and State thresholds for Tanner crab (*Chionoecetes bairdi* (Rathbun, 1924)) in: a, Prince William Sound, Northern and Hinchinbrook; b, Prince William Sound, Other; c, Yakutat; d, Southeast Alaska.

nine areas and red king crab is the target species, while the fall survey covers six areas and Tanner crab is the target species. Similar methods and gear are employed during both surveys (Bednarski et al., 2008). Pot survey and commercial harvest data are used as input to a 3-stage catch-survey model (Zheng et al., 2006) to produce annual estimates of MMB and legal male biomass (LMB) by survey area.

Survey areas are assigned a stock status based on the CPUE of four size and sex classes and on female reproductive potential (Siddon et al., 2009). The five stock status tiers correspond to exploitation rates of 0, 5, 10, 15, or 20% of MMB, or a maximum of 50% of LMB. Harvestable surplus is determined as the sum of the products of exploitation rate and MMB, expanded to account for 29% of harvest from unsurveyed areas (Siddon et al., 2009).

Harvest strategy

The harvest strategy for Tanner crab in Southeast Alaska includes a MSST of half the $B_{MSY}$, using mean MMB for 1997-2007 as a proxy, and "core" and "non-core" area definitions. A core area base season length of five days is established, with provisions for additional fishing time of up to 10 days, depending upon MMB and the number of pots registered. For non-core areas, season length is that of the core season plus an additional five days (table II). For the Yakutat Registration Area, a maximum harvest level of 454 t is in effect.

Harvest and population trends

The Tanner crab fishery in Southeast Alaska achieved a peak harvest of 1500 t in 1981/82 (Hebert et al., 2008), but by 2007/08 had declined to a modern low of 274 t (fig. 5); however, the fishery remains above threshold. The relatively short survey time series prohibits determination of long-term biomass trends (fig. 5).

Statewide grooved and triangle Tanner crabs

There is no formal stock assessment for the deepwater grooved or triangle Tanner crab in Alaska; however, observers have been routinely deployed to monitor CPUE, bycatch, and gear performance in the Westward Region (Byrne & Cross, 1997). Statewide harvest peaked at 630 t in 1994/95 (fig. 6). Due to low effort, management area-specific harvest is confidential, but 92% by

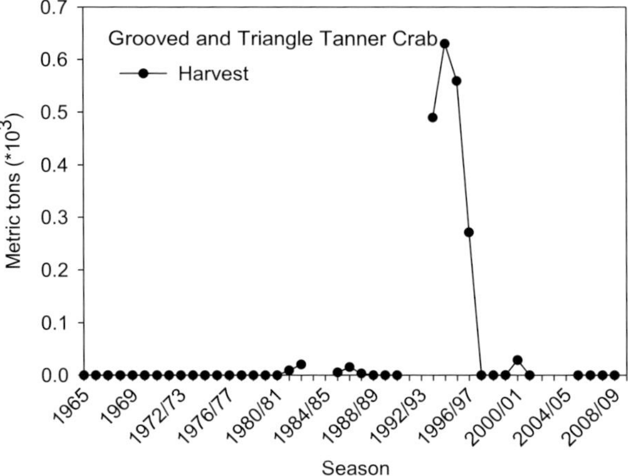


Fig. 6. Interannual trends in the harvest of grooved (*Chionoecetes tanneri* (Rathbun, 1893)) and triangle Tanner crab (*Chionoecetes angulatus* (Rathbun, 1893)) in Alaska; missing data points are due to confidential data where less than three permit holders made landings in a season.

weight of all Alaskan production has come from EBS, Eastern Aleutians, and South Peninsula combined; most harvest is of grooved Tanner crab, with incidental catch of triangle Tanner crab.

Grooved Tanner crab may be taken incidentally while fishing for golden king crab in Eastern Aleutians, and a guideline harvest range (GHR) of 23-91 t is in effect. Elsewhere in the state a commissioner's permit is required for directed fishing. Crab population size, catchability, product quality and markets all limit deepwater Tanner crab production in Alaska.


## PERSPECTIVE

Better benthic habitat mapping would improve stratification and reduce variability of population estimates.

Ontogenetic stage-specific and stock-specific parameter estimates are needed for modeling.

The effectiveness of proxy thresholds, and resiliency to environmental perturbation of current harvest strategies should be reviewed.


## ACKNOWLEDGEMENTS

This paper is PP # 255 of Commercial Fisheries Division of the Alaska Department of Fish and Game. The authors acknowledge numerous ADF&G

staff who collected and reported population and harvest data over the years, as well as those who assisted in the preparation of this paper, including John E. Clark, Chris Siddon, Ken Goldman, Charlie Trowbridge, Doug Pengilly, and Heather Fitch.

## REFERENCES

ALVERSON, D. L. & W. T. PEREYRA, 1969. Demersal fish explorations in the northeastern Pacific Ocean — An evaluation of exploratory fishing methods and analytical approaches to stock size and yield forecasts. Journal of the Fisheries Research Board of Canada, **26**: 1985-2001.

BECHTOL, W. R., 1999. A bottom trawl survey for crabs and groundfish in the Prince William Sound Management Area 16-26 August 1997. Alaska Department of Fish and Game, Division of Commercial Fisheries, Regional Information Report, **2A99-24**: 1-47.

— —, 2000. A bottom trawl survey for crabs in the Southern, Kamishak Bay, and Barren Islands districts of the Cook Inlet management area 8-12 June and 26 June–1 July 1997. Alaska Department of Fish and Game, Commercial Fisheries Management and Development Division, Regional Information Report, **2A00-21**: 1-50.

BECHTOL, W. R., C. TROWBRIDGE & N. SZARZI, 2002. Tanner and king crabs in the Cook Inlet management area, stock status and harvest strategies. Alaska Department of Fish and Game, Commercial Fisheries Division, Sport Fisheries Division, Regional Information Report, **2A02-07**: 1-38.

BEDNARSKI, J., G. BISHOP & C. SIDDON, 2008. Tanner crab pot survey methods for Southeast Alaska. Alaska Department of Fish and Game, Regional Information Report, **1J08-02**: 1-37.

BYRNE, L. C. & D. CROSS, 1997. Summary of special projects carried out by an observer onboard a crab boat in the 1996 western Aleutian area *Chionoecetes tanneri* fishery. Alaska Department of Fish and Game, Division of Commercial Fisheries, Regional Information Report, **4K97-17**: 1-23.

CHILTON, E. A., C. E. ARMISTEAD & R. J. FOY, 2009. The 2009 Eastern Bering Sea Continental Shelf Bottom Trawl Survey: results for commercial crab species. NOAA Fisheries, Alaska Fisheries Science Center, Kodiak Laboratory, NOAA Technical Memorandum, **NMFS-AFSC-201**: 1-111.

DONALDSON, B., 1988. Prince William Sound Tanner Crab Tagging and Index Survey, 1988. Alaska Department of Fish and Game, Division of Commercial Fisheries, Central Region, Regional Information Report, **2C89-02**: 1-32.

HEBERT, K., J. STRATMAN, K. BUSH, G. BISHOP, C. SIDDON, J. BEDNARSKI & A. MESSMER, 2008. 2009 Report to the Board of Fisheries on Region 1 Shrimp, Crab, and Scallop Fisheries. Alaska Department of Fish and Game, Fisheries Management Report, **08-62**: 1-210.

KIMKER, A. T., 1985. Shellfish Report to the Alaska Board of Fisheries. Alaska Department of Fish and Game, Division of Commercial Fisheries, Regional Information Report, **85-4**: 1-56.

OTTO, R. S., 1989. An overview of eastern Bering Sea king and Tanner crab fisheries. In: B. MELTEFF (ed.), Proceedings of the International Symposium on King and Tanner Crabs, **90-04**: 9-26. (Alaska Sea Grant College Program Report, Anchorage).

SIDDON, C., J. BEDNARSKI & G. H. BISHOP, 2009. Southeast Alaska Tanner crab 2006 stock assessment and recommendations for the 2007 commercial fishery. Alaska Department of Fish and Game, Fishery Data Series, **09-18**: 1-39.

SPALINGER, K. A. & M. E. CAVIN, 2004. Standard project operational plan: bottom trawl survey of crab and groundfish: Kodiak, Chignik, South Alaska Peninsula, and eastern Aleutian Areas. Alaska Department of Fish and Game, Division of Commercial Fisheries, Regional Information Report, **4K04-47**: 1-50.

TURNOCK, B. J. & L. RUGOLO, 2008. Stock assessment of eastern Bering Sea snow crab. In: Stock assessment and fishery evaluation report for the king and Tanner crab fisheries of the Bering Sea and Aleutian Islands Regions: 1-76. (North Pacific Fisheries Management Council, Crab Plan Team).

URBAN, D., D. PENGILLY, D. A. JACKSON & I. VINING, 1999. Tanner crab harvest strategy for Kodiak, Chignik, and the South Peninsula Districts, report to the board of fisheries. Alaska Department of Fish and Game, Commercial Fisheries Division, Regional Information Report, **4K99-21**: 1-15.

URBAN, D. & I. VINING, 1999. Reconstruction of historic abundances of Kodiak, Chignik, and South Peninsula Tanner crab, report to the Alaska Board of Fisheries. Alaska Department of Fish and Game, Commercial Fisheries Division, Regional Information Report, **4K99-17**: 1-24.

ZHENG, J., 2008. Temporal changes in size at maturity and their implications for fisheries management for eastern Bering Sea Tanner crab. Journal of Northwest Atlantic Fishery Science, **41**: 137-149.

ZHENG, J., J. M. RUMBLE & G. H. BISHOP, 2006. Estimating Southeast Alaska Tanner crab abundance using pot survey and commercial catch data. Alaska Fisheries Research Bulletin, **12**(2): 196-211.


First received 16 December 2009.
Final version accepted 20 February 2010.

THE CURRENT STATUS OF BIOLOGICAL KNOWLEDGE RELATING
TO THE MANAGEMENT OF FISHERIES FOR TANNER
(*CHIONOECETES BAIRDI* (RATHBUN, 1924)) AND SNOW CRABS
(*CHIONOECETES OPILIO* (FABRICIUS, 1788)) IN ALASKAN WATERS

BY


J. ZHENG[1,4]), L. M. SLATER[2]), J. WEBB[1]) and G. BISHOP[3])

[1]) Commercial Fisheries Division, Alaska Department of Fish and Game, P.O. Box 115526,
Juneau, AK 99811-5526, U.S.A.
[2]) Westward Regional Office, Commercial Fisheries Division, Alaska Department of Fish
and Game, 211 Mission Road, Kodiak, AK 99615, U.S.A.
[3]) Southeast Regional Office, Commercial Fisheries Division, Alaska Department of Fish
and Game, P.O. Box 110024, Juneau, AK 99811-0024, U.S.A.


## ABSTRACT


In Alaskan waters, male-only fisheries for Tanner crab (*Chionoecetes bairdi* (Rathbun, 1924)) and snow crab (*Chionoecetes opilio* (Fabricius, 1788)) are prosecuted and make important contributions to the economy. Concern over population declines has resulted in increased research efforts, which have yielded improvements to our understanding of biological and spatial patterns of these two species in Alaska. In summary, (1) male and female snow and Tanner crabs undergo a terminal molt to maturity, (2) the size at maturity shows large spatial variation in both Tanner and snow crabs and interannual trends are present for Tanner crab, (3) female sperm reserves are spatiotemporally variable and may be influenced by sex ratio and exploitation, (4) snow crab exhibit biennial spawning at low temperatures, (5) spatiotemporal trends in fecundity are dwarfed by effects of ontogeny for both species, and (6) snow crab in the eastern Bering Sea appear to undergo an ontogenetic migration towards the shelf edge. Continued research, especially related to reproductive potential, growth and age, natural mortality, and the roles of oceanography and predation in determining crab distribution and recruitment success, is increasingly vital for setting overfishing levels, refining harvest strategies, and making predictions about population responses to climate change.


## INTRODUCTION

Tanner crab (*Chionoecetes bairdi* (Rathbun, 1924)) and snow crab (*Chionoecetes opilio* (Fabricius, 1788)) are moderately long-lived, brachyuran crabs with a complex life history; they are sexually dimorphic, store sperm, and have

___
[4]) Corresponding author; e-mail: jie.zheng@alaska.gov

bipartite breeding behavior (Somerton, 1982). Snow crab have a circumpolar distribution (Macginitie, 1955), and Tanner crab are broadly distributed within the Northeast and Northwest Pacific Ocean (Slizkin, 1989). In Alaskan waters, male-only fisheries for Tanner crab are prosecuted from the eastern Bering Sea (EBS) to Southeastern Alaska, and for snow crab, in the EBS. As discussed in Bishop et al. (in press), fisheries for these species make important contributions to the economy of Alaska. Recent declines in these fisheries have catalyzed research efforts; hence, the purpose of this paper is to review some of the resulting biological advances in male terminal molt, size at maturity, reproductive potential, and snow crab migration and distribution.

## BIOLOGICAL KNOWLEDGE RELATING TO THE MANAGEMENT OF FISHERIES

### Terminal molt and size at maturity

Both male and female Tanner and snow crabs undertake a terminal molt when reaching morphometric maturity (Conan & Comeau, 1986; Tamone et al., 2005). Morphometrically mature males are distinguished from immature males by an increase in chela height relative to size (as measured by carapace width (CW)) (fig. 1), as further evidenced by changes in ecdysteroid levels (Tamone et al., 2005). For females, a prominent increase in the width of the abdominal flap indicates sexual maturity (Somerton, 1981). Thus, the determination of maturity status for the collection of data on size at maturity can be accomplished visually for female crab, whereas for male crab, measurements of CW and chela height must be taken and maturity status then determined through the relationship between size and chela height (fig. 1).

The mean size of mature females is considered a better measure of the median size at maturity than size at 50% maturity because females undergo their molt to maturity at similar ages and their mature size distribution can be assumed normal (Somerton, 1981). Size at 50% maturity for males is determined using logistic regression once maturity status is determined as described above (Conan & Comeau, 1986; Zheng, 2008). Data used to estimate sizes at maturity for males and females include survey data from 1975 to 2006 for Bering Sea female Tanner crab, from 1990 to 2006 for Bering Sea male Tanner crab, from 1989 to 1994 for Bering Sea snow crab, from 1996 to 2008 for Marmot Bay (Kodiak) Tanner crab, from 2007 to 2009 for Cook Inlet Tanner crab, and from 1997 to 2007 for southeastern Alaska Tanner crab. The results are from Zheng (2008) for EBS (consisting of Bristol Bay and Pribilof

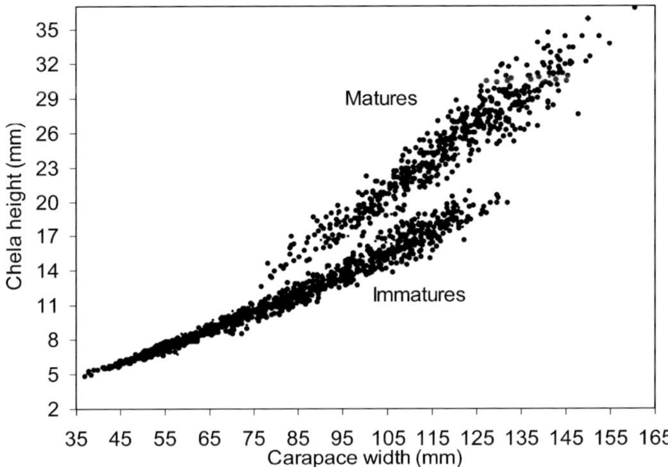


Fig. 1. Relationship of chela height to carapace width for male Tanner crab (*Chionoecetes bairdi* (Rathbun, 1924)) from waters around Kodiak Island, 2003. The visible separation of this relationship into two groups illustrates that males that have undergone their terminal molt to maturity have a larger chela at size (upper data points) in comparison to immature males (lower data points).

Islands) Tanner crab, from Otto (1998) for EBS snow crab, and from Siddon & Bednarski (in press) for southeastern Alaska female Tanner crab.

Estimated sizes at 50% maturity vary greatly among locations for male Tanner crab (table I). In southeastern Alaska, size at 50% maturity was largest in Stephens Passage and smallest in Glacier Bay (Siddon & Bednarski, in press). Cook Inlet and Kodiak (Marmot Bay) had similar sizes at 50% maturity, whereas the Pribilof Islands had the smallest size at 50% maturity for all Alaska Tanner crab stocks examined. Sizes at 50% maturity in Bristol Bay were found to have a downward trend from 1990 to 2006 (Zheng, 2008). Mean size at maturity for female Tanner crab also differed among locations and was largest in Port Camden, in southeastern Alaska, and smallest off the Pribilof Islands. Like Bristol Bay male Tanner crab, mean sizes at maturity for female Tanner crab in Bristol Bay and the Pribilof Islands also had a downward trend from 1975 to 2006 (Zheng, 2008). The size at 50% maturity for EBS male snow crab and the mean size at maturity for female snow crab were much smaller than those for Tanner crab stocks.

Declines in the size at maturity, whether fishery-induced, due to genetic effects, or environment-driven, have great consequences for fisheries management. When males mature at a smaller size, they increase the likelihood of mating prior to capture by commercial fisheries. Legal size of retention and harvest rates may need to be adjusted to maximize the yield and protect the

TABLE I

Summary of size at 50% maturity for Tanner crab (*Chionoecetes bairdi* (Rathbun, 1924)) and snow crab (*Chionoecetes opilio* (Fabricius, 1788)) by location in Alaska. Values estimated by authors unless otherwise noted

| Species | Location | Sublocation | Male | | Female | |
|---|---|---|---|---|---|---|
| | | | Mean CW (mm) | Range CW (mm) | Mean CW (mm) | Range CW (mm) |
| Tanner crab | Eastern Bering Sea[a] | Bristol Bay | 108.9 | 92.8-127.4 | 89.7 | 81.9-98.5 |
| | | Pribilof islands | 99.7 | 82.0-117.8 | 82.8 | 76.5-87.8 |
| | Southeastern Alaska[b] | Stephens Passage | 131.3 | | 93.3 | |
| | | Icy Strait | 125.7 | | 84.2 | |
| | | Glacier Bay | 109.2 | | 84.7 | |
| | | Port Camden | 123.1 | | 95.3 | |
| | | Thomas Bay | 111.1 | | 91.0 | |
| | | Holkham Bay | 121.1 | | 89.0 | |
| | Cook Inlet | Southern District | 116.4 | | 95.5 | |
| | Kodiak | Marmot Bay | 116.3 | | 83.5 | |
| Snow crab | Eastern Bering | Sea[c] | 75.0 | | 56.1 | |

Sources: [a]Zheng, 2008; [b]Siddon & Bednarski, in press; [c]Otto, 1998.

stock when the size at maturity changes (Zheng, 2008). Management options, such as reducing the legal size limit, targeting mature males by using a limit ratio of chela height against carapace width, or implementing a slot fishery (minimum and maximum size limits) may need to be considered to reduce the loss of harvest opportunity for small-sized mature crab and to avoid potential adverse genetic consequences to the stock (Zheng, 2008). However, before an appropriate management response can be determined, the relative importance of anthropogenic and environmental factors in producing the decline in size at maturity must be assessed.

## Sperm reserves

All crab stocks in Alaska are managed with large male-only harvest rules. As knowledge of Tanner and snow crabs' complex reproductive biology and mating systems has increased, relationships between large male-only harvest, environmental factors, reproductive biology, and the probability of sperm limitation in exploited crab stocks have received increased attention (Sainte-Marie et al., 2008). Females mate soon after their terminal molt to maturity and receive sperm that they store internally in paired spermatheca;

these females, after extruding the first egg clutch of ontogeny, are termed "primiparous". Female Tanner and snow crabs which have not mated, or which have inadequate sperm reserves, may fail to extrude a clutch and resorb eggs in their ovaries or alternatively extrude a partially or fully unfertilized egg clutch (Adams, 1985; Sainte-Marie & Carriere, 1995). Terminally molted females that have hatched at least one clutch may mate in the hard-shell condition to increase their sperm reserves or alternately may fertilize clutches with sperm stored from previous mating(s); these females, after extruding the second or greater egg clutch of ontogeny, are termed "multiparous". Two measures of sperm reserves are frequently used, "spermathecal load" or the weight of the contents of the spermathecae, and estimated "sperm cell count". Because spermathecal load was significantly higher in multiparous than primiparous Tanner crab, while sperm cell counts were similar between these life history stages, sperm cell counts are considered a more reliable estimate of viable sperm availability for multiparous females (Webb & Bednarski, in press).

To evaluate the effect of male-only fisheries on female sperm reserves, recent studies of Tanner and snow crabs in Alaska have focused on variability with life history, including ontogenetic mating frequency, and spatiotemporal factors, including operational sex ratios and fishing pressure. Webb (2009) compared sperm reserves of multiparous Tanner crab in southeastern Alaska collected before and after the mating season. Those captured before mating and held in the lab without males through hatching and extrusion of a new clutch were found to have successfully fertilized their new clutches and most had sufficient sperm reserves remaining for fertilization of a subsequent clutch. Those collected after the mating season with newly extruded clutches had increased frequency of grasping mark and significantly higher sperm reserves than those fertilizing clutches with stored sperm in the laboratory, indicating that nearly all multiparous females with access to males had mated. There are selective pressures both for and against multiparous females re-mating. Because males are larger than females, the females may be injured during mating. It may be beneficial for multiparous females with adequate sperm reserves to avoid mating and instead fertilize clutches with stored sperm. Alternatively, since sperm from the last male mated appears to be used first for fertilization and multiparous females have greater opportunity for male selection than nulliparous females, multiparous females may select to re-mate to increase not only the quantity, but possibly also the quality of their sperm reserves (Sainte-Marie et al., 2008). Recent mating, as evidenced by the presence of a white layer of ejaculate in the ventral portion of the spermathecae, was associated with three to five-fold increases in sperm cell counts among both primiparous and

TABLE II

Summary of mean spermathecal load (SL) for Tanner crab (*Chionoecetes bairdi* (Rathbun, 1924)) and snow crab (*Chionoecetes opilio* (Fabricius, 1788)) by location, ontogenetic life stage, and year

| Species | Location | Ontogenetic Life Stage | Year | N | Mean SL (g) | SE |
|---|---|---|---|---|---|---|
| Tanner crab | Eastern Bearing Sea | Primiparous | 2005[a] | 15 | 0.09 | 0.013 |
| | Southeastern | Primiparous | 2007[b] | 72 | 0.04 | 0.006 |
| | Alaska | Multiparous | 2006[c] | 21 | 0.44 | 0.026 |
| | | | 2007[b] | 52 | 0.20 | 0.016 |
| Snow crab | Eastern Bearing Sea | Primiparous | 2005[d] | 56 | 0.02 | 0.002 |
| | | | 2007[d] | 100 | 0.03 | 0.003 |
| | | | 2008[d] | 153 | 0.06 | 0.003 |

Sources: [a]Gravel & Pengilly, 2007; [b]Webb & Bednarski, in press; [c]Webb, 2009; [d]Slater et al., in press.

multiparous females (Webb & Bednarski, in press). Comparisons among five discrete stocks of Tanner crab in southeastern Alaska found that the highest levels of sperm reserves for primiparous females was correlated with the lowest exploitation rate, while the highest levels for multiparous females were associated with an increased sex ratio of large, old-shell males to multiparous females (Webb & Bednarski, in press).

Sperm reserves of primiparous female snow crab from the EBS from 2002 through 2008 were very low compared to those of conspecifics from the Gulf of St. Lawrence (Rugolo et al., 2005; Gravel & Pengilly, 2007; Sainte-Marie et al., 2008; Slater et al., in press). Within the overall low levels (table II), there is a trend of larger reserves further south in the distributional range (Rugolo et al., 2005; Slater et al., in press), which may be related to the size and maturity status of available males. Since variation in operational sex ratios due to both spatial and temporal factors affects sperm reserves in Canadian stocks (Sainte-Marie et al., 2008), continued monitoring is necessary to determine how these factors affect sperm reserves in the EBS. However, since limited research on sperm reserves was conducted in the EBS prior to the decline in snow crab abundance in the late 1990s, it will be difficult to determine if spatial variation in sex ratio, either due to natural processes or fishery removals of large males, is associated with variability in sperm reserves.

Fecundity

Early studies of fecundity of *Chionoecetes* crabs in Alaska have focused on establishing fecundity-at-size relationships for Tanner and snow crabs

(Haynes et al., 1976). More recent studies have taken a broader approach to investigating factors affecting egg production. Fecundity increases with female size for Tanner and snow crabs, and multiparous Tanner crab females have ~30 to 50% higher fecundity than primiparous females of similar size (Somerton & Meyers, 1983; Webb & Bednarski, in press). Ambient temperature affects the rate of egg development, and snow crab in the EBS switch from an annual to a biennial clutch production at temperatures of ~1.0°C (Rugolo et al., 2005), which is similar to that for conspecifics from eastern Canada. The extent of the cold pool (<2°C) over the EBS shelf is spatiotemporally variable and related to ice extent during the previous winter (Mueter & Litzow, 2008). Egg production is likely to decrease during periods when the cold pool covers much of the snow crab distribution in the EBS and a higher proportion of females produce clutches biennially.

Comparisons of fecundity-at-size relationships for each species suggest that ontogeny may be more important than spatiotemporal differences in determining variability in fecundity. For example, the mean fecundity of a 90 mm CW primiparous Tanner crab ranges from 100 000 in Southeastern Alaska to 115 000 in the EBS, whereas the mean fecundity of a multiparous Tanner crab of the same size ranges from 163 000 in southeastern Alaska to 169 000 in the EBS (fig. 2 (Gravel & Pengilly, 2007; Webb & Bednarski, in press)). Ontogenetic and spatiotemporally-driven differences in fecundity are not as well described for snow crab in the EBS. The mean fecundity of a mixed population of primiparous and multiparous female snow crab of size 55 mm CW ranges from 20 000 in the EBS (Haynes et al., 1976) to 23 000 in the Chukchi Sea (fig. 2 (Jewett, 1981; Paul et al., 1997)).

## Snow crab distribution and movement

The large spatial scale of the EBS and extensive seasonal sea ice cover make it difficult to assess distribution and movement of snow crab. This is further complicated by the approximately six-month lag between the annual National Oceanic and Atmospheric Administration survey and the fishery. During this period, the center of distribution of large males ($\geqslant 102$ mm CW) moves from the middle continental shelf to further south and west on the outer continental shelf (Orensanz et al., 2004). Since the number of crab harvested can exceed the localized population estimate in some areas, crab are either migrating seasonally into the fishery area or the survey catchability is less than assumed (area-swept estimates use a survey catchability of one). During the summer of 2005, the Alaska Department of Fish and Game initiated a

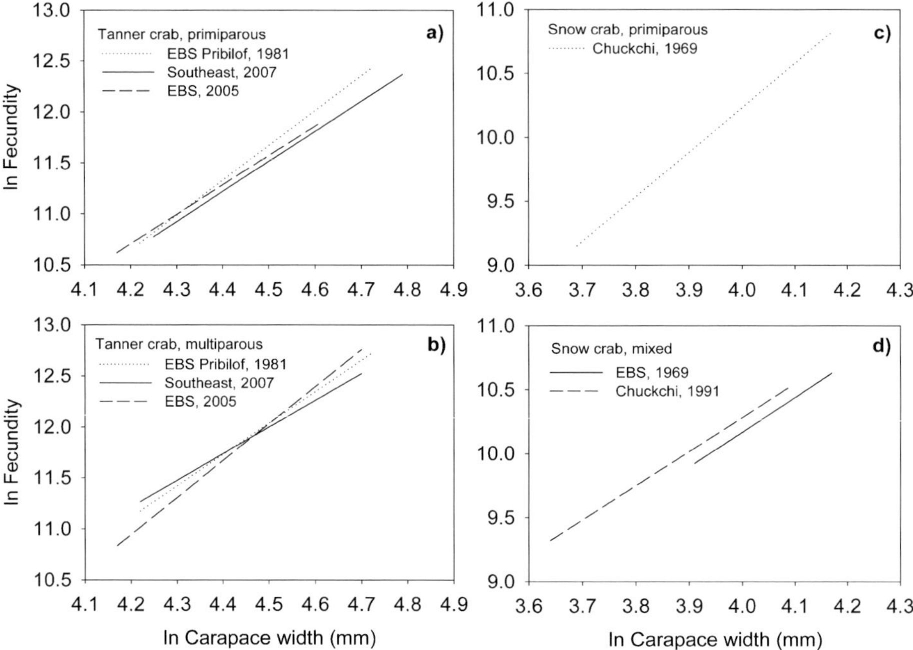

Fig. 2. Fecundity-at-size by location and year for: a, primiparous Tanner crab (*Chionoecetes bairdi* (Rathbun, 1924)); b, multiparous Tanner crab; c, primiparous snow crab (*Chionoecetes opilio* (Fabricius, 1788)); d, mixed primiparous and multiparous snow crab. Information compiled from Jewett (1981) for primiparous snow crab from the Chukchi Sea; Haynes et al. (1976) for mixed primiparous and multiparous snow crab from the eastern Bering Sea (EBS); Paul et al. (1997) for mixed primiparous and multiparous snow crab from the Chukchi Sea; Webb and Bednarski (in press) for primiparous and multiparous Tanner crab from southeastern Alaska; Gravel & Pengilly (2007) for primiparous and multiparous Tanner crab from the EBS; and Somerton & Meyers (1983) for primiparous and multiparous Tanner crab from the Pribilof Islands in the EBS.

study to assess the movement of snow crab in the EBS to better understand this spatial mismatch in seasonal distribution (Pengilly, 2006). On a broad scale, crab tagged and released closest to the fishing grounds were ten times more likely to be recaptured during the fishery than those released furthest from the fishing grounds. This indicates that crab captured outside the fishing grounds during the survey have a lower exploitation rate, suggesting limited migration. However, on a smaller spatial scale, there was very little difference in the probability of recapture for males released on the outer shelf versus the middle shelf. Since most harvest during the fishery occurs on the outer shelf, this suggests that seasonal migration of males from the middle to the outer shelf is occurring.

In addition to seasonal migration, a number of studies have indicated snow crab have an ontogenetic migration pattern. Zheng et al. (2001) described a change in distribution of successive age classes of snow crab in EBS survey data from the inner shelf in the northeast EBS to the deeper outer shelf to the southwest. This observation was further developed by Ernst et al. (2005) for mature females, suggesting that migration may be driven by bottom temperature and/or depth gradients.

Zheng et al. (2001) noted that small males and immature and primiparous females, which are presumably the least mobile component of the population, were more broadly distributed in years of high abundance and contracted to the north during years of low abundance. This pattern suggests a process other than movement is contributing to distributional changes over time. During the period 1978 to 1999, the center of distribution of the stock gradually shifted toward the northwest (Zheng et al., 2001), coinciding with a general population decline (Orensanz et al., 2004). The northward shift of snow crab has been estimated at 89 km for the whole stock (Mueter & Litzow, 2008) to 220 km for primiparous females and large males (Orensanz et al., 2004; Ernst et al., 2005). This shift in distribution has been variously attributed to a single factor, the reduction in winter sea ice cover (Mueter & Litzow, 2008), or to a succession of factors, proposed as the 'environmental ratchet hypothesis' (Orensanz et al., 2004). This hypothesis states that favorable conditions exist for snow crab larvae to grow in the middle shelf when factors such as the timing of the sea ice retreat and the location of the cold pool allow for post-larvae settlement to the cold bottom temperatures preferred by early juvenile instars (Dionne et al., 2003). Between 1982 and 2006, the average summer bottom temperature increased significantly in the EBS and the cold pool moved northward by up to 230 km (Mueter & Litzow, 2008). Once settlement and growth have shifted to the north, reestablishment of the stock in the south is difficult because larval transport is northward and predation pressure exerted by Pacific cod (*Gadus macrocephalus*), which have expanded northward to the southern range of the EBS, may prevent juvenile movement to the south (Orensanz et al., 2004; Mueter & Litzow, 2008).


## CONCLUSIONS

Concerns over the role of a male-only harvest strategy in population declines have resulted in large improvements to our understanding of reproductive biology of *Chionoecetes* crabs in Alaska. First, the sizes at maturity vary greatly among spatial stocks within a species. Due to a terminal molt to morphometric maturity for males, variability in size at maturity has implications for harvest strategy. Second, female sperm reserves may be reduced for stocks with high fishing pressure. Third, biennial spawning exists for snow crab in the EBS at temperatures of ~1.0°C. Fourth, ontogeny may be more important than spatiotemporal differences in determining variability in fecundity. Finally, snow crab in the EBS have an ontogenetic migration pattern, moving from the inner shelf in the northeast to the deeper outer shelf to the southwest and were more broadly distributed in years of high abundance and contracted to the north during years of low abundance.

Improved understanding of crab reproductive potential, crab distribution, and the role of oceanography and predation in determining crab distribution and recruitment success is increasingly vital for setting overfishing levels, refining harvest strategies and making predictions about population response to climate change. Future research needs include sperm limitation, the role of mature males on reproductive success, annual snow crab movement in the EBS, impacts of oceanography and predation on crab distributions, and factors influencing sizes at maturity and recruitment success.

## ACKNOWLEDGEMENTS

We would like to acknowledge Doug Pengilly for reviewing the early draft of this paper and Kally Spalinger, Richard Gustafson and Chris Siddon for providing data to estimate sizes at maturity for Kodiak, Cook Inlet and southeastern Alaska Tanner crab. This is contribution PP-259 from the Alaska Department of Fish and Game, Commercial Fisheries Division.

## REFERENCES

ADAMS, A. E., 1985. Some aspects of the reproductive biology of the crab *Chionoecetes bairdi*: final project report: 10 pp. (AK-SG-85-07, University of Alaska Sea Grant).

BISHOP, G., J. ZHENG, L. M. SLATER, K. SPALINGER & R. GUSTAFSON, 2011. The current status of the fisheries for *Chionoecetes* spp. (Decapoda, Oregoniidae) in Alaskan waters. Crustaceana Monographs, **15**: 109-126.

CONAN, G. Y. & M. COMEAU, 1986. Functional maturity and terminal molt of male snow crab, *Chionoecetes opilio*. Can. J. Fish. Aquat. Sci., **43**: 1710-1719.

DIONNE, M., B. SAINTE-MARIE, E. BOURGET & D. GILBERT, 2003. Distribution and habitat selection of early benthic stages of snow crab, *Chionoecetes opilio*. Mar. Ecol. Prog. Ser., **259**: 117-128.

ERNST, B., J. M. ORENSANZ & D. A. ARMSTRONG, 2005. Spatial dynamics of female snow crab (*Chionoecetes opilio*) in the eastern Bering Sea. Can. J. Fish. Aquat. Sci., **62**: 250-268.

GRAVEL, K. A. & D. PENGILLY, 2007. Investigations on reproductive potential of snow and Tanner crab females from the eastern Bering Sea in 2005. Alaska Department of Fish and Game, Fishery Data Series, **07-23**: 43 pp.

HAYNES, E., J. F. KARINEN, J. WATSON & D. J. HOPSON, 1976. Relation of number of eggs and egg length to carapace width in the brachyuran crabs *Chionoecetes bairdi* and *C. opilio* from the southeastern Bering Sea and *C. opilio* from the Gulf of St. Lawrence. J. Fish. Res. Board Can., **33**(11): 2592-2595.

JEWETT, S. C., 1981. Variations in some reproductive aspects of female snow crabs *Chionoecetes opilio*. J. Shellfish Res., **1**(1): 95-99.

MACGINITIE, G. E., 1955. Distribution and ecology of the marine invertebrates of Point Barrow, Alaska. Smithsonian Miscellaneous Collections, **128**(9): 1-201.

MUETER, F. J. & M. A. LITZOW, 2008. Sea ice retreat alters the biogeography of the Bering Sea continental shelf. Ecol. Appl., **18**(2): 309-320.

ORENSANZ, J., B. ERNST, D. A. ARMSTRONG, P. STABENO & P. LIVINGSTON, 2004. Contraction of the geographic range of distribution of snow crab (*Chionoecetes opilio*) in the eastern Bering Sea: an environmental rachet? CalCOFI Report, **45**: 65-79.

OTTO, R. S., 1998. Assessment of the eastern Bering Sea snow crab, *Chionoecetes opilio*, stock under the terminal molting hypothesis, In: G. S. JAMIESON & A. CAMPBELL (eds.), Proceedings of the North Pacific Symposium on Invertebrate Stock Assessment and Management. Can. Spec. Publ. Fish. Aquat. Sci., **125**: 109-124.

PAUL, J. M., A. J. PAUL & W. E. BARBER, 1997. Reproductive biology and distribution of the snow crab from the northeastern Chukchi Sea, In: J. B. REYNOLDS (ed.), Fish ecology in Arctic North America. American Fisheries Society Symposium, **19**: 287-294.

PENGILLY, D., 2006. Bering Sea crab research (IV): final comprehensive performance report. NOAA Cooperative Agreement, **NA04NMF4370175**: 1-115.

RUGOLO, L., D. PENGILLY, R. MACINTOSH & K. GRAVEL, 2005. Reproductive potential and life history of snow crabs in the eastern Bering Sea. In: D. PENGILLY (ed.), Bering Sea snow crab fishery restoration research: final comprehensive performance report. NOAA Cooperative Agreement, **NA04NMF4370175**: 57-324.

SAINTE-MARIE, B. & C. CARRIERE, 1995. Fertilization of the second clutch of eggs of snow crab, *Chionoecetes opilio*, from females mated once or twice after their molt to maturity. Fish. Bull., **93**(4): 759-764.

SAINTE-MARIE, B., T. GOSSELIN, J.-M. SEVIGNY & N. URBANI, 2008. The snow crab mating system: opportunity for natural and unnatural selection in a changing environment. Bull. Mar. Sci., **83**(1): 1-31.

SIDDON, C. E. & J. A. BEDNARSKI, in press. Variation in size at maturity of Tanner crab in southeastern Alaska, In: G. H. KRUSE, G. L. ECKERT, R. J. FOY, R. N. LIPCIUS, B. SAINTE-MARIE, D. L. STRAM & D. WOODBY (eds.), Biology and management of exploited crab populations under climate change. (Alaska Sea Grant College Program, University of Alaska Fairbanks).

SLATER, L. M., K. A. MACTAVISH & D. PENGILLY, in press. Preliminary analysis of spermathecal load of primiparous snow crabs (*Chionoecetes opilio*) from the eastern Bering Sea, 2005-2008, In: G. H. KRUSE, G. L. ECKERT, R. J. FOY, R. N. LIPCIUS, B. SAINTE-MARIE, D. L. STRAM & D. WOODBY (eds.), Biology and management of exploited crab populations under climate change. (Alaska Sea Grant College Program, University of Alaska Fairbanks).

SLIZKIN, A. G., 1989. Tanner crabs (*Chionoecetes opilio, C. bairdi*) of the Northwest Pacific: distribution, biological peculiarities, and population structure, In: B. MELTEFF (ed.), Proceedings of the International Symposium on King and Tanner Crabs: 27-33. (AK-SG-90-4, Alaska Sea Grant College Program, University of Alaska Fairbanks).

SOMERTON, D. A., 1981. Regional variation in the size of maturity of two species of Tanner crab (*Chionoecetes bairdi* and *C. opilio*) in the eastern Bering Sea, and its use in defining management subareas. Can. J. Fish. Aquat. Sci., **38**: 163-174.

SOMERTON, D. A., 1982. Bipartite breeding: a hypothesis of the reproductive pattern in Tanner crabs, In: Proceedings of the International Symposium on the Genus *Chionoecetes*: 283-289. (AK-SG-82-10, Alaska Sea Grant College Program, University of Alaska Fairbanks).

SOMERTON, D. A. & W. S. MEYERS, 1983. Fecundity differences between primiparous and multiparous female Alaskan Tanner crab (*Chionoecetes bairdi*). J. Crust. Biol., **3**(2): 183-186.

TAMONE, S. L., M. M. ADAMS & J. M. DUTTON, 2005. Effect of eyestalk-ablation on circulating ecdysteroids in hemolymph of snow crabs, *Chionoecetes opilio*: physiological evidence for a terminal molt. Integrated Comparative Biology, **45**: 166-171.

WEBB, J. & J. BEDNARSKI, in press. Variability in reproductive potential among exploited stocks of Tanner crab, *Chionoecetes bairdi*, in southeast Alaska, In: G. H. KRUSE, G. L. ECKERT, R. J. FOY, R. N. LIPCIUS, B. SAINTE-MARIE, D. L. STRAM & D. WOODBY (eds.), Biology and management of exploited crab populations under climate change. (Alaska Sea Grant College Program, University of Alaska Fairbanks).

WEBB, J. B., 2009. Reproductive success of multiparous female Tanner crab (*Chionoecetes bairdi*) fertilizing eggs with or without recent access to males. J. Northw. Atl. Fish. Sci., **41**: 163-172.

ZHENG, J., 2008. Temporal changes in size at maturity and their implications for fisheries management for eastern Bering Sea Tanner crab. J. Northw. Atl. Fish. Sci., **41**: 137-149.

ZHENG, J., G. H. KRUSE & D. R. ACKLEY, 2001. Spatial distribution and recruitment patterns of snow crabs in the eastern Bering Sea. In: G. H. KRUSE, N. BEZ, A. BOOTH, M. W. DORN, S. HILLS, R. N. LIPCIUS, D. PELLETIER, C. ROY, S. J. SMITH & D. WITHERELL (eds.), Spatial processes and management of marine populations: 233-255. (AK-SG-01-02, Alaska Sea Grant College Program, University of Alaska Fairbanks).


First received 16 December 2009.
Final version accepted 22 May 2010.

# RECORD OF A MALE SNOW CRAB, *CHIONOECETES OPILIO* WITH TWO EXTRA FINGERS ON THE LEFT CHELA


BY

HAJIME MATSUBARA[1]

Laboratory of Aquatic Genome Science, Department of Aquatic Biology, 196 Yasaka, Abashiri, Hokkaido 099-2493, Japan



## ABSTRACT

A male snow crab (*Chionoecetes opilio*) with two extra fingers on the left chela was caught from Abashiri, Hokkaido, Japan. These extra fingers arose from the inner proximal portion of the original immovable finger. Morphology of these extra fingers is quite similar to those of the normal fixed finger of the propodus and the dactyl. Previously, such abnormal cheliped has not been recorded in this crab species.


## INTRODUCTION

Naturally occurring abnormal chelae have been reported in the lobsters *Homarus americanus* (cf. Faxon, 1881; Cole, 1910; Przibram, 1921), *Nephrops norvegicus* (cf. Shelton et al., 1981), the crabs *Cancer pagurus* (cf. Przibram, 1921), *Geryon affinis granulatus* (cf. Okamoto, 1991), the mud crab *Scylla* spp. (cf. Fuseya & Watanabe, 1999), the intertidal mud crab *Macrophthalmus japonicus* (cf. Suzuki, 1963), the Japanese swimming crab *Charybdis japonica* (cf. Nakatani & Matsuno, 2004), the hair crab *Erimacrus isenbeckii* (cf. Suzuki & Odawara, 1971), the crayfish *Astacus fluviatilis* (cf. Bateson, 1894; Przibram, 1921) and the American crayfish *Procambarus clarkii* (cf. Nakatani et al., 1992, 1997). Recently, a male snow crab (*Chionoecetes opilio*) with two extra fingers on the left chela was caught. Such shape has previously not been recorded, and its morphology is herein described.

---

[1] e-mail: h3matsub@bioindustry.nodai.ac.jp




New frontiers in crustacean biology: 139-143
">

## MATERIAL AND METHODS

A snow crab with the extra fingers was caught from Abashiri, Hokkaido, Japan (44°20′N 144°20′E, 500 m) on 22 April 2009. The crab was weighed, and the carapace length, carapace width, movable finger length and fixed finger length were measured.

## RESULTS AND DISCUSSION

The carapace length, carapace width and body weight of the specimen are 120.75 mm, 123.26 mm and 676 g, respectively (fig. 1). The two extra fingers arise from the inner proximal portion of the original fixed finger (fig. 2). The length of the normal movable finger (dactyl) is 64.2 mm in the left (fig. 2a-b) and 66.89 mm in the right finger (fig. 2i-j). The length of the normal fixed finger is 50.94 mm in the left (fig. 2c-d) and 53.29 mm in the right finger (fig. 2k-l). In contrast, the extra movable finger (fig. 2e-f) measures 48.85 mm in length and the extra fixed finger (fig. 2g-h) 40.56 mm.

Fig. 1. Dorsal view of a male snow crab (*Chionoecetes opilio*) with an extra chela on the left cheliped. Scale bar: 5 cm.

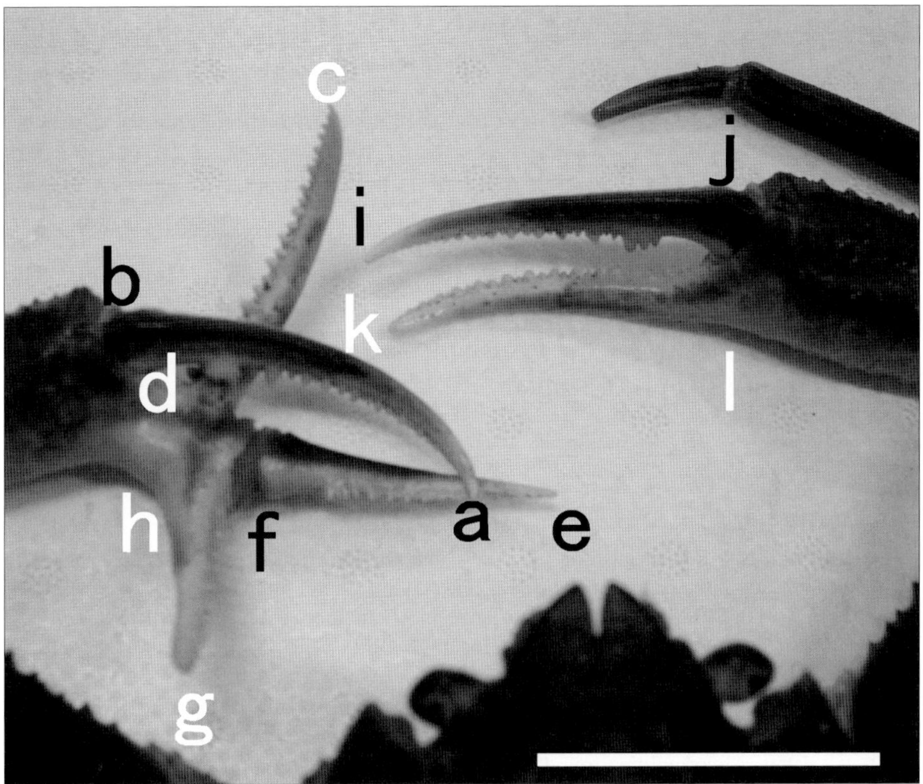


Fig. 2. The enlarged figure of the chelae. a-b, normal dactyl of left chela; c-d, normal immovable finger of left chela; e-f, extra dactyl of left chela; g-h, extra immovable finger of left chela; i-j, normal dactyl of right chela; k-l, normal immovable finger of right chela. Scale bar: 5 cm.

Interestingly, these extra movable and fixed fingers were functional and have a quite similar shape of the cutting edges as in normal ones (fig. 2). *Chionoecetes* spp. with an extra dactyl or extra fixed finger on the cheliped have been reported (Itoh, 1956, 1960, 1965, 1967; Mitsuhashi, 1993; Mizusawa and Satoh, 1965; Motoh, 1971, 1972, 2002; Motoh & Toyoda, 2003; Suzuki & Odawara, 1971; Tange & Iwasa, 1991), however, these pincers were not functional. Functional extra dactyls have been observed in the Japanese swimming crab *Charybdis japonica* (cf. Nakatani & Matsuno, 2004). In addition, Murayama et al. (1994) reported lateral outgrowths on the first cheliped of crayfish *Procambarus clarkii* (Girard) that can be induced by wounding. Moreover, these induced extra claws on the chelipeds of crayfish had nerve fibers (Nakatani et al., 1997, 1998). Therefore, I also hypothesize that the two extra fingers on the present *Chionoecetes opilio* specimen occurred by hypertypic regeneration of the wounded dactyl. Albeit probably rare, this male snow crab had two extra fingers on the left chela.

## ACKNOWLEDGEMENTS

The author thanks Mr. Masaharu Hara (Abashiri Fisheries Co.) for the supply of the male snow crab with the two extra fingers on the left chela. The author thanks Dr. Chris Norman (Japan Scientific Texts) for critical reading of the manuscript. This research was supported by grants for young scintists from the Tokyo University of Agriculture.

## REFERENCES

BATESON, W., 1894. Materials for the study of variation, treated with especial regard to discontinuity in the origin of species: i-xvi, 1-598. (Macmillan, London, New York).

COLE, L. J., 1910. Description of an abnormal lobster cheliped. Biol. Bull., **18**: 252-268.

FAXON, W., 1881. On some crustacean deformities. Bull. Mar. Comp. Zool. at Harvard, **8**: 257-274.

FUSEYA, R. & S. WATANABE, 1999. Abnormalities specimens of the mud crab genus *Scylla*. Cancer, **8**: 17-20. [In Japanese.]

ITOH, K., 1956. Abnormalities found in the snow crab, *Chionoecetes opilio*. Sampling and rearing, **18**(11): 347. [In Japanese.]

— —, 1960. Abnormalities re-found in the snow crab, *Chionoecetes opilio*. Sampling and rearing, **22**(4): 123-125. [In Japanese.]

— —, 1965. The two abnormal chela of snow crab, *Chionoecetes opilio*. Rep. Japan Sea National Fisheries Research Institute, **14**: 91-93. [In Japanese.]

— —, 1967. The three abnormal chela of snow crab, *Chionoecetes opilio*. Rep. Japan Sea National Fisheries Research Institute, **17**: 141-142. [In Japanese.]

MITSUHASHI, T., 1993. Abnormalities found in the Japanese edible crab, *Chionoecetes japonicus*. Let. Hokkaido Fisheries Experiment Station, **23**: 21. [In Japanese.]

MIZUSAWA, R. & Y. SATOH, 1965. Abnormalities found in Japanese edible crab, *Chionoecetes japonicus*. Sampling and rearing, **27**(9): 340-341. [In Japanese.]

MOTOH, H., 1971. Abnormalities found in the left cheliped of Japanese edible crab, *Chionoecetes japonicus*. Res. Crust., **4/5**: 184-190. [In Japanese.]

— —, 1972. Abnormalities found in the right cheliped of Japanese edible crab, *Chionoecetes japonicus*. Let. Hokkaido fisheries experiment station, **2**: 21-27. [In Japanese.]

— —, 2002. The two abnormal snow crab, *Chionoecetes opilio*. Cancer, **11**: 3-6. [In Japanese.]

MOTOH, H. & K. TOYOTA, 2003. Abnormal cheliped found on the third maxilliped of a male crab, *Chionoecetes opilio*. Cancer, **12**: 19-22. [In Japanese.]

MURAYAMA, O., I. NAKATAN & M. NISHITA, 1994. Induction of lateral outgrowths on the chelae of the crayfish, *Procambarus clarkii* (Girard). Crust. Res., **23**: 69-73.

NAKATANI, I. & S. MATSUNO, 2004. The Japanese swimming crab, *Charybdis japonica*, with two extra dactyls on the chela. Cancer, **13**: 17-18.

NAKATANI, I., Y. OKADA & T. KITAHARA, 1998. Induction of extra claws on the chelipeds of a crayfish, *Procambarus clarkii*. Biol. Bull., **195**: 52-59.

NAKATANI, I., Y. OKADA & T. YAMAGUCHI, 1997. An extra claw on the first and on the third cheliped of the crayfish, *Procambarus clarkii* (Decapoda, Cambaridae). Crustaceana, **70**: 788-798.

NAKATANI, I., K. YAMAUCHI & O. MURAYAMA, 1992. Abnormalities found in the chela of the crayfish, *Procambarus clarkii* (Girard). Res. Crust., **21**: 207-209.

OKAMOTO, K., 1991. Abnormality found in the cheliped of *Geryon affinis granulatus* Sakai. Res. Crust., **20**: 63-65.

PRZIBRAM, H., 1921. Die Bruchdreifachbildung im Tiemche. Wilhelm Roux Arch. Entw-Meck Org., **48**: 205-4.44.

SHELTON, P. M. J., P. R. TRUBY & R. G. J. SHELTON, 1981. Naturally occurring abnormalities (Bruchdreifachbildungen) in the chelae of three species of Crustacea (Decapoda) and a possible explanation. J. Embryol. Exp. Morphol., **63**: 285-304.

SUZUKI, H., 1963. Abnormalities found in the cheliped of the intertidal mud crab *Macrophthalmus japonicus*. Res. Crust., **1**: 51-53. [In Japanese.]

SUZUKI, H. & T. ODAWARA, 1971. Malformation found in the chelipeds of two edible crabs. Res. Crust., **4/5**: 191-195.

TANGE, K. & T. IWASA, 1991. Abnormalities found in the right cheliped of Japanese edible crab, *Chionoecetes japonicus* Rathbun. Let. Hyogo fisheries experiment station, **29**: 73-76. [In Japanese.]

First received 1 December 2009.
Final version accepted 21 December 2009.

# AMPHIDROMY AND MIGRATIONS OF FRESHWATER SHRIMPS. I. COSTS, BENEFITS, EVOLUTIONARY ORIGINS, AND AN UNUSUAL CASE OF AMPHIDROMY

BY

RAYMOND T. BAUER[1])

Department of Biology, University of Louisiana at Lafayette, Lafayette,
Louisiana 70504-2451, U.S.A.

## ABSTRACT

Many freshwater shrimps (Decapoda, Caridea) have amphidromous life histories, with extended planktonic larval development in the sea. Larvae either are hatched upstream to drift down to the sea or are carried and released there by females. After development, postlarvae (juveniles) must migrate back up to their adult freshwater habitat. An amphidromous life cycle thus involves long distance migrations between marine and fresh waters. Other freshwater shrimps have abbreviated (or direct) larval development (ALD) with a completely freshwater life cycle and without such migrations. The question of which life history pattern (amphidromy versus ALD) is ancestral in freshwater shrimps is discussed in terms of costs, benefits, and phylogeny. Competing hypotheses are presented to explain the unusual distribution of *Macrobrachium ohione*, an amphidromous shrimp with recently abundant populations located very far ($>1500$ km) from the sea.

## INTRODUCTION

Although the majority of caridean shrimps are marine, approximately 25% of the 3200 species occur in fresh water habitats (De Grave et al., 2007, 2009). The life history of some of these species is completely adapted to fresh water in that all stages of the life cycle occur there. The extended planktonic development of most marine species is abbreviated in these freshwater species, with hatching as advanced larvae and few subsequent larval stages, or is direct, with the embryo hatching out as a postlarva (small juvenile) (Hayashi &

---

[1]) e-mail: rtbauer@louisiana.edu



New frontiers in crustacean biology: 145-156

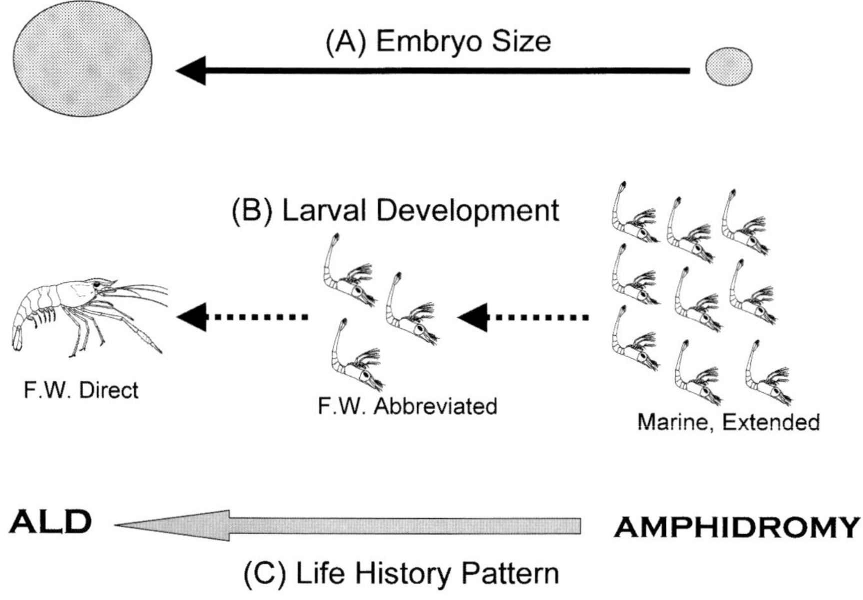


Fig. 1. The relationship of embryo (egg) size, larval development and life history pattern in freshwater shrimps. ALD, abbreviated larval and direct development. F.W., fresh water.

Hamano, 1984; Jalihal et al., 1993). To sustain extended incubation and embryonic development before hatching in these species, mature oocytes (eggs) must contain considerable yolk. Thus, females with abbreviated larval development (ALD) spawn relatively few, large eggs (fig. 1). At the other extreme in the life history spectrum of freshwater shrimps are amphidromous species, whose larvae require development in saline waters. Larval development occurs in the brackish water of estuaries and coastal bays or in the open sea. In amphidromous species, females spawn many small eggs which hatch at a much less advanced larval stage than those of species with abbreviated or direct development, and larval development is extended, with several stages (fig. 1) and is marine planktonic (Bauer, 2004).

Amphidromy is defined as a life history cycle in which there are recurring migrations between fresh water and the sea for purposes other than reproduction (McDowall, 2007). In amphidromous freshwater shrimps, principally found in the caridean families Atyidae, Xiphocarididae, and Palaemonidae (primarily *Macrobrachium* spp.), females live, breed, and spawn in fresh water but the larvae must go to the sea for development. Larvae may simply be hatched and released in the upstream habitat, using river flow to drift to the sea. Alternately, females migrate downstream, carrying their brooded embryos nearer to or into coastal bays and estuaries where hatching of embryos to larvae takes place. After larval development in coastal or open ocean environments, the newly metamorphosed postlarvae (juveniles) must find and enter the mouth of a coastal river or stream and migrate up to the adult freshwater habitat, sometimes considerable distances from the sea. Amphidromous shrimps have received ever increasing attention in the last 2-3 decades, with studies stimulated both by basic interest in life cycles and by human impacts on their lotic habitats, especially dam construction (Holmquist et al., 1998) and diversion of stream water (March et al., 2003), which impede or completely stop the downstream transport of larvae to the sea or the subsequent "return" upstream migration of juveniles.

## COSTS, BENEFITS, AND ORIGINS OF AMPHIDROMY IN SHRIMPS

Life histories of freshwater shrimps exhibit a continuum between two extremes: (a) amphidromy, with extended larval development in salt water versus (b) ALD in fresh water, with a reduction of larval stages, sometimes with direct development, with hatching as a benthic postlarva or juvenile (e.g., Magalhães & Walker, 1988; Jalihal, 1993). Amphidromy involves long distance movements or migrations while ALD does not. Hatching stages of ALD species are benthic or nearly so and can remain in or close to the adult habitat (Magalhães & Walker, 1988). What are the selective pressures that favor (benefits or advantages) or disfavor (costs or disadvantages) amphidromy and ALD? Obviously, ALD must evolve in freshwater species whose larvae are blocked from access to the sea or they become extinct. Such species are those which occur or become land-locked in inland waters or in which the distances to the sea are great, beyond the capacity of the lecithotrophic, non-feeding Stage-1 larva to survive until reaching the larval habitat downstream.

McDowall (2007), referring especially to fishes, made a comprehensive list of the advantages and disadvantages of amphidromy, which can be viewed in terms of shrimp biology. A major disadvantage is that the delicate hatching larvae leave the adult habitat to make a long dangerous trip down rivers or streams, sometimes in rapidly flowing, turbulent waters, to the marine environment. After development, the resulting juveniles must then run the gauntlet in reverse to get to back to the adult habitat. Mortality during these migrations is obviously high, given the large number of larvae that amphidromous species produce. However, amphidromy allows larvae access to the abundant planktonic food supply of estuarine and marine habitats. Magalhães & Walker (1988), working on a group of Amazonian freshwater shrimps, showed that ALD occurred in species living in inland nutrient-poor or poorly illuminated creeks and lakes in which larval food supply (plankton) was poor or lacking.

The initial invasion of freshwater habitats by carideans may have been due a complex of selective pressures: invasion of an empty or under-occupied freshwater niche, as well as escape from marine competition and predation. The stream systems of the tropical rainforests in which many freshwater shrimps reside, such as those of the island of Puerto Rico, have a detritus-based food web with a primary organic input of leaf litter, twigs and fruit (Covich & McDowall, 1996). Atyid shrimps, with their unique scraping and filtering chela brushes, are important harvesters of detritus and periphyton. *Xiphocaris elongata* is a somewhat more generalized consumer (primarily a leaf-shredder) and, at a higher tropic level, *Macrobrachium* spp. are omnivorous scavengers and predators.

The hypothesis of "escape from fish predation" as a selective pressure which favored invasion of fresh water by shrimps has some support. McDowall (2007) suggested that the freshwater fish fauna (including predators) is highly impoverished, at least on island streams where amphidromous species are abundant. Covich et al. (2009) have shown that upstream migration by the amphidromous *Atya lanipes* and *Xiphocaris elongata* allows them access to headwater refugia which cannot be reached by stream-fish predators.

A major advantage of amphidromy is dispersal (Hunte, 1978; McDowall, 2007). As larvae develop in estuaries or the open sea, dispersal to adjacent and sometimes distant streams and rivers, sometimes on other land masses, can occur. A variety of studies on populations of amphidromous species show gene flow among populations from stream systems on the same or other land masses (Page et al., 2005, 2007, 2008; Cook et al., 2006; Mashiko & Shy, 2008). Amphidromy allows colonization of new but similar habitats as well as recruitment back into the stream of larval origin (Hunte, 1978). The biogeography of Caribbean and Pacific atyid shrimps appears, in large part, to be a product of larval dispersal (Page et al., 2008). "Estuary hopping" (larval movement between nearby estuaries) or limited dispersal in the open sea has allowed gene flow among Indo-Australian populations of *Macrobrachium rosenbergii* (de Bruyn & Mather, 2007). The broader distributions of amphidromous *Macrobrachium* spp. on western Pacific islands relative to related ALD species has been attributed to larval dispersal at sea (Mashiko & Shy, 2008).

A question which has generated some controversy is: which is the plesiomorphic (ancestral) life history, amphidromy or ALD (fig. 2). The issue

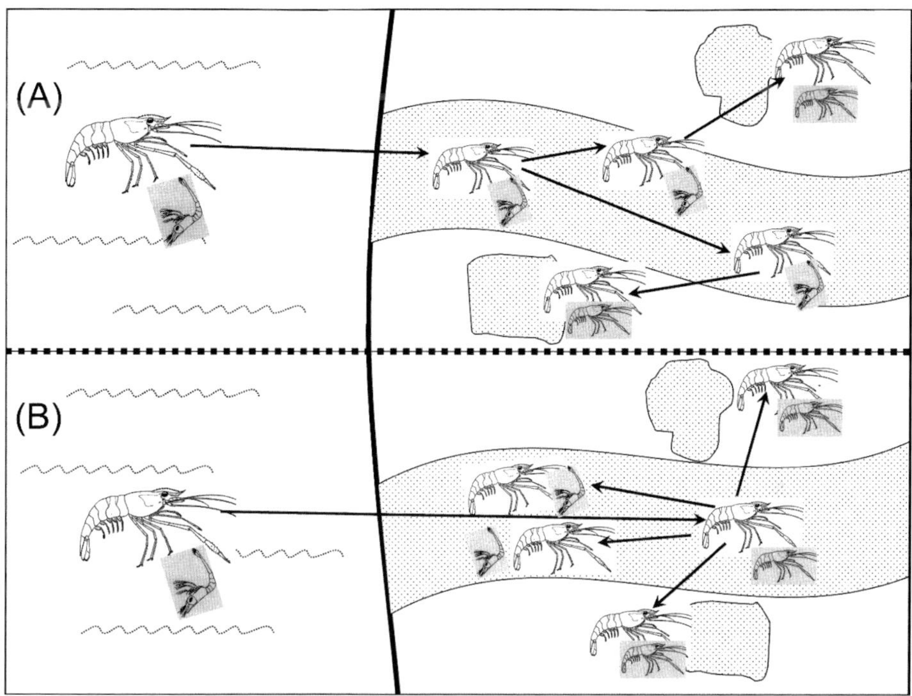



Fig. 2. Two hypotheses on the origin of amphidromy and ALD in freshwater shrimps. A, ampidromous species are derived from marine species which invade fresh water (stippled rivers and lakes to right of shoreline) and give rise to both ALD and amphidromous descendants; B, the initially marine shrimp invaders evolve ALD upon entry into freshwater habitats and then give rise to both amphidromous and other ALD species. Species with extended larval development (marine, amphidromous) are represented by an adult plus larva (shaded), ALD species by adult and miniature adult (shaded).

must be addressed separately in the Atyidae and in the palaemonid genus *Macrobrachium*, in which the full range of amphidromy to ALD (or the reverse) is found. Of 41 atyid genera (De Grave et al., 2009), only 3 occur in specialized saline habitats (Bauer, 2004). These genera aside, the atyids are strictly freshwater shrimps as adults, with life histories ranging from amphidromy to direct development. Ortmann (1894) believed that atyid ancestors entered fresh water at "a very early geological period" (perhaps the Jurassic; Ortmann, 1902). He based this on their "exclusively" freshwater habits and their primitive morphology, very closely allied to that of the marine "Acanthephyridae" (now termed Oplophoridae), which he considered the most primitive caridean family (see Bauer, 2004 and Bracken et al., 2009 for more recent views on caridean phylogeny). He regarded the freshwater habits of the family as the "original manner of living". However, he was apparently unaware of the estuarine or marine larval development of many atyids.

Other students of the Atyidae also supposed that the group radiated into fresh water at an early geological age but they suspected or knew that many atyids have marine larvae. Chace & Hobbs (1969) wrote of the "primeval *Atya* with its presumed marine larvae" while Hobbs & Hart (1982) stated, based on a morphological phylogeny and the biogeography of the genus, that an *Atya*-like ancestor existed by the late Mesozoic (early Jurassic). Carpenter (1977) hypothesized that atyids probably originated in a shallow-water Tethys Sea in the Cretaceous. Recently, the divergence time from the most recent ancestor has been estimated independently, using molecular "clocks", as the early Jurassic for the Xiphocarididae and the mid-Jurassic for its sister group, the Atyidae (Bracken et al., in press). This view is concordant with that hypothesized by the biogeographical/morphological phylogeny studies cited above.

Pereira & Garcia (1995) proposed the opposite evolutionary history for the other major amphidromous group, *Macrobrachium* spp. (Palaemonidae). In their view, the *Macrobrachium* ancestor was a completely freshwater shrimp with ALD which then gave rise to both ALD and amphidromous descendents. This view has received little support. One objection has to do with larval development. Amphidromous *Macrobrachium* spp. and atyids have extended planktonic development, as do most marine carideans. Nothing in amphidromous atyid and palaemonid larval development suggests that it is somehow significantly different from that of other marine shrimps. Extended larval development derived secondarily from ALD would presumably show some unique, recognizable features. No such features have been reported, although many descriptive studies on caridean larvae have been published. As Williamson (1982) stated in his review of decapod larval morphology and diversity, "abbreviation of larval development may certainly be regarded as a departure from the ancestral condition in Decapoda". Given the abundance, diversity, and fossil record of decapod taxa with known or presumed extended, marine, planktonic larval development, there is no reason to suppose that this is not the ancestral condition in the Decapoda including the Caridea, a primarily marine group.

Phylogenies based on gene sequences allow for an independent test of the alternative hypotheses that amphidromy or ALD is ancestral in freshwater shrimps. Some recent studies have mapped amphidromy or ALD on gene-based phylogenies. Cook et al. (2006), using the COI mtDNA gene and biogeographical data on various populations of *Paratya australiensis*, showed that the amphidromy to freshwater transition has occurred several times in this species on Australia. Page et al. (2005) used molecular phylogeny to show that amphidromy is plesiomorphic to ALD in this genus *Paratya*. Murphy & Austin (2004) analyzed the phylogeography of *Macrobrachium* from a sample of 30 species (of 238 worldwide; De Grave et al., 2009). When amphidromy and ALD were mapped on the phylogeny, 5 primarily amphidromous lineages contained derived ALD species, supporting the "amphidromy as primitive" view. However, the most basal lineages in the overall tree were ALD species, supporting the Pereira & Garcia (1995) hypothesis. Fortunately, the hypothesis was further tested in a recent analysis of phylogeny (based on several genes) and life history evolution of 45 Asian *Macrobrachium* spp. by Wowor et al. (2009). The mapping of ALD and amphidromy on their phylogenetic tree clearly shows that, as in Atyidae and Xiphocarididae, amphidromy is the primitive life history trait in the genus *Macrobrachium*.


## *MACROBRACHIUM OHIONE*, AN UNUSUAL AMPHIDROMOUS SPECIES

*Macrobrachium ohione* is exceptional in a number of ways that illustrate various aspects of amphidromous life-history migrations. It is one of six *Macrobrachium* species that inhabit coastal river systems emptying into the Gulf of Mexico and along the southeastern Atlantic coast of the United States (Bowles et al., 2000). All of these species have been assumed to be amphidromous (Bowles et al., 2000), primarily because of their geographic distribution and requirement of salt water for larval development (Dugan et al., 1975). Observations on one species, *M. ohione*, show that it is amphidromous, with both a female downstream hatching migration and an upstream juvenile migration after marine development (Bauer & Delahoussaye, 2008). This species is extraordinary among amphidromous species around the world in the distances from the sea that upstream populations are now or were formerly abundant. Instead of maximum upstream distances of several to a few hundred kilometers from the sea, as in most amphidromous shrimps, substantial reproductive *M. ohione* populations (with embryo-bearing females) were found as recently as the 1930's and 1940's as far north as 1500-2000 km from the sea (Gulf of Mexico) in the Mississippi/Ohio River System (fig. 3) (Bowles et al., 2000; Bauer & Delahoussaye, 2008).

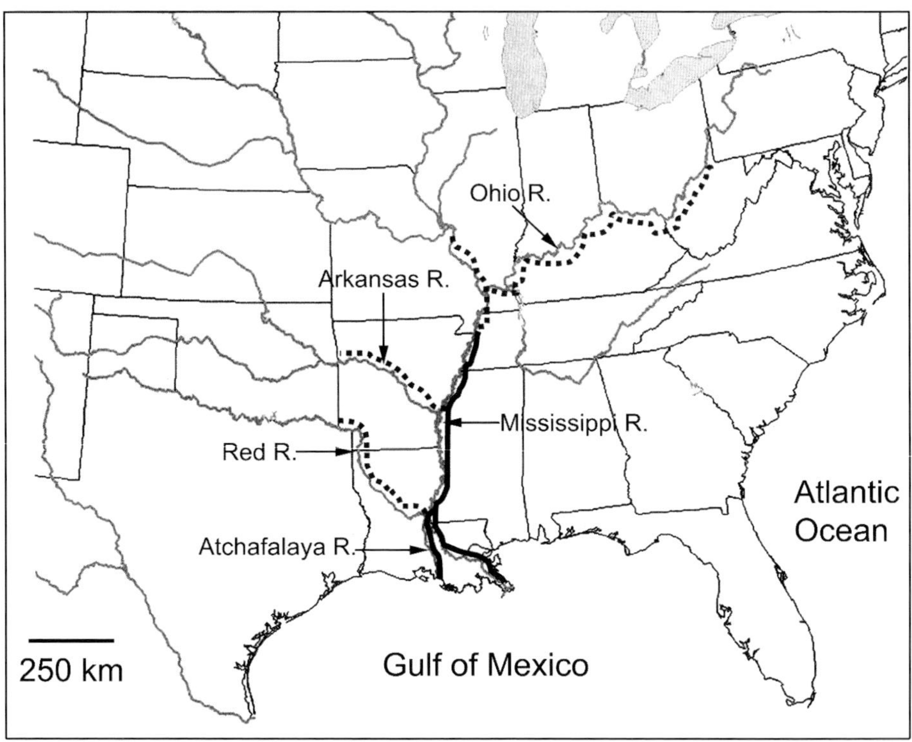


Fig. 3. The distribution *Macrobrachium ohione*, past (dotted lines) and present (solid dark lines), in the Mississippi River System, North America. Major rivers in the system are labeled (arrows).

Such a far-ranging distribution away from the sea in an amphidromous species (*M. ohione*) presents a puzzle. Is it possible that females can migrate 1500 km or more to release larvae in or within larval drifting distance of coastal estuaries? Various hypotheses may be proposed. One is that females from far northern populations do migrate down to the sea and that juveniles migrate back up again. An estimate of $\sim 1$ km hr$^{-1}$ upstream swimming speed of migrating juveniles was made by Bauer & Delahoussaye, 2008. Assuming that the nocturnally active juveniles swim upstream for 8 hr d$^{-1}$ at this speed, it would take 6 months to cover 1500 km. Juveniles do grow while migrating and could mature reproductively after arrival at far-upstream locations in their first year, as they do in coastal populations, and then migrate down again almost immediately, perhaps more quickly as reproductive adults swimming downstream. As the embryo incubation period at 22-23°C is 2-3 weeks (Bauer & Delahoussaye, 2008), embryos would hatch far upstream before the females reach the sea, and not within the maximum $\sim 150$ km distance that hatching (Stage-1) larvae can drift before safely reaching saline waters to continue development (Rome et al., 2009). Thus, a downstream female migration and return upstream juvenile migration seems implausible as a life history strategy for these far-upstream populations if life span is similar to those estimated from coastal populations (1-2 years; Truesdale & Mermilliod, 1979).

Another hypothesis that might explain the former abundance of large populations far from the sea is that larval development is abbreviated or occurs in fresh water, eliminating the need for migrations. Larval development was not, unfortunately, studied when far-northern populations were abundant and females were readily available. However, embryo size is a good indicator of the type of larval development in shrimps (Bauer, 2004). Of five females from upstream populations (museum collections) observed, all had early embryos in the same small size range as those from a coastal amphidromous population (pers. obs.). These few observations indicate full planktonic development in far-upstream populations, like that of coastal populations. However, whether such development requires salt water needs to be tested experimentally using larvae hatched from far-upstream females.

One intriguing possibility is that larval development formerly occurred upstream in low salinity larval nurseries. Along the Ohio River, in the upper Mississippi River, the Red River, and other rivers inhabited by *M. ohione* in eastern North America are ancient and extensive salt deposits and salt springs (fig. 1 in Brown, 1980). Their original influence upon rivers is now greatly diminished by human impact, e.g., prevention of salt-spring flow into river waters extracted for human use. Formerly, overflow during the spring flood, when *M. ohione* larval release takes place, may have inundated adjacent salt deposits or salt springs to create temporary low salinity areas in which amphidromous larvae could develop.

A final hypothesis is that far-upstream populations of *M. ohione* are (were) "population sinks" (McDowall, 2007), i.e., stocked solely by immigration of juveniles produced by downstream coastal populations. Individuals in upstream populations might produce embryos and larvae but these would not survive to contribute to the next generation. Such population sinks have been demonstrated in amphidromous fishes in Hawaiian streams (McRae, 2007). In *M. ohione*, it is not known what factors might induce upstream migrating juveniles to stop and remain at a particular location or continue moving upstream. Perhaps the first juveniles that come in from the sea recruit into downstream populations. Later-arriving juveniles find the downstream habitats occupied and continue upriver until finding a location in which the local density is not too high to settle into. In this way, the latest arrivals may be the juveniles that continue far upstream and keep (kept) the far-upstream populations stocked with recruits. Such a situation might be evolutionarily stable if there is no opportunity for selection to operate on the timing at which juveniles begin their upstream migration.

## ACKNOWLEDGEMENTS

I am grateful to editor Akira Asakura for organizing the 2009 Tokyo Crustacean meetings at which the symposium on migration of freshwater shrimps, which stimulated this paper, was presented. This research was supported by NOAA grant No. NA06OAR4170022 (R/SA-04) to RTB and Louisiana State University. This is Contribution number 138 of the University of Louisiana Laboratory for Crustacean Research.

## REFERENCES

BAUER, R. T., 2004. Remarkable shrimps: adaptations and natural history of the carideans. (University of Oklahoma Press, Norman).

BAUER, R. T. & J. DELAHOUSSAYE, 2008. Life history migrations of the amphidromous river shrimp *Macrobrachium ohione* from a continental large river system. J. Crust. Biol., **28**: 622-632.

BOWLES, D. E., K. AZIZ & C. L. KNIGHT, 2000. *Macrobrachium* (Decapoda: Caridea: Palaemonidae) in the contiguous United States: a review of the species and assessment of threats to their survival. J. Crust. Biol., **20**: 158-171.

BRACKEN, H. D., S. DE GRAVE & D. L. FELDER, 2009. Phylogeny of the infraorder Caridea based on mitochondrial and nuclear genes (Crustacea: Decapoda). Crust. Issues, **18**: 281-308.

BRACKEN, H. D., S. DE GRAVE, A. TOON, D. L. FELDER & K. A. CRANDALL, 2009. Phylogenetic position, systematic status, and divergence time of the Procarididea (Crustacea: Decapoda). Zoologica Scripta, **39**(2): 198-212.

BROWN, I. W., 1980. Salt and the eastern North American Indian: an archaeological study. Lower Mississippi Survey, Peabody Mus., Harvard Univ., Bull., **6**: 1-106.

CARPENTER, A., 1977. Zoogeography of the New Zealand freshwater Decapoda: a review. Tuatara, **23**: 41-48.

CHACE, F. A., JR. & H. H. HOBBS, JR., 1969. The freshwater and terrestrial decapod crustaceans of the West Indies with special reference to Dominica. Bull. United States Nat. Mus., **292**: 1-258.

COOK, B. D., A. W. BAKER, T. J. PAGE, S. C. GRANT, J. H. FAWCETT, D. A. HURWOOD & J. M. HUGHES, 2006. Biogeographic history of an Australian freshwater shrimp, *Paratya australiensis* (Atyidae): the role life history transition in phylogeographic diversification. Mol. Ecol., **15**: 1083-1093.

COVICH, A. P., T. A. CROWL, C. L. HEIN, M. J. TOWNSHEND & W. H. McDOWALL, 2009. Predator-prey in river networks: comparing spatial refugia in two drainage systems. Fresh. Biol., **54**: 450-465.

COVICH, A. P. & W. H. MCDOWALL, 1996. The stream community. In: D. REAGAN & R. WAIDE (eds.), The food web of a tropical rain forest community: 433-459. (University of Chicago Press, Chicago).

DE BRUYN, M. & P. B. MATHER, 2007. Molecular signatures of Pleistocene sea-level changes that affected connectivity among freshwater shrimp in Indo-Australian waters. Mol. Ecol., **16**: 4295-4307.

DE GRAVE, S. Y., A. CAI & A. ANKER, 2007. Global diversity of shrimps (Decapoda: Caridea) in freshwater. Hydrobiologia, **595**: 287-293.

DE GRAVE, S., N. D. PENTCHEFF, S. T. AHYONG, T.-Y. CHAN, K. A. CRANDALL, P. C. DWORSCHAK, D. L. FELDER, R. M. FELDMANN, C. H. J. M. FRANSEN, L. Y. D. GOULDING, R. LEMAITRE, M. E. Y. LOW, J. W. MARTIN, P. K. L. NG, C. E. SCHWEITZER, S. H. TAN, D. TSHUDY & R. WETZER, 2009. A classification of living and fossil genera of decapod crustaceans. Raffles Bull. Zool., (Supplement) **21**: 1-114.

DUGAN, C. C., R. W. HAGOOD & T. A. FRAKES, 1975. Development of spawning and mass larval rearing techniques for brackish-freshwater shrimps of the genus *Macrobrachium* (Decapoda: Palaemonidae). Florida. Res. Publ., **12**: 1-28.

FRYER, G., 1977. Studies on the functional morphology and ecology of atyid prawns in Dominica. Phil. Trans. Roy. Soc. London, (B, Biol. Sci.) **277**: 57-129.

HAYASHI, K.-I. & T. HAMANO, 1984. The complete larval development of *Caridina japonica* De Man reared in the laboratory. Zool. Sci., **1**: 571-589.

HOBBS, H. H., JR. & C. W. HART, JR., 1982. The shrimp genus *Atya* (Decapoda: Atyidae). Smith. Cont. Zool., **364**: 1-143.

HOLMQUIST, J. G., J. M. SCHMIDT-GENGENBACH & B. BUCHANAN-YOSHIOKA, 1998. High dams and marine-freshwater linkages: effects on native and introduced fauna in the Caribbean. Cons. Biol., **12**: 621-630.

HUNTE, W., 1978. The distribution of freshwater shrimps (Atyidae and Palaemonidae) in Jamaica. Zool. J. Linn. Soc., **64**: 35-150.

JALIHAL, D. R., K. N. SANKOLLI & S. SHENOY, 1993. Evolution of larval developmental patterns and the process of freshwaterization in the prawn genus *Macrobrachium*. Crustaceana, **65**: 365-376.

MAGALHÃES, C. & I. WALKER, 1988. Larval development and ecological distribution of central amazonian palaemonid shrimps (Decapoda, Caridea). Crustaceana, **55**: 279-292.

MARCH, J. G., J. P. BENSTEAD, C. M. PRINGLE & F. N. SCATENA, 2003. Damming tropical island streams: problems, solutions, alternatives. Bioscience, **53**: 1069-1078.

MASHIKO, K. & J.-Y. SHY, 2008. Derivation of four morphologically affiliated species of *Macrobrachium* (Caridea, Palaemonidae) with divergent reproductive characteristics in northeastern Asia. J. Crust. Biol., **28**: 370-377.

MCDOWALL, R. M., 2007. On amphidromy, a distinct form of diadromy in aquatic organisms. Fish & Fisheries, **8**: 1-13.

MCRAE, M. G., 2007. The potential for source-sink population dynamics in Hawaii's amphidromous fishes. Bishop Mus. Bull. Cult. Environ. Stud., **3**: 87-98.

MURPHY, N. P. & C. M. AUSTIN, 2005. Phylogenetic relationships of the globally distributed freshwater prawn genus *Macrobrachium* (Crustacea: Decapoda: Palaemonidae): biogeography, taxonomy and the convergent evolution of abbreviated larval development. Zool. Scr., **34**: 187-197.

ORTMANN, A. E., 1894. A study of systematic and geographical distribution of the decapod family Atyidae Kingsley. Proc. Nat. Sci. Philiadelphia, **1894**: 397-416.

——, 1902. The geographical distribution of freshwater decapods and its bearing upon ancient geography. Proc. American Phil. Soc., **41**: 267-400.

PAGE, T. J., A. M. BAKER, B. D. COOK & J. M. HUGHES, 2005. Historical transoceanic dispersal of a freshwater shrimp: the colonization of the South Pacific by the genus *Paratya* (Atyidae). J. Biogeogr., **32**: 581-593.

PAGE, T. J., B. D. COOK, T. VON RINTELEN, K. VON RINTELEN & J. M. HUGHES, 2008. Evolutionary relationships of atyid shrimp imply both ancient Caribbean radiations and common marine dispersals. J. North American Benthol.Soc., **27**: 68-83.

PAGE, T. J., K. VON RINTELEN & J. M. HUGHES, 2007. An island in the stream: Australia's place in the cosmopolitan world of Indo-West Pacific freshwater shrimp (Decapoda: Atyidae: *Caridina*). Mol. Phylog. Evol., **43**: 645-659.

PEREIRA, G. A. & J. V. GARCÍA, 1995. Larval development of *Macrobrachium reyesi* Pereira (Decapoda: Palaemonidae), with a discussion on the origin of abbreviated development in palaemonids. J. Crust. Biol., **15**: 117-133.

ROME, N., S. L. CONNER & R. T. BAUER, 2009. Delivery of hatching larvae to estuaries by an amphidromous river shrimp: tests of hypotheses based on larval moulting and distribution. Freshw. Biol., **54**: 1924-1932.

TRUESDALE, F. M. & W. J. MERMILLIOD, 1979. The river shrimp *Macrobrachium ohione* (Smith) (Decapoda, Palaemonidae): its abundance, reproduction, and growth in the Atchafalaya basin of Louisiana, U.S.A. Crustaceana, **32**: 216-220.

WILLIAMSON, D. I., 1982. Larval morphology and diversity. In: D. E. BLISS (ed.), The biology of the Crustacea. Volume 2. Embryology, morphology, and genetics: 43-110. (Academic Press Inc., New York).

WOWOR, D., V. MUTHUS, R. MEIER, M. BALKE, Y. CAI & P. K. L. NG, 2009. Evolution of life history traits in Asian freshwater prawns of the genus *Macrobrachium* (Crustacea: Decapoda: Palaemonidae) based on multilocus molecular phylogenetic analysis. Mol. Phyl. Evol., **52**: 340-350.

First received 16 December 2009.
Final version accepted 13 January 2010.

# AMPHIDROMY AND MIGRATIONS OF FRESHWATER SHRIMPS. II. DELIVERY OF HATCHING LARVAE TO THE SEA, RETURN JUVENILE UPSTREAM MIGRATION, AND HUMAN IMPACTS


BY

RAYMOND T. BAUER[1])

Department of Biology, University of Louisiana at Lafayette, Lafayette,
Louisiana 70504-2451, U.S.A.



## ABSTRACT

Hatching (Stage-1) larvae of amphidromous shrimps do not feed and must reach salt water within a few days to molt to Stage 2, the first feeding instar. Stage-1 larvae are transported from to the sea after upstream hatching by drifting in stream flow or are carried to estuaries for hatching by females migrating downstream. Hatching usually occurs during seasons or periods of high stream flow. After development in the sea, newly metamorphosed benthic postlarvae (juveniles) must find stream mouths and migrate upstream to the adult freshwater habitat. Such migrations are striking, occurring during periods of low but continuous flow, with many juveniles walking or swimming alongside the shore at night. The migratory behavior is a positive rheotaxis, with downstream river flow the directional cue. Juveniles are capable of climbing over or around low obstacles in their path provided that there is some downstream flow. Both larval drift and juvenile migrations are blocked by high dams without passageways and by the reservoirs behind them. Water extraction from streams is a significant source of larval mortality. Human impacts can be mitigated by appropriate conservation measures, e.g., restriction of water extraction during periods of larval abundance, and construction of passageways up and around dams and reservoirs to allow juvenile migration to upstream habitats.


## INTRODUCTION

Amphidromy is a life history pattern defined by diadromous migrations (Bauer, 2011, this volume). Here, I address two major aspects of amphidromy, the delivery of larvae to the sea and the return upstream juvenile migration, as well as human impacts on these migrations.

---

[1]) e-mail: rtbauer@louisiana.edu



New frontiers in crustacean biology: 157-168

## DELIVERY OF LARVAE TO THE SEA

Research on amphidromous shrimps has long indicated that females hatch their larvae into stream flow, with larvae drifting more or less passively to downstream estuarine or marine habitats (fig. 1) (Hunte, 1978; Hamano & Hayashi, 1992; March et al., 1998, 2003; Benstead et al., 1999, 2000; Bauer & Delahoussaye, 2008). Many amphidromous species inhabit streams in which distances from the adult habitat to the sea are relatively short, a few to dozens of kilometers, e.g., Caribbean islands, Japan, Taiwan, Costa Rica). Stage-1 (hatching) larvae of amphidromous species are lecithotrophic, i.e., do not feed, instead utilizing yolk remaining from embryonic development. Such larvae must molt to Stage 2 (first feeding stage) before their food stores are used up or face starvation (Rome et al., 2009). Thus, Stage-1 larvae have a limited period, usually a few days, to drift to the saline waters which will trigger molting to Stage 2. In amphidromous species in streams with a 1-3 days drifting distance to the sea, larvae can simply be released upstream to drift to the sea.

However, in river systems on large land masses, distances from the adult habitat to the sea may be hundreds of kilometers or more (Bauer & Delahoussaye, 2008). Such distances may be well beyond the drifting capacity of Stage-1 larvae. Females may have to assist larval delivery by migrating downstream to or near coastal waters where hatching then occurs. Various observations have indicated such migrations in different *Macrobrachium* species on continental land masses, e.g., *M. rosenbergii* (cf. Ling, 1969), *M. malcomsonii* (cf. Ibrahim, 1962), and *M. ohione* (cf. Bauer & Delahoussaye, 2008). Females of the palaemonid *Cryphiops caementarius* migrate from as much as 100 km upstream to enter brackish water to hatch embryos for larval development in coastal waters (Hartmann, 1958). In such species, how long (far) can a non-feeding (Stage-1) safely drift in fresh water until reaching the sea and still molt successfully to Stage 2, the first feeding stage? This question was addressed with a factorial experiment on larval development in *M. ohione* by Rome et al. (2009). High survival and molting occurred in treatments in which larvae were maintained in fresh water 1 or 3 d before transfer to saline water of 6 or 10 (but not 2) ppt. On the other hand, larvae maintained 5 d in fresh water before transfer had poor survival and molting at all salinities. Thus, *M. ohione* larvae hatched within or very near the estuary have the greatest chance of continuing larval development.

In many amphidromous species, release of larvae coincides with high stream flows which facilitate both female downstream migration or rapid larval drift to the sea (fig. 2A). In palaemonid species in continental large

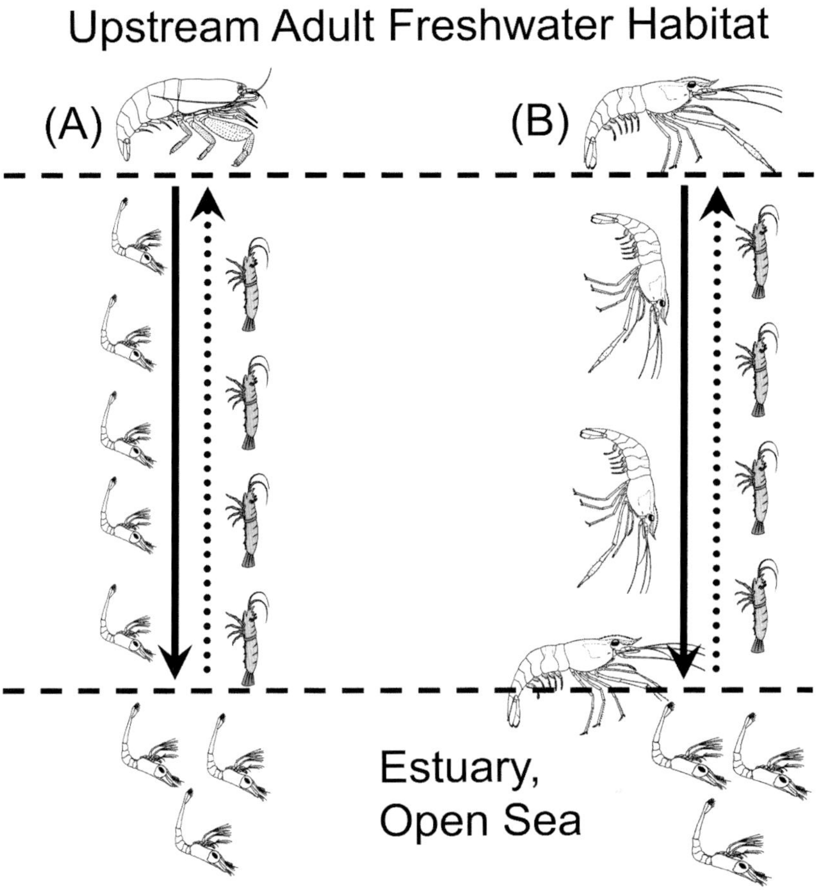

## Upstream Adult Freshwater Habitat

(A)

(B)

### Estuary, Open Sea

Fig. 1. Migrations of amphidromous shrimps. A, upstream females hatch larvae (upside-down swimmers) which then drift in stream flow down to the sea; B, females incubating embryos migrate down to river mouths to hatch larvae. In both A and B, after planktonic development in salt water, larvae metamorphose to benthic postlarvae which then migrate as young juveniles upstream to the adult freshwater habitat.

rivers systems, female migration to or near estuaries occurs during the river's seasonal flood (Hartmann, 1958; Ibrahim, 1962; Bauer & Delahoussaye, 2008). In amphidromous species which depend only on river flow to deliver larvae to the sea, hatching by upstream females usually occurs during periods or seasons of high stream flow. In Central America, distances to the sea are relatively short, and hatching and larval drift apparently occur during the rainy season, when stream flows are high (Ingo Wehrtmann and Luis Rólier, pers. comm.). Likewise, freshwater shrimps in high gradient streams on the mountainous island of Puerto Rico tend to have their peak reproduction when

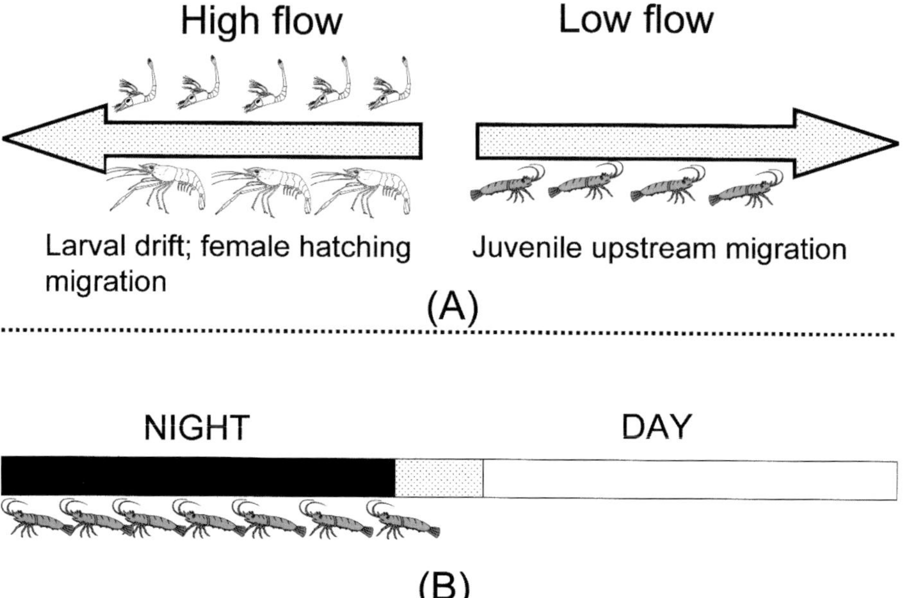


Fig. 2. Major factors affecting the timing of amphidromous shrimp migrations. A, larval release and drift to the sea, as well as female downstream hatching migrations in species which have them, tend to occur during seasonal periods of high downstream flow; juvenile upstream migrations take place during low stream flow, when flow resistance to upstream movement is lower; B, juvenile migrations occur at night in the relative absence of light; migrating juveniles will avoid (move away from) strong illumination (e.g., floodlights) on shore structures and bridges.

river flow is high (Johnson et al., 1998). On Miyako-jima Island (Ryukyus, Japan), two amphidromous carideans from an anchialine cave habitat release larvae when freshwater levels of small cave pools rise sufficiently, due to seasonal precipitation, to allow larval exit from the caves into the sea for development (Yoshihisa Fujita, pers. comm.).

## RETURN UPSTREAM MIGRATION BY JUVENILES

After larval development, the newly-metamorphosed individual must find the mouth of a stream and migrate back up to the adult habitat (fig. 1). In carideans, the zoeal larva swims with natatory thoracic exopods; in the last larval (decapodid) stage, the young shrimp has functional pleopods (swimerets) but retains natatory exopods. When the latter are lost completely, the individual is a juvenile; transitional stages in which these degenerate are postlarvae (Anger, 2001). Young individuals migrating upstream in *M*.

*rosenbergii* and *M. ohione* are juveniles (Ling, 1969; pers. obs., respectively) as is likely in other amphidromous species.

Juveniles migrate upstream from the sea at night (fig. 2B) (Ibrahim, 1962; Hamano & Hayashi, 1992; Benstead et al., 1999; Bauer & Delahoussaye, 2008; Kikkert et al., 2009). The ultimate cause of nocturnal migration is avoidance of predation by visually hunting fish and birds (e.g., Kikkert et al., 2009), with reduction in light intensity the proximate factor. However, Kikkert et al. (2009) analyzed the influence of cloud cover and moonlight on juvenile migrations of three species (from 3 families) and did not always find the expected positive or negative effects. During the day, migrating juveniles may be resting, feeding, and molting in protected habitat along the river bank. The latter is suggested by the increase in size (growth) with increasing distance upstream from the sea observed in migrating juveniles of various amphidromous species (Hartmann, 1958; Bauer & Delahoussaye, 2008; Kikkert et al., 2009; Ingo Wehrtmann and Luis Rólier, pers. comm.).

Migrating juveniles are usually found along the stream bank in very shallow water or in the splash zone, often with their bodies partially or completely out of the water (e.g., Hamano & Honke, 1997; Benstead et al., 1999). They move upstream by a combination of swimming, walking, and crawling along the bottom. Juveniles of various amphidromous species have been observed crawling up vertical or near-vertical natural barriers such as low waterfalls and brush piles as well as artificial barriers such as low weirs and dams (e.g., Ibrahim, 1962; Ling, 1969; Hamano & Hayashi, 1992; Benstead et al., 1999; Kikkert et al., 2009). When juveniles encounter an obstacle, they can crawl up or around it along the wet edges of the obstacle (Benstead et al., 1999). There must be some flow over the barrier for movement to occur (e.g., Hamano & Hayashi, 1992; Benstead et al., 1999, Fièvet, 1999; March et al., 2003). On the other hand, in *Macrobrachium ohione*, which occurs in large deep rivers, migrating juveniles swim near the surface in a band or swarm within 1-2 m of the river bank, sometimes right along the edge of the water (Bauer & Delahoussaye, 2008). The unidirectional flow of water downstream is the probable cue that stimulates a positive rheotaxis in migrating juveniles, whether they are crawling or swimming.

An obvious hypothesis to explain juvenile migrations along the stream edge is that water velocity is lowest there (fig. 3). Downstream flow (the directional cue) is present, but less energy is required to move against it. When encountering an obstacle, juveniles seek areas of low flow to climb up or around it (Benstead et al., 1999). Perhaps for the same reason, juvenile migrations generally

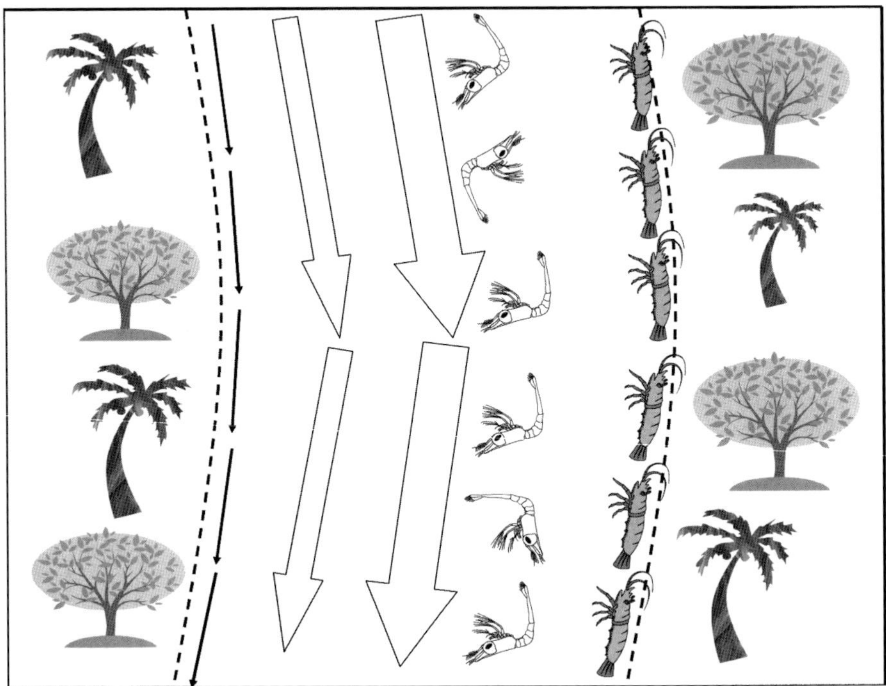

Fig. 3. Relationship between the midchannel to shore stream velocity and location of larval and juvenile migrations of amphidromous shrimps. River velocity (width of downstream pointing arrows) is greatest towards midstream and diminishes towards the shore where it is minimal. Larvae drift downstream in the bulk flow of the stream; juveniles migrate along the shore where river flow offers the least resistance to upstream movement while still retaining a directional cue for juveniles to follow.

take place when stream flows are seasonally low (but not absent) (fig. 2A). In *Macrobrachium malcolmsonii*, the migration takes place in the River Godavari from August to February, when the river is lower and water velocity is slowing from highs of the previous June–September monsoon flood (Ibrahim, 1962). Similarly, the upstream migration of *Cryphiops caementarius* occurs during the low flow periods in Peruvian coastal streams from June–September (austral winter) (Hartmann, 1958). Peak juvenile migrations of *M. ohione* in the Atchafalaya River coincide with decreasing water velocity that occurs during the summer in the lower Mississippi River system (Bauer & Delahoussaye, 2008). Juveniles of various *Macrobrachium* spp. on the Pacific coast generally migrate upstream during the seasonal dry season, when stream flow was reduced (Ingo Wehrtmann and Luis Rólier, pers. comm.).

Differences among species in migratory response to stream velocities are related to differences in body morphology and degree of resistance to high flows. Kikkert et al. (2009) analyzed juvenile migrations of amphidromous species from three different genera and families in Puerto Rico. Migrations of two of them (*Xiphocaris elongata*; *Macrobrachium* spp.) were negatively correlated with high flows, as might be expected, but not those of a third (*Atya* spp.). *Xiphocaris elongata* is a slender shrimp whose body is held high off the substratum by long slender legs (Fryer, 1977) and thus is most easily displaced downstream by high flows. *Atya* spp. have a much stouter, heavier body which hangs down close to the substratum between robust, short legs better adapted for clinging. The juveniles of *Macrobrachium* spp. are intermediate in overall morphology and climbing behavior.

## OCCURRENCE AND REDUCTION OF HUMAN IMPACTS

The most dramatic and significant human alteration of amphidromous shrimp habitat is the blocking of migratory routes by high dams (spillway height > 15 m; March et al., 2003) (fig. 4). Headwaters above high dams without any spillway discharge or fishway (fish ladder, passageway, ramp) completely lack amphidromous shrimps, which were present in equivalent streams without dams (e.g., Holmquist et al., 1998). Horne & Besser (1977) trapped *Macrobrachium* spp. at different points along the San Marcos and Guadalupe Rivers in Texas. Several high bottom-release dams had been built along the 325 km length of the river, and 3 of 4 *Macrobrachium* spp. now occur only downstream of the dam nearest the river mouth. Only 1 species, *M. carcinus*, which apparently can crawl around dams, occurs throughout the length of the river system (Horne & Besser, 1977).

Juveniles are capable of climbing low-incline, man-made passageways with water flow (see below). Although juveniles can surmount low dams with flow, the latter are still a partial impediment to migration. The juveniles encountering an obstacle tend to accumulate below it, attracting predators such as birds and fishes, including migrating predatory fishes which are blocked from moving upstream (Benstead et al., 1999) (fig. 4).

Although no construction of new dams and elimination of unnecessary ones is the best alternative to blockage of amphidromous migrations, passage around such barriers is possible. Various studies have shown that juvenile shrimps migrating upstream will climb up fish ladders or other suitable constructed ramps (Hamano et al., 1995; Hamano & Honke, 1997; Benstead et al., 1999; Fièvet, 1999). Japanese researchers have conducted experimental studies showing that the ideal "shrimp ladder" is a ramp with an inclination

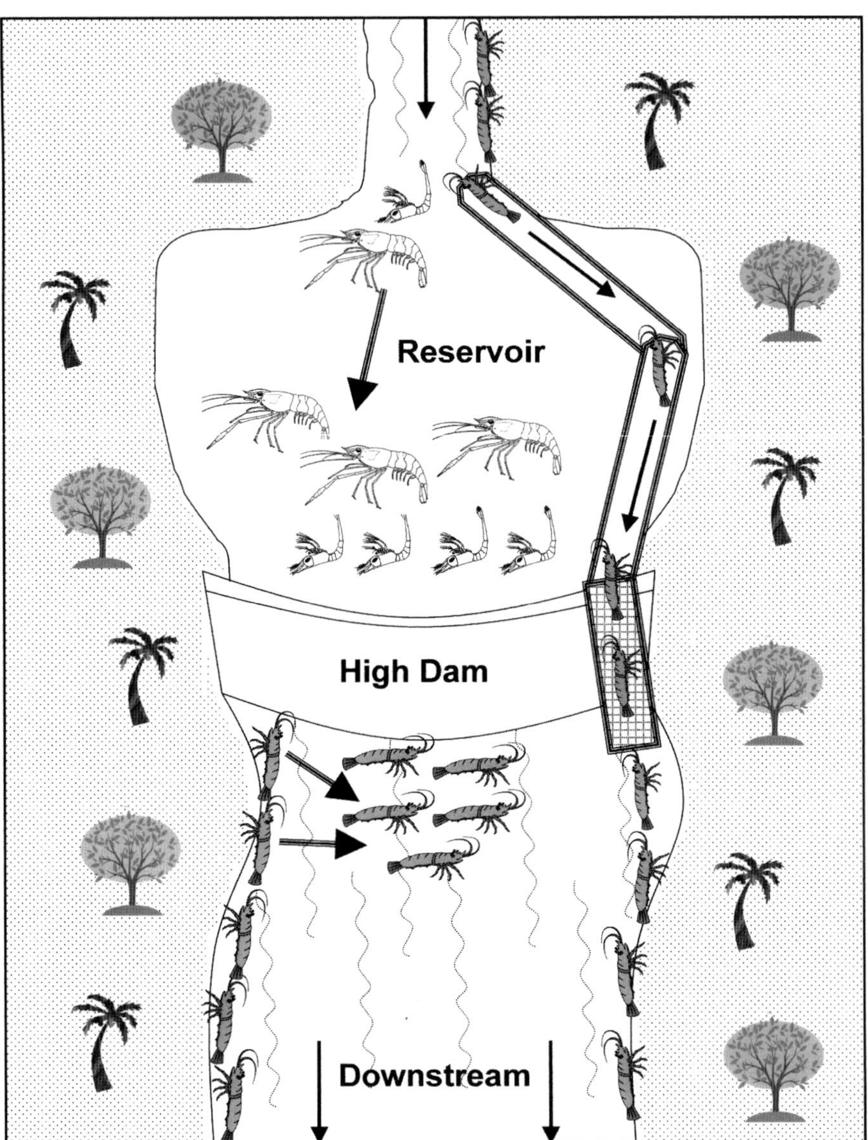


Fig. 4. Human impacts on amphidromous shrimp migrations. Stage-1 larvae (upside-down swimmers) released in upstream headwaters (top of figure) by females in one species, and adult females (unshaded, upright) of another species migrating downstream to hatch in coastal waters, are trapped (arrow with double line) in the reservoir upstream of the high dam. Juveniles (shaded) of both species migrating upstream after larval development are blocked (left stream bank) and accumulate (arrows with double lines) downstream of the dam; on the right side of the stream, a shrimp ramp (rectangle with mesh fill) allows juveniles to climb up and over the dam. If a channel with directional flow (solid arrows) is provided, juveniles may be able to bypass the still waters of the reservoir and move into the headwaters upstream of the reservoir.

of $\leqslant 50°$, a flow of water at speeds of $\leqslant 65$ cm sec$^{-1}$, and a flooring with sufficient purchase for the tips of the shrimps' walking leg ($\sim$0.5 mm mesh, e.g., lined with artificial sponge scrubber mesh, or constructed with cellular concrete). Hamano & Honke (1997) showed how floodlight illumination of one bank can be used to direct migrating shrimps, which avoid such light, to the opposite bank below a dam equipped with a fishway. Pompeu et al. (2006) reported that juvenile migraters enter and are transported to the upstream side of power plant dam with a fish lift (elevator). If dams or other obstacles are low enough, continual or periodic flow over the structure will stimulate juvenile movement over them. No studies, however, have addressed the issue of how adult females moving downstream to release larvae, in those species which do so, might be able to continue downstream past dams. Whether or not they would be able to find and migrate down shrimp or fish ladders is unknown.

Other structures along the bank, such as wharves, jetties, revetments, wing dikes and other river control structures may block or interrupt the migration route of juveniles. Flow patterns downstream of such structures may be complex and confuse the directional response of migrating juveniles. The decline in the once-abundant populations of the amphidromous *M. ohione* in the upper Mississippi and lower Ohio Rivers might due to such interruption of juvenile recruitment to upstream populations (Bauer & Delahoussaye, 2008; Bauer, 2010). The actual effect of along-bank structures on juvenile migrations needs to be tested.

The reservoirs behind high dams are also a problem for amphidromous shrimps. Even if juvenile migraters pass by a high dam via a "shrimp ladder", the lack of directional flow in the reservoir may confuse them and prevent further movement upstream. For this reason, Holmquist et al. (1998) recommended the construction of side channels between the shrimp ladder and upstream flow to circumvent the reservoir (fig. 4). Many reservoirs are stocked for recreational fishing with species predatory on shrimps (Holmquist et al., 1998). Reservoirs of high dams without frequent spillway discharge must also act as traps for shrimp larvae drifting down from upstream, preventing the larvae from continuing on to the sea (fig. 4). Although likely, this source of significant larval mortality has not been studied (March et al., 2003). In species in which females migrate downstream to hatch larvae, the reservoir-dam complex may block the migration (fig. 4).

Reservoirs behind dams are often the site for extracting water to use in municipal water systems and agriculture. The large volumes of water removed contain large numbers of larvae (Benstead et al., 1999; March et al., 2003).

Intake screens to keep out large fish and debris do and can not have a mesh small enough to block tiny shrimp larvae from entering. However, various measures can be taken to greatly reduce this source of larval mortality (Benstead et al., 1999; March et al., 2003). In tropical streams, at least, females release larvae in the early evening, i.e., in the $\sim$3 h after sunset (March et al., 1998). Limiting extraction of water from a reservoir above a dam for 3-5 h in the early evening would greatly reduce larval mortality (March et al., 2003). A knowledge of the species reproductive season would make this limitation necessary only during that period of the year. Reduction of all water extraction by water conservation measures and elimination of wasteful water usage would further limit larval mortality (March et al., 2003).

Amphidromous shrimps are important components of the ecosystems in which they occur. Within the freshwater (juvenile and adult) portion of their life cycle, they serve both as primary and secondary consumers. In tropical island streams, biomass of these shrimps is significant. Although the ecology of amphidromous shrimps in continental river systems is much less studied, given their often high abundance and use in artisanal fisheries, they must also have important ecological roles in their habitats. Larvae delivered to and developing in estuaries and nearshore coastal waters must represent a measurable and possibly significant energy transfer from the freshwater to the marine environment. Having grown in size and energy content during development in the sea, the upstream migrating juveniles must likewise represent an important energy input from marine habitats into freshwater streams, their adult habitat. Amphidromous shrimps, as adults, are often the focus of local artisanal fisheries. For these reasons, the considerable human impacts on amphidromous species should be reduced as much as possible. An understanding of their migrations is essential in identifying and mitigating human impacts on these species.

## ACKNOWLEDGEMENTS

I am grateful to editor Akira Asakura for organizing the 2009 Tokyo Crustacean meetings at which the symposium on migration of freshwater shrimps, which stimulated this paper, was presented. This research was supported by NOAA grant No. NA06OAR4170022 (R/SA-04) to RTB and Louisiana State University. This is Contribution number 139 of the University of Louisiana Laboratory for Crustacean Research.

# REFERENCES

ANGER, K., 2001. The biology of decapod crustacean larvae. Crust. Issues, **14**: 1-420. (A.A. Balkema, Lisse).

BAUER, R. T., 2011. Amphidromy and migrations of freshwater shrimps. I. Costs, benefits, evolutionary origins, and an unusual case of amphidromy. In: A. ASAKURA (ed.), New frontiers in crustacean biology: Proceedings of the TCS Summer Meeting, Tokyo, 20-24 September 2009. Crustaceana Monographs, **15**: 145-156. (Brill, Leiden).

BAUER, R. T. & J. DELAHOUSSAYE, 2008. Life history migrations of the amphidromous river shrimp *Macrobrachium ohione* from a continental large river system. J. Crust. Biol., **28**: 622-632.

BENSTEAD, J. P., J. G. MARCH & C. M. PRINGLE, 2000. Estuarine larval development and upstream post-larval migration of freshwater shrimps in two tropical rivers of Puerto Rico. Biotropica, **32**: 545-548.

BENSTEAD, J. P., J. G. MARCH, C. M. PRINGLE & F. N. SCATENA, 1999. Effects of a low-head dam and water abstraction on migratory tropical stream biota. Ecol. Appl., **9**: 656-668.

FIÈVET, É., 1999. An experimental survey of freshwater shrimp migration in an unimpounded stream of Guadeloupe Island, Lesser Antilles. Archive. für Biol., **144**: 339-355.

FRYER, G., 1977. Studies on the functional morphology and ecology of atyid prawns in Dominica. Phil. Trans. Roy. Soc. London, (B, Biol. Sci.) **277**: 57-129.

HAMANO, T. & K. I. HAYASHI, 1992. Ecology of an atyid shrimp *Caridina japonica* (De Man, 1892) migrating to upstream habitats in the Shiwagi Rivulet, Tokushima prefecture. Res. Crust. Carcinol. Soc. Japan, **21**: 1-13.

HAMANO, T. & K. HONKE, 1997. Control of the migrating course of freshwater amphidromous shrimps by lighting. Crust. Res., **26**: 162-171.

HAMANO, T., K. YOSHIMI, K.-I. HAYASHI, H. KAKIMOTO & S. SHOKITA, 1995. Experiments on fishways for freshwater amphidromous shrimps. Nippon Suisan Gakkaishi, **61**: 171-178.

HARTMANN, G., 1958. Apuntes sobre la biología de camarón de río, *Cryphiops caementarius* (Molina) Palaemonidae, Decapoda. Pesca y Casa, **8**: 15-28.

HOLMQUIST, J. G., J. M. SCHMIDT-GENGENBACH & B. BUCHANAN-YOSHIOKA, 1998. High dams and marine-freshwater linkages: effects on native and introduced fauna in the Caribbean. Cons. Biol., **12**: 621-630.

HORNE, F. & S. BESSER, 1977. Distribution of river shrimp in the Guadalupe and San Marcos Rivers of central Texas, U.S.A. (Decapoda, Caridea). Crustaceana, **33**: 56-60.

HUNTE, W., 1978. The distribution of freshwater shrimps (Atyidae and Palaemonidae) in Jamaica. Zool. Jour. Lin. Soc., **64**: 35-150.

IBRAHIM, K. H., 1962. Observations on the fishery and biology of the freshwater prawn *Macrobrachium malcomsonii* Milne Edwards in River Godvari. Indian J. Fish., **9**: 433-467.

JOHNSON, S. L., A. P. COVICH, T. A. CROWL, A. ESTRADA-PINTO, J. BITHORN & W. A. WURSTBAUGH, 1998. Do seasonality and disturbance influence reproduction in freshwater atyid shrimp in headwater streams, Puerto Rico? Verhand. Internat. Ver. Limnol., **26**: 2076-2081.

KIKKERT, D. A., T. A. CROWL & A. P. COVICH, 2009. Upstream migration of amphidromous shrimp in the Luquillo Experimental Forest, Puerto Rico: temporal patterns and environmental cues. J. North American Benthol. Soc., **28**: 233-246.

LING, S. W., 1969. The general biology and development of *Macrobrachium rosenbergii* (De Man). FAO Fish. Rep., **57**: 589-606.

MARCH, J. G., J. P. BENSTEAD, C. M. PRINGLE & F. N. SCATENA, 1998. Migratory drift of larval freshwater shrimps in two tropical streams, Puerto Rico. Fresh. Biol., **40**: 261-273.

— —, — —, — — & — —, 2003. Damming tropical island streams: problems, solutions, alternatives. Bioscience, **53**: 1069-1078.

POMPEU, P. DOS S., F. VIEIRA & C. B. MARTINEZ, 2006. Utilização do mecanismo de transposição de peixes da Usina Hidreléctrica Santa Clara por camarões (Palaemonidae), bacia do rio Mucuri, Minas Gerais, Brasil. Rev. Brasileira Zool., **23**: 293-297.

ROME, N., S. L. CONNER & R. T. BAUER, 2009. Delivery of hatching larvae to estuaries by an amphidromous river shrimp: tests of hypotheses based on larval moulting and distribution. Freshw. Biol., **54**: 1924-1932.

First received 16 December 2009.
Final version accepted 13 January 2010.

# A MIGRATORY SHRIMP'S PERSPECTIVE ON HABITAT FRAGMENTATION IN THE NEOTROPICS: EXTENDING OUR KNOWLEDGE FROM PUERTO RICO

BY


MARCIA N. SNYDER[1,3]), ELIZABETH P. ANDERSON[2,4])
and CATHERINE M. PRINGLE[1,5])

[1]) Odum School of Ecology, University of Georgia, Athens, Georgia 30602, U.S.A.
[2]) Global Water for Sustainability Program, Florida International University, Miami, FL 33199, U.S.A.


## ABSTRACT


Migratory freshwater fauna depend on longitudinal connectivity of rivers throughout their life cycles. Amphidromous shrimps spend their adult life in freshwater but their larvae develop into juveniles in salt water. River fragmentation resulting from pollution, land use change, damming and water withdrawals can impede dispersal and colonization of larval shrimps. Here we review current knowledge of river fragmentation effects on freshwater amphidromous shrimp in the Neotropics, with a focus on Puerto Rico and Costa Rica. In Puerto Rico, many studies have contributed to our knowledge of the natural history and ecological role of migratory neotropical shrimps, whereas in Costa Rica, studies of freshwater migratory shrimp have just begun. Here we examine research findings from Puerto Rico and the applicability of those findings to continental Costa Rica. Puerto Rico has a relatively large number of existing dams and water withdrawals, which have heavily fragmented rivers. The effects of fragmentation on migratory shrimps' distribution have been documented on the landscape-scale in Puerto Rico. Over the last decade, dams for hydropower production have been constructed on rivers throughout Costa Rica. In both countries, large dams restrict shrimps from riverine habitat in central highland regions; in Puerto Rico 27% of stream kilometers are upstream of large dams while in Costa Rica 10% of stream kilometers are upstream of dams. Research about amphidromy specific to non-island shrimps is increasingly important in light of decreasing hydrologic connectivity.


---


[3]) e-mail: msnyder@uga.edu
[4]) e-mail: epanders@fiu.edu
[5]) e-mail: cpringle@uga.edu




New frontiers in crustacean biology: 169-182

## INTRODUCTION

Migratory shrimps are an important component of the aquatic fauna in many regions of the new and old world tropics (Pringle et al., 1993; Crowl et al., 2001). Many migratory shrimp are amphidromous, living primarily in freshwater but dependent on saltwater for parts of their life cycle (Chace & Hobbs, 1969). Adult shrimp spawn in freshwaters, and then larval shrimp passively drift from upstream freshwater reaches to salt water where they develop into juveniles. They then migrate back upstream, where they spend the majority of their lifetime. Migratory shrimps play an important part in stream food webs and ecosystem function, particularly as organic matter processors (i.e., leaf "shredders" and algal consumers). In addition, they are conduits for movement of energy and matter between marine and freshwater systems.

Freshwater migratory shrimps can be negatively affected by river fragmentation, which occurs when rivers lose hydrological connectivity where dams, water withdrawals or water pollution create un-passable stream reaches for downstream drift of shrimp larvae or upstream juvenile migration (Holmquist et al., 1998; Pringle & Scatena, 1999b). Research on effects of fragmentation on migratory stream biota has been heavily concentrated in temperate regions and overwhelmingly biased towards fishes (Pringle et al., 2000; March et al., 2003). Nevertheless, tropical rivers are becoming increasingly fragmented, with losses in connectivity threatening the long-term persistence of migratory shrimps (Greathouse et al., 2006a). The relatively little knowledge of general shrimp ecology and limited information on the effects of river fragmentation in the tropics makes predicting shrimp response to river fragmentation more challenging. The exception to this case is Puerto Rico, where decades of research has examined the ecology of freshwater migratory shrimps and documented the impacts of fragmentation on their distribution and abundance (fig. 1; Pringle et al., 1993; Holmquist et al., 1998; Crowl et al., 2001).

We suggest that lessons from Puerto Rico may provide insights for other Neotropical regions inhabited by migratory shrimps that are under similar increasing pressure for river development. Here, we review what is known about the ecology and migratory behavior of shrimps in Puerto Rico and summarize effects of river fragmentation. We then use Costa Rica as a case study to test the broad applicability of this knowledge to other tropical countries.

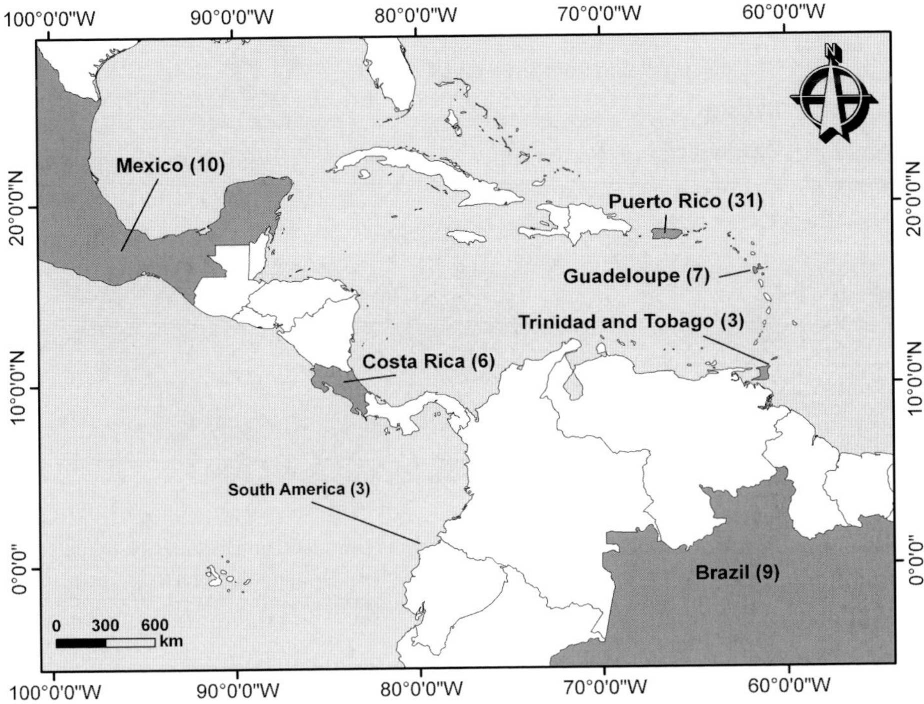


Fig. 1. Number of scientific publications on freshwater shrimp per region through 2009, found using the Web of Science search engine which did not include strictly regional journals.

## HABITAT FRAGMENTATION AND SHRIMP: LESSONS FROM PUERTO RICO

Dams and water withdrawals have fragmented nearly all rivers in Puerto Rico. Contrary to most other tropical regions, over the period (1940-50) Puerto Rico exhibited an accelerated shift from an agricultural to an industrial based economy because of its association as a commonwealth of the United States (Grau et al., 2003). This rapid industrialization was accompanied by hydropower dam construction, which peaked in the 1950s (Pringle & Scatena, 1999a). Today, Puerto Rico has three times the number of large dams per unit land area as the continental United States and more dams than any other Caribbean island (Greathouse et al., 2006a). On the basis of a 1:100 000 scale stream network, we estimate that 27% of stream kilometers are upstream from large dams in Puerto Rico (fig. 2). Puerto Rico also has many small dams associated with water withdrawal for human consumption. For example the 11 269 hectare Caribbean National Forest (CNF), which is managed as a multi-usage area, contains 34 small dams and associated water withdrawal structures.

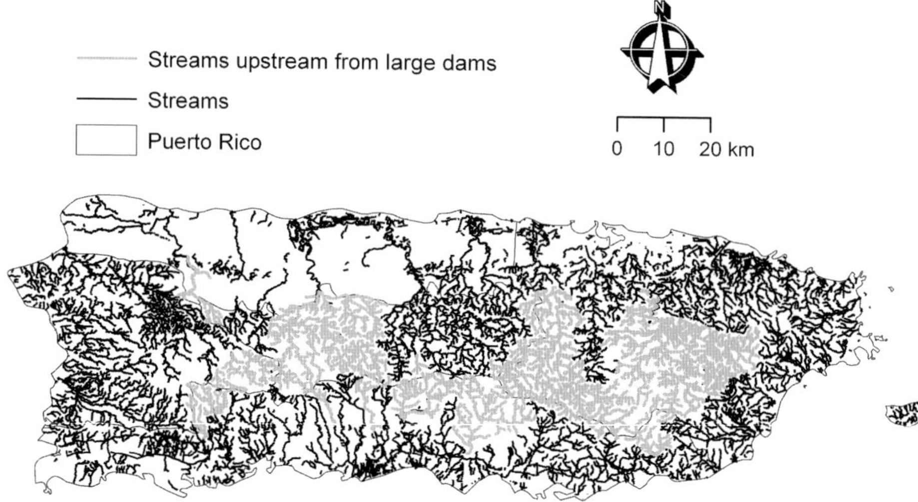



Fig. 2. Using 1:100 000 scale maps of rivers and streams from the United States National Atlas and a published map of large dams we calculate that 27% of stream kilometers are upstream of large dams (Greathouse, 2006a).

Both large and small dams in Puerto Rico affect distribution and abundance of shrimps, with 15 known species (table I; Holmquist et al., 1998). Dams impede upstream migration of shrimp juveniles and downstream drift of shrimp larvae to the estuary (Holmquist et al., 1998; Benstead et al., 1999; Greathouse et al., 2006a). They can also decrease successful recruitment of post-larval shrimp that rely on the signature of freshwater for migratory clues (March et al., 2003). Holmquist et al. (1998) found that dams >15 m tall, without spillway discharge (a structure which allows a limited quantity of water to flow over the dam), completely impeded upstream migration by juvenile shrimp, effectively extirpating shrimp populations above the dam. In contrast, dams >15 m tall, with intermittent spillway discharge, still allowed some shrimp to persist upstream from the dam (Greathouse et al., 2006a).

Dams <15 m tall directly affect in-stream biota by reducing suitable habitat, blocking access to suitable habitat, and causing direct and indirect mortality (March et al., 2003). Smaller dams are typically formed by a concrete wall (low-head dam) that creates a reservoir from which water is directly removed. The other less common structure is an "in-channel" water withdrawal system which pumps water out through risers and alters the physical environment much less than the typical low-head dam (March et al., 2003). Small low-head dams can impede upstream migration of juvenile shrimp when there is no spillway discharge, whereas in-channel intakes are less likely to cause

TABLE I

The decapod species of Puerto Rico (Holmquist 1998)

| Decapod species |
| --- |
| *Macrobrachium faustinum* |
| *Micratya poeyi* |
| *Atya lanipes* |
| *Atya scabra* |
| *Xiphocaris elongata* |
| *Macrobrachium heterochirus* |
| *Potimirim mexicana* |
| *Atya innocuous* |
| *Potimirim americana* |
| *Epilobocera sinuatifrons* |
| *Macrobrachium acanthurus* |
| *Macrobrachium carcinus* |
| *Potimirim glabra* |
| *Macrobrachium crenulatum* |
| *Palaemon pandaliformis* |

mortality (March et al., 2003). According to a hydrological budget created for the CNF, on a typical day, water intakes collectively divert 70% of the mean flow generated within the CNF before it reaches the ocean (Crook et al., 2007). These water intakes reduce suitable habitat and many streams are dry during part of the year. In some cases, wastewater discharged below the water intakes is the only flowing water in streams during dry periods (Hunter & Arbona, 1995; Pringle & Scatena, 1999a).

Cumulative effects of dams in highly fragmented systems can impede the migration of shrimp to varying degrees. Approaches that consider cumulative effects of multiple barriers on connectivity at the river scale are rare but important for management of migratory fauna that spend most of their lives in freshwater. One example is the Index of Longitudinal Riverine Connectivity (ILRC) which was created to evaluate the relative level of connectivity or alteration of the streams in the Caribbean National Forest of Puerto Rico, through estimation of the probability that an individual shrimp will be able to migrate downstream to the estuary and return to the reach where it was released as a larvae (Crook et al., 2009). The CNF includes nine watersheds and at least thirty-four water intakes. The ILRC characterized nine of the streams as "low" connectivity, three of the streams as "moderate" connectivity and five streams as "high" connectivity. During drought years, some CNF streams are dry for most of the year and are more likely to become impermeable barriers to shrimp migration. When there is no water flowing over the dam, mortality of juvenile shrimps migrating upstream is 100%. If some minimum flow is maintained, the probability of upstream migration increases from "zero" to some probability. The ILRC illustrated that the maintenance of some minimum flow is the most important factor to ensure shrimp passage. Other studies have gone into further detail about ways to mitigate the impact of dams on shrimp migration by maintaining fish ladders, reducing abstraction during peak migration times and altering dam structures (Benstead et al., 1999; March et al., 2003).

Thus far we have only considered connectivity within the freshwater part of the shrimp life cycle. By having larvae that are dispersed through the marine system, shrimp have the ability to reach wider distributional ranges, and to re-colonize freshwater streams following disturbance events (McDowall, 2004; Covich, 2006). Genetic analysis of mitochondrial DNA of seven amphidro-mous shrimp species (*Atya lanipes*, *Atya scabra*, *Atya innocous*, *Micratya poeyi*, *Micratya* sp., *Xiphocaris elongata*, and *Macrobrachium faustinum*) showed high gene flow in shrimp within and between drainages, highlight-ing the role of marine dispersal (Cook et al., 2009). Marine dispersal may be the determining component in the genetic structure and connectivity of shrimp species island-wide and underscores the advantages of obligate amphidromy. The majority of river fragments upstream from large dams are concentrated in the central highland region of Puerto Rico (fig. 2); if flow were restored, shrimps would have the potential to recolonize streams where they had been extirpated because of habitat fragmentation.

Understanding impacts of fragmentation on shrimps, beyond just impeding movement and hindering migration, requires an appreciation of the important ecological role played by shrimps. In Puerto Rico, shrimps are highly abun-dant, regulate stream ecosystem processes, and structure the stream food web (Covich et al., 1999; Pringle et al., 1999; Crowl et al., 2001; March et al., 2001). For example, Pringle (1996) found that atyid shrimps both significantly reduced the algal standing crop and altered the species composition of algae. Atyid and xiphocarid shrimps have also been found to reduce the fine particu-late organic matter and increase the nutrient quality of the epilithon (Pringle et al., 1999). Shrimps in Puerto Rican streams have also been shown to increase leaf decomposition rates (Crowl et al., 2001; March et al., 2001). Shrimps not only directly affect primary production and decomposition, they also af-fect the macroinvertebrate community by depressing populations of several taxa including chironomids (Pringle et al., 1993). In high densities shrimp can be a key component in maintaining stability of stream ecosystem processes (Pringle et al., 1999; Crowl et al., 2001). In areas where shrimp have been extirpated (i.e., upstream from dams >15 m with no spillway discharge), in-stream ecosystem processes change significantly. The consequences of shrimp extirpation include greater algal standing crop, and more fine benthic organic and inorganic matter of a lesser quality (Greathouse et al., 2006a).

## COSTA RICA CASE STUDY

In this section we examine whether insights that we have gained from Puerto Rico about the ecology of migratory shrimps and the effects of river fragmentation are applicable to other Neotropical shrimp populations. Freshwater migratory shrimps in other Neotropical countries are increasingly facing some of the same threats to their long-term viability as the shrimp fauna in Puerto Rico (Greathouse et al., 2006a). Costa Rica provides a case study on the basis of the fact that (1) there is good general knowledge of ecology of Costa Rican rivers and (2) the country is in the midst of a wave of hydropower development that has fragmented rivers that harbor migratory shrimps.

Local environmental differences can alter the life history of shrimp species and pose different conservation concerns. Two differences in the natural history and ecology of shrimps between Puerto Rico and Costa Rica are noteworthy. First, the Costa Rican shrimp fauna is composed of 17 shrimp species, the distribution of which relates to the country's central volcanic mountain range (table II). All shrimp species in both Puerto Rico and Costa Rica are known to be amphidromous (Chace & Hobbs, 1969). Of the 15 shrimp species found in Puerto Rico, six also occur in Costa Rica. Four of these (*M. acanthurus*, *M. heterochirus*, *M. crenulatum*, and *M. carcinus*) have only been found on the Caribbean slopes. The other two (*Atya scabra* and *A. innocuous*) have a more universal distribution and occur on both the Pacific and Caribbean slopes.

Second, in Puerto Rican streams, shrimps reach such high densities as to be considered the dominant macrofauna, whereas in Costa Rican streams, shrimps occur at a lower density and fish are considered to be dominant. The occurrence and density of shrimp fauna in Puerto Rico and Costa Rica is variable, depending on the size of the stream, the distance to the estuary and other factors such as density of predators. In Puerto Rico, in predator-free stream reaches upstream of waterfalls (which serve as barriers to predatory fish but not shrimp), the density of *Xiphocaris* shrimp can reach $10/m^2$ and *Atya* shrimp $16/m^2$; below waterfalls (in predator-rich reaches) the mean density of shrimps has been estimated at $0.2/m^2$ for *Xiphocaris* and $0.3/m^2$ for *Atya*

TABLE II

The freshwater shrimp species that occur on the Atlantic, Pacific or both slopes of Costa Rica (Obregon, 1986). * Indicate species that also occur in Puerto Rico

| Atlantic | Pacific | Both |
| --- | --- | --- |
| *Macrobrachium acanthurus** | *Macrobrachium tenellum* | *Atya scabra** |
| *Macrobrachium heterochirus** | *Macrobrachium occidentale* | *Atya crassa* |
| *Macrobrachium olfersi* | *Macrobrachium digueti* | *Atya innocous** |
| *Macrobrachium carcinus** | *Macrobrachium hancocki* | |
| *Macrobrachium amazonicum* | *Macrobrachium americanum* | |
| *Macrobrachium crenulatum** | *Macrobrachium Panamense* | |
| | *Atya margaritacea* | |
| | *Palaemon gracilis* | |

(Covich et al., 2009). *Macrobrachium* spp. can vary in density from 1-5 per pool, with mean pool area being 75.7 m$^2$ (Covich et al., 2006). These lower densities are comparable to estimates of *Macrobrachium* spp. populations (0.2/m$^2$) from lowland stream reaches in Costa Rica (Snyder, unpub. data).

Rivers in Costa Rica, as well as other Central and South American countries, are being targeted for new dam construction (Anderson et al., 2006a). Since 1990, >30 small to medium sized hydropower plants have been built in Costa Rica (Anderson et al., 2006b). Many hydropower plants operate as water diversion dams, where water is diverted from the river to an off-channel reservoir and then through a turbine house, before it is discharged again into the stream (Anderson et al., 2006a). Operation leaves a reach with highly reduced flow downstream from the diversion dam. The size of the dam and geographic relief of the landscape are factors that affect the length of the reach, but the dewatered reach in general ranges from 1-7 km for many Costa Rican diversion dams (Anderson et al., 2006b).


## APPLICABILITY OF LESSONS LEARNED FROM PUERTO RICO

Lesson 1. — On the basis of studies from Puerto Rico, whether water passes over a dam and how frequently it does is important to connectivity from a shrimp's perspective. Large dams without spillway discharge are making streams impassable to migratory shrimp. Maintaining some minimum flow is the most important variable for maintaining connectivity in shrimp populations in Puerto Rico. A few studies suggest this lesson may be true for shrimp in Costa Rica as well. In the Sarapiquí River drainage on the northern Atlantic slope, Costa Rica, shrimp populations were quantified with shocking and snorkel surveys near two small dams in 2001 during the wet and dry seasons (E.P.A., unpublished data; Anderson et al., 2006a). One dam, on the Puerto Viejo River, is 8 m tall and has a 4 km dewatered reach directly downstream of the dam which is maintained with a minimum flow of $\sim$10% of the average annual river discharge. The other dam is $\sim$3 m and is located on a tributary to the Puerto Viejo River, with a dewatered reach of 2 km. The mean annual discharge of the Puerto Viejo River and Quebradon stream at the site of the dam was 8.5 m$^3$/s and 1 m$^3$/s respectively, based on historical records over the period 1960-1990 (Anderson et al., 2006a). Both of these dams get overtopped during periods of high discharge during frequent storm events; this happens more frequently at the smaller dam. At the smaller dam (3 m), there was no difference in shrimp abundance above or below the dam, whereas at the taller dam (8 m) far fewer shrimp were found upstream from the dam than below the dam (fig. 3). The downstream end of the dewatered reach contained an order of magnitude more shrimp than other reaches of the stream.

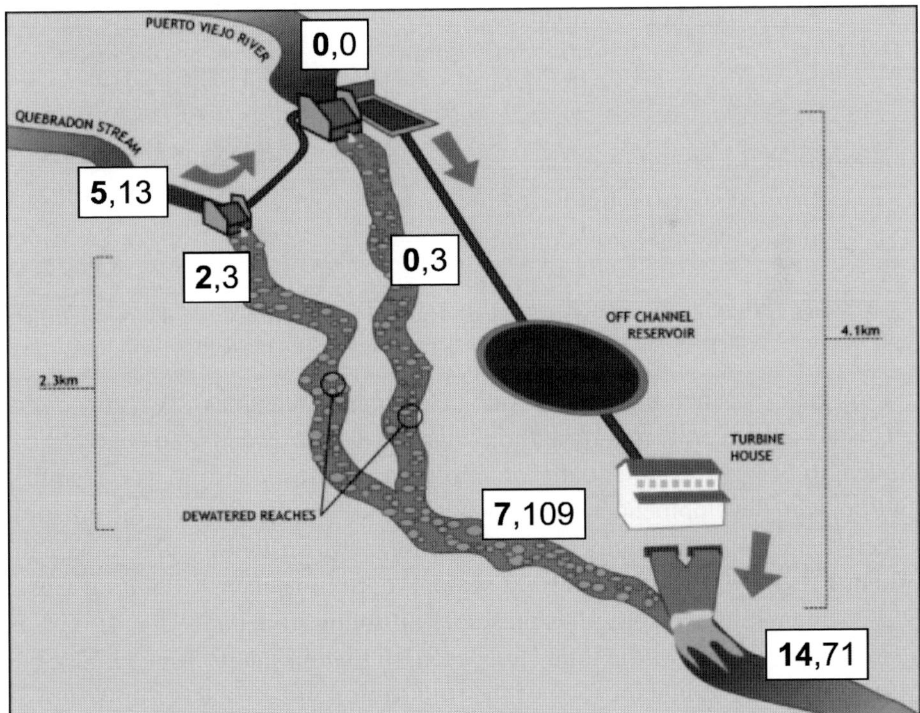


Fig. 3. Density of *Atya* and *Macrobrachium* shrimps collected by snorkeling and electrofishing surveys above and below the Quebradon (3 m) and Dona Julia (8 m) dams on the Quebradon stream and Puerto Viejo River in Sarapiqui, Costa Rica. Black boxes represent *Macrobrachium* spp. and grey boxes indicate *Atya* spp.

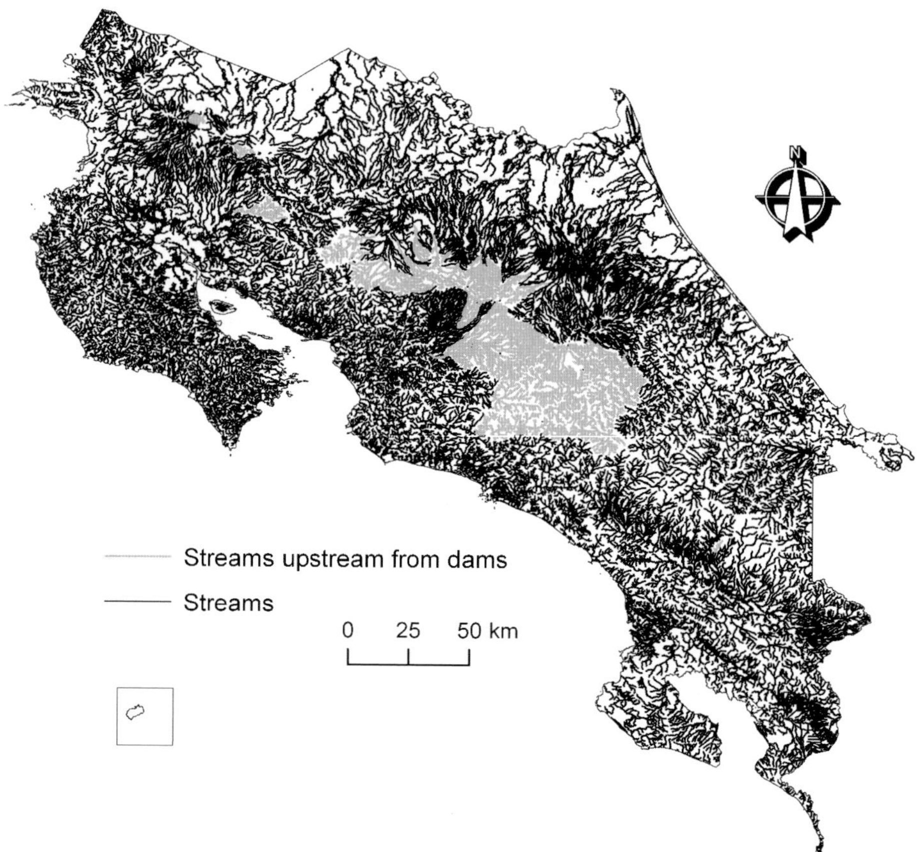


Fig. 4. Using 1:50 000 scale maps of rivers and streams from the Atlas Costa Rica we calculate that 10% of stream kilometers are upstream of large dams.

Lesson 2. — Much like Puerto Rico (7% in protected areas), very few rivers in Costa Rica are protected from the estuary to the headwaters. To characterize the length of streams potentially off-limits to migratory shrimp as a result of hydropower development in Costa Rica, we used a geographic information system to measure the number of stream kilometers upstream and downstream of dams nationwide at the 1:50 000 scale. Approximately 10% of total Costa Rican stream kilometer length is presently upstream from dams (fig. 4). This is a smaller proportion of the total stream length than in Puerto Rico (27%), but in both areas the spatial location of the fragmented stream reaches upstream of large dams is in the central highlands. To further examine river fragmentation in Costa Rica, we quantified the number of stream kilometers within national parks that are protected from hydropower development. Twenty-seven percent of Costa Rican land is located in protected areas but most of this land is

TABLE III

The percentage of stream kilometers in national parks in seven elevational bands in Costa Rica

| Elevation (m) | % in park | % of total stream length |
|---|---|---|
| 0-500 | 19 | 44.57561937 |
| 501-1000 | 42 | 24.95530437 |
| 1001-1500 | 49 | 15.52169264 |
| 1501-2000 | 64 | 9.103736272 |
| 2001-2500 | 82 | 5.205147832 |
| 2501-3000 | 0 | 0.217288801 |
| 3001-3700 | 81 | 0.421210713 |

at higher elevations which leaves lowland rivers and streams relatively more vulnerable to fragmentation than those at higher elevations (table III).

The relationship between marine dispersal and population connectivity between watersheds has yet to be examined in Costa Rica. Compared to Puerto Rico's almost uniformity of short, steep streams, Costa Rica has a high diversity of river types, including steep mountain streams and meandering lowland rivers. On average, the migratory distances a shrimp would need to travel upstream to reach a fish-barrier waterfall is much longer in Costa Rica than in Puerto Rico ($\sim$14 km in Puerto Rico vs. >60 km in Costa Rica). Continental estuaries are larger than those of Caribbean islands, which might influence the estuary retention time and create longer travel distances of amphidromous species. The levels of hydrologic connectivity at which viable shrimp populations can be maintained are likely to be different among species and between island and mainland populations. In Costa Rica, maintaining the same level of integrity of shrimp populations could require more free-flowing protected rivers than in Puerto Rico.

Lesson 3. — Shrimps in Puerto Rico are key components of ecosystems, and ecosystem processes are altered in areas upstream of dams where shrimp no longer persist (Pringle et al., 1993; Covich et al., 1999; Greathouse et al., 2006b). In Costa Rica, the impact of shrimps' absence on general ecosystem processes may be less severe than in Puerto Rico. Although shrimps in Costa Rica have been shown to reduce inorganic sediment mass, organic ash free dry mass, and densities of macroinvertebrate insects (Pringle & Hamazaki, 1998), they are less abundant than shrimps in Puerto Rico and part of a much more diverse freshwater fauna. Including introduced species, 174 fishes have been reported from Costa Rican freshwater ecosystems (Froese & Pauly, 2009). While the native fish fauna of Puerto Rican (n $=$ 70) is composed of all migratory fish, the Costa Rican fish fauna contains many non-migratory species and potentially predatory species (Greathouse et al., 2006b).

## FUTURE RESEARCH DIRECTIONS

We propose four potential future research directions that could significantly increase our understanding of shrimp populations and their viability in the face of increasing river fragmentation. First, what are the biological factors that affect the seasonal and daily migration patterns of upstream and downstream migration of different species? Second, how do other sources of river fragmentation such as harvesting and decreased water quality from wastewater or agricultural run-off affect shrimp populations? Third, there is a need for greater understanding of cumulative effects of river fragmentation on shrimp populations. Fourth, we recommend use of population genetic techniques to examine the extent that marine dispersal is determining the genetic make-up of continental fauna and the connectivity of shrimp populations across watersheds.

## ACKNOWLEDGEMENTS

We thank the organizers of the 2009 Annual Meeting of the Crustacean Society. We are grateful to the Pringle lab for helpful feedback that improved this manuscript. The authors received support from the National Science Foundation, Long-Term Studies in Environmental Biology program (DEB 9528434, DEB 0075339, DEB 0545463), and the Luquillo LTER (DEB-0218039 and DEB-0620910).

## REFERENCES

ANDERSON, E. P., M. C. FREEMAN & C. M. PRINGLE, 2006a. Ecological consequences of hydropower development in Central America: impacts of small dams and water diversion on neotropical stream fish assemblages. River Research and Applications, **22**: 397-411.

ANDERSON, E. P., C. M. PRINGLE & M. ROJAS, 2006b. Transforming tropical rivers: an environmental perspective on hydropower development in Costa Rica. Aquatic Conservation-Marine and Freshwater Ecosystems, **16**: 679-693.

BENSTEAD, J. P., J. G. MARCH, C. M. PRINGLE & F. N. SCATENA, 1999. Effects of a low-head dam and water abstraction on migratory tropical stream biota. Ecological Applications, **9**: 656-668.

CHACE, F. A. & H. H. HOBBS, 1969. The freshwater and terrestrial decapod crustaceans of the West Indies with special reference to Dominica. United States National Museum Bulletin, **292**: i-vi, 1-258. (Smithsonian Institution Press, Washington, D.C.).

COOK, B. D., S. BERNAYS, C. M. PRINGLE & J. M. HUGHES, 2009. Marine dispersal determines the genetic population structure of migratory stream fauna of Puerto Rico: evidence for island-scale population recovery processes. Journal of the North American Benthological Society, **28**: 709-718.

COVICH, A. P., 2006. Dispersal-limited biodiversity of tropical insular streams. Polish Journal of Ecology, **54**: 523-547.

COVICH, A. P., T. A. CROWL & T. HEARTSILL-SCALLY, 2006. Effects of drought and hurricane disturbances on headwater distributions of palaemonid river shrimp (*Macrobrachium* spp.) in the Luquillo Mountains, Puerto Rico. Journal of the North American Benthological Society, **25**: 99-107.

COVICH, A. P., T. A. CROWL, C. L. HEIN, M. J. TOWNSEND & W. H. MCDOWELL, 2009. Predator-prey interactions in river networks: comparing shrimp spatial refugia in two drainage basins. Freshwater Biology, **54**: 450-465.

COVICH, A. P., M. A. PALMER & T. A. CROWL, 1999. The role of benthic invertebrate species in freshwater ecosystems — Zoobenthic species influence energy flows and nutrient cycling. Bioscience, **49**: 119-127.

CROOK, K. E., C. M. PRINGLE & M. C. FREEMAN, 2009. A method to assess longitudinal riverine connectivity in tropical streams dominated by migratory biota. Aquatic Conservation-Marine and Freshwater Ecosystems, **19**: 714-723.

CROOK, K. E., F. N. SCATENA & C. M. PRINGLE, 2007. Water withdrawn from the Luquillo Experimental Forest, 2004. (United States Department of Agriculture).

CROWL, T. A., W. H. MCDOWELL, A. P. COVICH & S. L. JOHNSON, 2001. Freshwater shrimp effects on detrital processing and nutrients in a tropical headwater stream. Ecology, **82**: 775-783.

GRAU, H. R., T. M. AIDE, J. K. ZIMMERMAN, J. R. THOMLINSON, X. ZOU & E. HELMER, 2003. The ecological consequences of socioeconomic and land-use changes in postagriculture Puerto Rico. Bioscience, **53**: 1159-1168.

GREATHOUSE, E. A., C. M. PRINGLE & J. G. HOLMQUIST, 2006a. Conservation and management of migratory fauna: dams in tropical streams of Puerto Rico. Aquatic Conservation-Marine and Freshwater Ecosystems, **16**: 695-712.

GREATHOUSE, E. A., C. M. PRINGLE, W. H. MCDOWELL & J. G. HOLMQUIST, 2006b. Indirect upstream effects of dams: consequences of migratory consumer extirpation in Puerto Rico. Ecological Applications, **16**: 339-352.

HOLMQUIST, J. G., J. M. SCHMIDT-GENGENBACH & B. B. YOSHIOKA, 1998. High dams and marine-freshwater linkages: effects on native and introduced fauna in the Caribbean. Conservation Biology, **12**: 621-630.

HUNTER, J. M. & S. I. ARBONA, 1995. Paradise lost: an introduction to the geography of water pollution in Puerto Rico. Social Science & Medicine, **40**: 1331-1355.

MARCH, J. G., J. P. BENSTEAD, C. M. PRINGLE & M. W. RUEBEL, 2001. Linking shrip assemblages with rates of detrital processing along an elevational gradient in a tropical stream. Canadian Journal of Fisheries & Aquatic Sciences, **58**: 470-478.

MARCH, J. G., J. P. BENSTEAD, C. M. PRINGLE & F. N. SCATENA, 2003. Damming tropical island streams: problems, solutions, and alternatives. Bioscience, **53**: 1069-1078.

MCDOWALL, R. M., 2004. Ancestry and amphidromy in island freshwater fish faunas. Fish & Fisheries **5**: 75-84.

PRINGLE, C. M., 1996. Atyid shrimps (Decapoda: Atyidae) influence the spatial heterogeneity of algal communities over different scales in tropical montane streams, Puerto Rico. Freshwater Biology, **35**: 125.
PRINGLE, C. M., G. A. BLAKE, A. P. COVICH, K. M. BUZBY & A. FINLEY, 1993. Effects of omnivorous shrimp in a montane tropical stream: sediment removal, disturbance of sessile invertebrates and enhancement of understory algal biomass. Oecologia, **93**: 1-11.
RINGLE, C. M., M. C. FREEMAN & B. J. FREEMAN, 2000. Regional effects of hydrologic alterations on riverine macrobiota in the New World: tropical-temperate comparisons. Bioscience, **50**: 807.
PRINGLE, C. M. & T. HAMAZAKI, 1998. The role of omnivory in a neotropical stream: separating diurnal and nocturnal effects. Ecology, **79**: 269-280.
PRINGLE, C. M., N. HEMPHILL, W. H. MCDOWELL, A. BEDNAREK & J. G. MARCH, 1999. Linking species and ecosystems: different biotic assemblages cause interstream differences in organic matter. Ecology, **80**: 1860-1872.
PRINGLE, C. M. & F. N. SCATENA, 1999a. Freshwater resource development: case studies from Puerto Rico and Costa Rica. In: L. U. HATCH & M. E. SWISHER (eds.), Managed ecosystems: the mesoamerican experience: 292. (Oxford University Press, New York).
— — & — —, 1999b. Aquatic ecosystem deterioration in Latin America and the Caribbean. In: L. U. HATCH & M. E. SWISHER (eds.), Managed ecosystems: the mesoamerican experience: 292. (Oxford University Press, New York).

First received 2 December 2009.
Final version accepted 4 May 2010.

# COMPARISON OF GENETIC POPULATION STRUCTURES BETWEEN THE LANDLOCKED SHRIMP, *NEOCARIDINA DENTICULATA DENTICULATA*, AND THE AMPHIDROMOUS SHRIMP, *CARIDINA LEUCOSTICTA* (DECAPODA, ATYIDAE) AS INFERRED FROM MITOCHONDRIAL DNA SEQUENCES

BY


JUNTA FUJITA[1,3]), KOUJI NAKAYAMA[2]), YOSHIAKI KAI[1]), MASAHIRO UENO[1])
and YOH YAMASHITA[1])

[1]) Maizuru Fisheries Research Station, Field Science Education and Research Center,
Kyoto University, Nagahama, Maizuru, Kyoto 625-0086, Japan
[2]) Laboratory of Estuarine Ecology, Field Science Education and Research Center,
Kyoto University, Kyoto 606-8502, Japan


## ABSTRACT


The genetic population structures of the landlocked shrimp, *Neocaridina denticulata denticulata*, and the amphidromous shrimp, *Caridina leucosticta*, were compared in order to evaluate the effect of dispersal ability on the genetic structure. We examined the mitochondrial DNA (mtDNA) sequence variations of the region extending from the NADH dehydrogenase subunit 2 (ND2) gene to the tryptophan transfer RNA (tRNA$^{Trp}$) gene and the region encoding NADH dehydrogenase subunit 5 (ND5) gene in 120 specimens of *N. d. denticulata* from six locations and 160 specimens of *C. leucosticta* from eight locations. The neighbour joining tree showed several distinct clades roughly corresponding to the local populations within *N. d. denticulata*, but no distinct clades within *C. leucosticta*. Pairwise $\Phi_{ST}$ and AMOVA also indicated a significantly strong genetic differentiation among local populations of *N. d. denticulata*, but no or little genetic differentiation among local populations of *C. leucosticta*. These results suggest the absence of gene flow between local populations in *N. d. denticulata*, but high gene flow in *C. leucosticta* owing to its dispersal in the sea. Overall, these data suggested that the pattern of genetic structures exhibited by each species is closely correlated with their dispersal abilities.


## INTRODUCTION

Comparative phylogeography has contributed to assessing the roles of life history traits in shaping the genetic structure (Dawson et al., 2002; Ayre et

---


[3]) Corresponding author; e-mail: junta9@kais.kyoto-u.ac.jp





New frontiers in crustacean biology: 183-196

al., 2009; Steele et al., 2009). This multispecies approach enables broader conclusions to be drawn than those generated from species-specific case studies, because patterns of gene flow unique to a species would suggest that species-specific biological characteristics have impacted the species (Hickey et al., 2009). To properly evaluate the effect of dispersal ability on genetic structuring, comparative studies should be performed in similar distributional patterns on closely related species with clear differences in dispersal abilities (Dawson et al., 2002; Hickey et al., 2009; Steele et al., 2009).

*Neocaridina denticulata denticulata* (De Haan, 1844) and *Caridina leucosticta* (Stimpson, 1860) meet these requirements. Both shrimps have nearly the same adult size (ca. 8 mm carapace length, Hamano et al., 2000), generation time (ca. one year, Shokita, 1979; Niwa & Hamano, 1990) and habitat (riverside with slow velocity under dense vegetation, Niwa & Hamano, 1990; Hamano et al., 2000). However, these shrimps have different life history traits influencing dispersal abilities. *Neocaridina d. denticulata* is a landlocked shrimp showing direct development with no pelagic larval stages and spawns a small number (100 to 140 eggs) of large (1.6 mm by 1.05 mm) eggs (Shokita, 1981). Ovigerous females of the species are observed from May to September, and juveniles are observed from June to September (Niwa & Hamano, 1990). In contrast, *C. leucosticta* has an amphidromous life history, which spawns a large number (average 1109 eggs) of small (0.51 mm by 0.31 mm) eggs (Shokita, 1981) and spends six to seven zoeal stages for at least 12 days in the sea (Nakahara et al., 2005). Ovigerous females and drifting larvae of the species have been observed from June to October (Yamahira et al., 2007 and Hamano et al., 2005, respectively), and juveniles observed from August to November (Yamahira et al., 2007). *Neocaridina d. denticulata* showing the landlocked life history has no or little dispersal ability because they are commonly distributed among drainages isolated from one another by marine and terrestrial habitats, while *C. leucosticta* showing the amphidromy has a high dispersal ability because they can disperse among rivers during their larval stages in the sea. These two shrimps are co-distributed in the western part of Japan. *Neocaridina d. denticulata* has a relatively restricted distributional area of the western part of the Japanese mainland (Kamita, 1970), while *C. leucosticta* is widely distributed in the Ryukyu Islands, Japan and western part of the Japanese mainland (Kamita, 1970; Shokita, 1979) (fig. 1). Furthermore, recent molecular phylogenetic analyses have shown that the genus *Neocaridina* and the genus *Caridina* are very closely related within the family Atyidae (Page et al., 2007, 2008). These two species, therefore, will be a good example for a

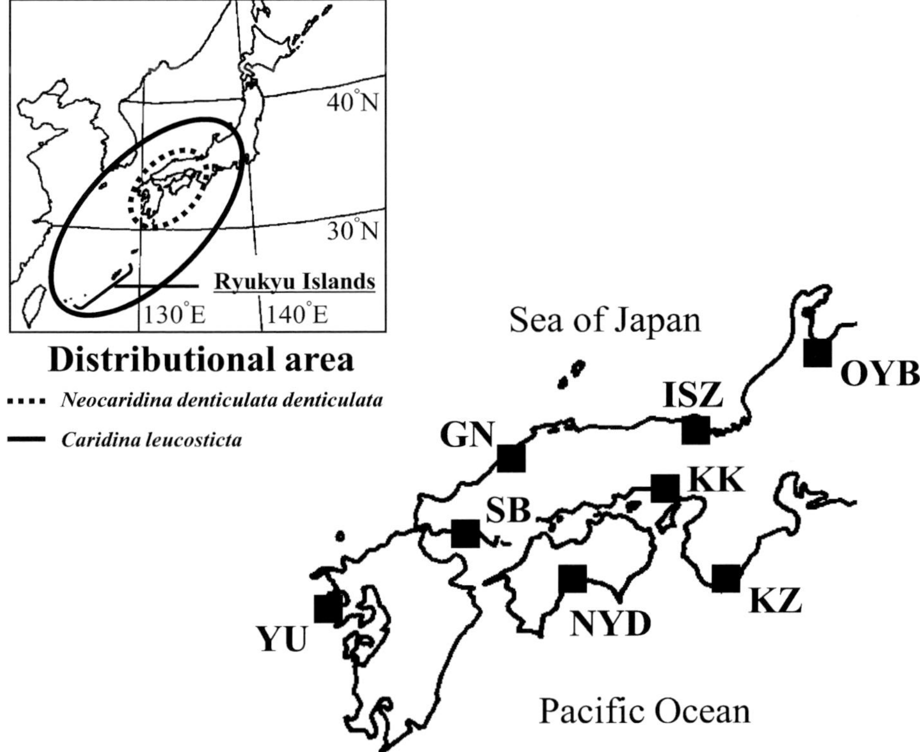


Fig. 1. Distributional area for each species and a map showing the sampling locations from the western part of Japan. Letters refer to collection locations, OYB, Oyabe River; ISZ, Isazu River; GN, Gono River; YU, Yukinoura River; SB, Saba River; KK, Kako River; NYD, Niyodo River; KZ, Koza River. Only *C. leucosticta* were collected in OYB and KZ.

comparative population genetic study. In this paper, we evaluated the effect of dispersal ability on the genetic population structure by using these two species.

## MATERIAL AND METHODS

### Sampling design

A total of 120 specimens of *N. d. denticulata* and 160 of *C. leucosticta* were collected in the western part of Japan with a dip net from spring to summer of 2008 (fig. 1). Both species were collected from the same six rivers: Isazu River (ISZ), Gono River (GN), Yukinoura River (YU), Saba River (SB), Kako River (KK) and Niyodo River (NYD). In addition, *C. leucosticta* was also collected from two other rivers, Oyabe River (OYB) and Koza River (KZ).

Twenty specimens for each river in both species were preserved in 99.5% ethanol for genetic analyses.

## DNA extraction, PCR and sequencing

Total DNA was extracted using the DNeasy Tissue Kit (Qiagen, Hilden, Germany) from abdominal tissue according to the manufacturer's protocol. The region extending from NADH dehydrogenase subunit 2 (ND2) gene to the tryptophan transfer RNA (tRNA[Trp]) gene and the region encoding NADH dehydrogenase subunit 5 (ND5) gene were amplified by means of polymerase chain reaction (PCR) using Ex Taq polymerase (Takara, Shiga, Japan). In the ND2-tRNA[Trp], PCR was performed on a model 9700 Thermal Cycler (Applied Biosystems, Foster City, CA, U.S.A.) using newly designed species-specific primers (ND2F-Neoden and ND2R-Neoden for *N. d. denticulata*; ND2F-Carileu and ND2R-Carileu for *C. leucosticta*, table I) with an initial denaturation step of 94°C for 5 min, 35 cycles of 94°C for 30 s, 55°C for 30 s, 72°C for 30 s and a final extension step of 72°C for 7 min. Partial sequences of ND5 gene were also amplified using newly designed primers (ND5F-Atyidae and ND5R-Atyidae, table I), under the same thermal cycling conditions as PCR for ND2-tRNA[Trp] (except for annealing at 47°C or 50°C). The PCR products, purified with ExoSAP-IT (GE Healthcare UK Ltd.), were sequenced with the BigDye Terminator Cycle Sequencing Kit

TABLE I

Primer information used in this study. Y and R denote T or C, and A or G, respectively. Positions refer to the corresponding complete mitochondrial sequences *Halocaridina rubra* (DDBJ/EMBL/GenBank accession number: DQ917432)

| Gene | Name | Sequence | Position | Direction |
|---|---|---|---|---|
| ND2-tRNA[Trp] | ND2F-Neoden | 5′-GTTTAYGYGGTT GTTTCCTCTTCAG-3′ | 15435-15459 | Forward |
| | ND2R-Neoden | 5′-CTCTTATRGGAAA CTTTGAAGGCTAC-3′ | 15879-15904 | Reverse |
| | ND2F-Carileu | 5′-TTAACCTATGCT GTAGTATCTTCATCAG-3′ | 15444-15471 | Forward |
| | ND2R-Carileu | 5′-TCTTATAAGAAACT TTGAAGGCTGCTAG-3′ | 15876-15903 | Reverse |
| ND5 | ND5F-Atyidae | 5′-CCCCCTATTAT YCGGATATCCTG-3′ | 5670-5692 | Forward |
| | ND5R-Atyidae | 5′-GCGGCTATAA CTAARAGRGC-3′ | 6072-6091 | Reverse |

ver. 1.1 (Applied Biosystems) and an automated sequencer (ABI310; Applied Biosystems) using each forward primer. Alignments were checked by eye in BioEdit (Hall, 1999).

## Data analyses

We used MEGA ver. 4 (Tamura et al., 2007) to generate an unrooted tree based on the neighbour joining (NJ) method using best fit model of substitution suggested by MODELTEST (Posada & Crandall, 1998). The models selected were: *N. d. denticulata*, Tamura-Nei (TrN) with a gamma distribution ($\Gamma = 0.2401$); *C. leucosticta*, TrN + $\Gamma$ ($\Gamma = 0.2825$). Bootstrap values were generated from a resampling of 1000 neighbour joining trees in order to evaluate the robustness of each node (Felsenstein, 1985).

Recently, the invasion of the Chinese shrimp, *Neocaridina denticulata sinensis* (Kemp, 1918), into Japanese rivers has been reported (Niwa et al., 2005). Because sequences of cytochrome oxidase subunit I (COI) gene for *N. d. sinensis* have been deposited in the DDBJ (Shih & Cai, 2007), we sequenced COI of several specimens randomly selected from each clade inferred from ND2-tRNA$^{\mathrm{Trp}}$ + ND5 sequences, and excluded those identified as *N. d. sinensis* from our analyses.

Haplotype ($h$) and nucleotide ($\pi$) diversities were calculated for each species in DnaSP ver. 5 (Librado & Rozas, 2009) to obtain measures of molecular diversity in each species. The number of net nucleotide substitutions per site within and between populations ($D_{\mathrm{A}}$) was calculated according to Nei (1987) using DnaSP ver. 5. Levels of genetic differentiation ($\Phi_{\mathrm{ST}}$) between pairs of populations were generated by Arlequin ver. 3.11 (Excoffier et al., 2005). We tested the significance of genetic differentiation by permuting 100 000 times. Finally, analyses of molecular variance (AMOVA; Excoffier et al., 1992) was used to determine the proportion of genetic variations (based on $\Phi_{\mathrm{ST}}$) that could be explained by partitioning of mitochondrial DNA diversities among and within populations using Arlequin ver. 3.11.

# RESULTS

## DNA sequences and NJ tree

Partial sequences of the ND2-tRNA$^{\mathrm{Trp}}$ (390 base pairs, bp) and ND5 (354 bp) were obtained for both species. ND2-tRNA$^{\mathrm{Trp}}$ was composed of 346 bp for posterior part of ND2 gene and 44 bp for anterior part of tRNA$^{\mathrm{Trp}}$

gene. Nucleotide sequences from both ND2-tRNA$^{Trp}$ and ND5 were of equal length with no insertions or deletions (indels) between the two species and could therefore be aligned unambiguously with each other. We generated 744 bp of ND2-tRNA$^{Trp}$ + ND5 sequences describing 31 unique haplotypes for *N. d. denticulata* (DDBJ/EMBL/GenBank Accession No. AB524916-AB524940 for ND2-tRNA$^{Trp}$ and AB524941-AB524963 for ND5) and 59 unique haplotypes for *C. leucosticta* (DDBJ/EMBL/GenBank Accession No. AB524974-AB525010 for ND2-tRNA$^{Trp}$ and AB525011-AB525038 for ND5).

The unrooted NJ tree inferred from haplotypes of the landlocked shrimp, *N. d. denticulata*, formed five distinct clades roughly corresponding to the local populations, and the robustness of all five clades were supported by high bootstrap supports (97 to 100%) (fig. 2). In contrast, NJ tree of the amphidromous shrimp, *C. leucosticta*, formed no distinct clades, and specimens of each location appeared to be placed randomly (fig. 2).

As mentioned in the "Material and Methods" section, we distinguished *N. d. sinensis* from *N. d. denticulata* based on the COI sequences (392 bp). NJ tree inferred from the COI sequences of our specimens (DDBJ/EMBL/GenBank Accession No. AB524964-AB524973) and previously determined sequences deposited in the DNA database showed that specimens from Clade V of the tree inferred from ND2-tRNA$^{Trp}$ + ND5 was closely related to sequences of *N. d. sinensis* identified by Shih & Cai (2007) (data not shown). Therefore, all the specimens of Clade V were assumed to be *N. d. sinensis* (fig. 2), and excluded from our analyses. Other clades (Clade I to IV) were assumed to be Japan-endemic species. When genetic differences between ND2-tRNA$^{Trp}$ + ND5 and COI were compared by using the same individuals, sequence divergences among putative Japan-endemic clades (Clade I to IV) showed nearly the same values (7.8-12.6% for ND2-tRNA$^{Trp}$ + ND5, 7.7-12.2% for COI).

## Comparative population genetic analyses

In ND2-tRNA$^{Trp}$, 68 sites (17.4%) for *N. d. denticulata* and 37 sites (9.5%) for *C. leucosticta* were variable, and 12 haplotypes were recognized for the former and 37 haplotypes for the latter. In ND5, 53 sites (15.0%) for *N. d. denticulata* and 27 sites (7.6%) for *C. leucosticta* were variable, and 12 haplotypes were recognized for the former and 28 haplotypes for the latter. The number of net nucleotide substitutions per site ($D_A$) within populations showed low values for both species ($-0.00150$ to $0.00000$), while average $D_A$ between populations showed a high value ($0.05788 \pm 0.00501$) for *N. d. denticulata*

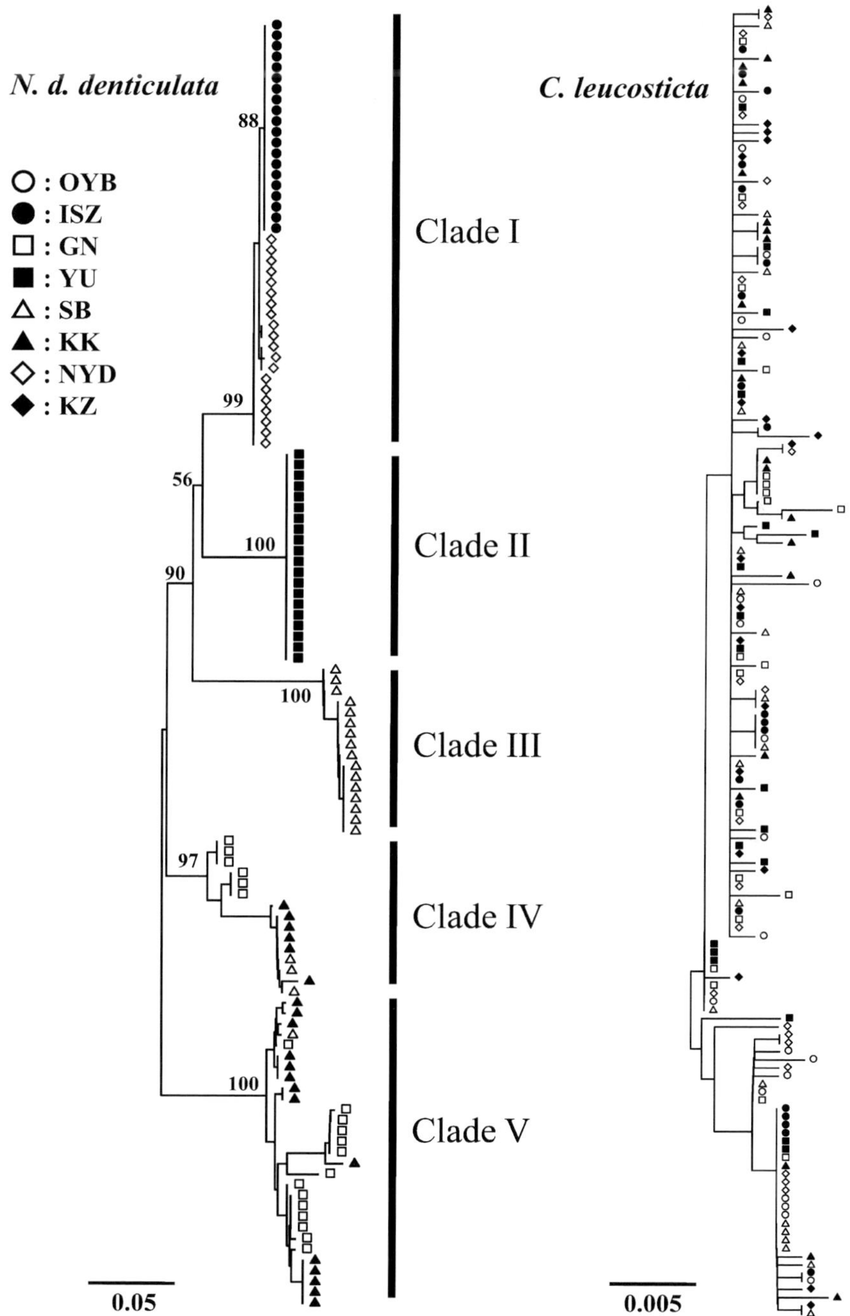


Fig. 2. Unrooted NJ tree for each species based on partial sequences of ND2-tRNA$^{Trp}$ + ND5. Scores as percentages of 1000 bootstraps above 50 are shown above or below associated nodes. Letters refer to collection locations are shown in fig. 1.

and a low ($0.00008 \pm 0.00003$) value for *C. leucosticta* (table II). *Neocaridina d. denticulata* had high levels of haplotype and nucleotide diversities, while *C. leucosticta* had a high level of haplotype diversity and a low level of nucleotide diversity (table III). However, *N. d. denticulata* tended to have low levels of haplotype and nucleotide diversities within populations (table III).

Pairwise $\Phi_{ST}$ values were relatively high and significant among all the local populations of *N. d. denticulata* under Bonferroni correction ($P < 0.001$ in almost all population-pairs), whereas relatively low and not significant among all the local populations of *C. leucosticta* (table IV). The result of AMOVA showed that the genetic variation in *N. d. denticulata* was distributed principally among populations (% of variation: 88.42%, $\Phi_{ST}$: 0.884, $P < 0.001$) and to a considerably lesser extent within populations (% of variation: 11.58%) (table V). In contrast, genetic variation in *C. leucosticta* was not significant among populations (% of variation: 2.08%, $\Phi_{ST}$: 0.021, $P = 0.086$) and distributed principally at the intrapopulation level (% of variation: 97.92%) (table V).

## DISCUSSION

Phylogeographic and population genetic studies in freshwater shrimps have been conducted by several authors, such as Page & Hughes (2007a) and Hurwood & Hughes (2001) for landlocked shrimps, and Cook et al. (2008) for amphidromous shrimps. Page & Hughes (2007a) reported that the phylogenetic tree inferred using the Australian landlocked shrimp, *Caridina indistincta*, showed distinct clades clearly corresponding to local populations. Hurwood & Hughes (2001) also reported that high levels of genetic differentiation within rivers and between adjacent rivers were shown in the Australian landlocked shrimp, *Caridina zebra*. In contrast, genetic differentiation among rivers by using pairwise $\Phi_{ST}$ analyses for Caribbean amphidromous shrimps found no significant values after Bonferroni correction (Cook et al., 2008).

Such relationships between life history traits and genetic population structures can be applied in the present comparative study, which simultaneously used sympatric closely related species with limited vs. greater dispersal abilities (Steele et al., 2009). *Neocaridina d. denticulata* is a landlocked shrimp and has a low dispersal ability, and this is reflected in its high level of genetic structure (fig. 2, tables IV and V). In the NJ tree, four distinct clades (except for the *N. d. sinensis* clade) roughly corresponded to the local populations (fig. 2).

TABLE II

The number of net nucleotide substitutions per site ($D_A$) within and between populations for the two species. Abbreviations of collection locations are shown in fig. 1. *: The data excluding the putative *N. d. sinensis* clade. na = not analyzed

| | $D_A$ within each population | | | | | | | | Average $D_A$ between populations |
|---|---|---|---|---|---|---|---|---|---|
| | OYB | ISZ | GN | YU | SB | KK | NYD | KZ | |
| *N. d. denticulata* | na | 0.00000 | −0.00134* | 0.00000 | −0.00150* | −0.00049* | −0.00010 | na | 0.05788 ± 0.00501* |
| *C. leucosticta* | −0.00024 | −0.00018 | −0.00014 | −0.00017 | −0.00021 | −0.00020 | −0.00021 | −0.00018 | 0.00008 ± 0.00003 |

## TABLE III

Sample size ($n$), number of haplotypes ($N_{hap}$), haplotype diversity ($h$) and nucleotide diversity ($\pi$) for the two species. Abbreviations of collection locations are shown in fig. 1. *: The data excluding the putative *N. d. sinensis* clade. na = not analyzed

| Location | *N. d. denticulata* | | | | *C. leucosticta* | | | |
|---|---|---|---|---|---|---|---|---|
| | $n$ | $N_{hap}$ | $h$ | $\pi$ | $n$ | $N_{hap}$ | $h$ | $\pi$ |
| OYB | na | na | na | na | 20 | 14 | 0.932 | 0.00482 |
| ISZ | 20 | 1 | 0.000 | 0.00000 | 20 | 7 | 0.763 | 0.00355 |
| GN | 6* | 2* | 0.600* | 0.00806* | 20 | 10 | 0.832 | 0.00284 |
| YU | 20 | 1 | 0.000 | 0.00000 | 20 | 11 | 0.868 | 0.00332 |
| SB | 19* | 4* | 0.754* | 0.02853* | 20 | 12 | 0.889 | 0.00428 |
| KK | 6* | 3* | 0.600* | 0.00296* | 20 | 12 | 0.900 | 0.00396 |
| NYD | 20 | 5 | 0.732 | 0.00204 | 20 | 10 | 0.832 | 0.00429 |
| KZ | na | na | na | na | 20 | 13 | 0.853 | 0.00353 |
| Overall | 91* | 15* | 0.879* | 0.05202* | 160 | 59 | 0.857 | 0.00390 |

## TABLE IV

Pairwise $\Phi_{ST}$ values between populations for *N. d. denticulata* (below diagonal) and *C. leucosticta* (above diagonal), excluding the putative *N. d. sinensis* clade. Letters refer to collection locations are shown in fig. 1. (*): $P < 0.05$ without Bonferroni correction. *, ** and ***: $P < 0.05$, $0.01$ and $0.001$ under Bonferroni correction, respectively, na = not analyzed

| | OYB | ISZ | GN | YU | SB | KK | NYD | KZ |
|---|---|---|---|---|---|---|---|---|
| OYB | – | 0.0028 | 0.0989(*) | 0.0671 | −0.0296 | 0.0671 | −0.0218 | 0.0919(*) |
| ISZ | na | – | 0.0223 | 0.0058 | −0.0187 | 0.0016 | −0.0153 | 0.0135 |
| GN | na | 0.9751*** | – | −0.0041 | 0.0683 | −0.0222 | 0.0430 | −0.0168 |
| YU | na | 1.0000*** | 0.9784*** | – | 0.0440 | 0.0006 | 0.0223 | −0.0151 |
| SB | na | 0.8223*** | 0.6786** | 0.8245*** | – | 0.0406 | −0.0324 | 0.0634 |
| KK | na | 0.9933*** | 0.8531* | 0.9937*** | 0.7322** | – | 0.0237 | −0.0099 |
| NYD | na | 0.7480*** | 0.9494*** | 0.9836*** | 0.8030*** | 0.9747*** | −0.0416 | |
| KZ | na | na | na | na | na | na | na | – |

## TABLE V

Results of the analyses of molecular variance (AMOVA) for *N. d. denticulata* and *C. leucosticta* displaying the percentage of variation and associated $\Phi_{ST}$, based on data excluding the putative *N. d. sinensis* clade

| Species | Source of variation | d.f. | Sum of squares | Variance components | % of variation | $\Phi_{ST}$ | $P$ value |
|---|---|---|---|---|---|---|---|
| *N. d. denticulata* | Among populations | 5 | 1766.38 | 23.96 | 88.42 | 0.884 | <0.001 |
| | Within populations | 85 | 266.88 | 3.14 | 11.58 | | |
| | Total | 90 | 2033.26 | 27.1 | | | |
| *C. leucosticta* | Among populations | 7 | 14.38 | 0.03 | 2.08 | 0.021 | 0.086 |
| | Within populations | 152 | 219.04 | 1.44 | 97.92 | | |
| | Total | 159 | 233.42 | | | | |

However, SB specimens were divided into Clade III and Clade IV. This poly-phyletic status may be due to human-mediated transfer from other rivers. On the other hand, *C. leucosticta* has a relatively long pelagic larval duration (at least 12 days), and the NJ tree for *C. leucosticta* showed no distinct clades (fig. 2). In addition, pairwise $\Phi_{ST}$ and AMOVA also indicated little genetic differentiation among populations for *C. leucosticta* (tables IV and V). The amphidromous life history of *C. leucosticta*, hence, appears to facilitate effective genetic exchange among rivers as a result of the larval dispersal via the marine environment.

The difference in the number of net nucleotide substitutions per site ($D_A$) and molecular diversity indices between species was also reflected by their different life history traits. $D_A$ showed low values within populations and high values between populations for *N. d. denticulata* (table II), suggesting that a close genetic relationship occurs within populations and a high genetic divergence occurs among populations. Furthermore, *N. d. denticulata* had high haplotype and nucleotide diversities as a species, but tended to have low haplotype and nucleotide diversities for each population (table III). Accordingly, the landlocked life history of *N. d. denticulata* yielded little dispersal ability among populations, thereby producing small population size, increased genetic drift, elevated inbreeding, and fixed mutations in each population. In contrast, *C. leucosticta* had a low $D_A$ within and between populations (table II), and high haplotype diversity and low nucleotide diversity at both levels of species and populations (table III). These results suggested that the amphidromous life history for *C. leucosticta* yielded high dispersal ability among populations, and thereby producing genetically homogenized populations linked to the other populations via the sea.

Page & Hughes (2007b) compared the phylogeographic structures of four cryptic species from the *Caridina indistincta* complex on the basis of their egg sizes and distributional areas: species with the large eggs showed high intraspecific divergence and smaller distribution, while those with the smaller eggs showed the least divergence and larger distribution. The results from our study were matched to the relationships proposed by Page & Hughes (2007b). *Neocaridina d. denticulata* with the landlocked life history shows large-sized eggs, strong population structures (fig. 2), and is distributed only in the Japanese mainland (Hayashi, 2007; fig. 1). Unlike the *Caridina indistincta* complex (Page et al., 2005), each clade of *N. d. denticulata* may not be significant at species level, because COI sequence divergences between clades (7.7-12.2%) are slightly less than mean divergence between congeneric Crustacea species (15.4%, Hebert et al., 2003). Comparing their morphological and ecological characters in detail will be needed in order to assign clades resulted from our study to their proper taxonomic status. In contrast, *C. leucosticta* with the amphidromous life history shows small eggs, little population structures (fig. 2), and is distributed not only in the Japanese mainland, but also in the Ryukyu Islands and even in Korea (Hayashi, 2007; fig. 1).

Overall, these data clearly indicated that the pattern of genetic structure exhibited by each species (*N. d. denticulata* and *C. leucosticta*) correlates with life history traits influencing dispersal abilities. In Japan, many freshwater shrimps with differences in the distributional area or other ecological traits occur (Shokita, 1979). Therefore, molecular ecological researches by using these species will be needed in order to resolve the effect of these ecological traits on the genetic structure patterns.

## ACKNOWLEDGEMENTS

We are grateful to M. Sakamoto (Kyoto University) for useful discussion, R. Ohata, K. Miwa, N. Muto, K. Torikoshi (Kyoto University), K. Sakai (Noto Marine Center, Ishikawa) and H. Matsuura (Kochi Prefectual Fisheries Experimental Station) for the help in collecting specimens. Chris Norman (University of Plymouth) read the manuscript and offered helpful comments. We are also grateful to anonymous reviewers for constructive comments on earlier versions of our manuscript.

## REFERENCES

AYRE, D. J., T. E. MINCHINTON & C. PERRIN, 2009. Does life history predict past and current connectivity for rocky intertidal invertebrates across a marine biogeographic barrier? Mol. Ecol., **18**: 1887-1903.

COOK, B. D., C. M. PRINGLE & J. M. HUGHES, 2008. Molecular evidence for sequential colonization and taxon cycling in freshwater decapod shrimps on a Caribbean island. Mol. Ecol., **17**: 1066-1075.

DAWSON, M. N., K. D. LOUIE, M. BARLOW, D. K. JACOBS & C. C. SWIFT, 2002. Comparative phylogeography of sympatric sister species, *Clevelandia ios* and *Eucyclogobius newberryi* (Teleostei, Gobiidae), across the California Transition Zone. Mol. Ecol., **11**: 1065-1075.

EXCOFFIER, L., G. LAVAL & S. SCHNEIDER, 2005. Arlequin ver. 3.0: an integrated software package for population genetics data analysis. Evol. Bioinfor. Onl., **1**: 47-50.

EXCOFFIER, L., P. E. SMAUSE & J. M. QUARTO, 1992. Analysis of molecular variance inferred from metric distances among DNA haplotypes: application to human mitochondrial DNA restriction data. Genetics, **131**: 479-491.

FELSENSTEIN, J., 1985. Confidence limits on phylogenies: an approach using the bootstrap. Evolution, **39**: 783-791.

HALL, T. A., 1999. Bioedit: a user-friendly biological sequence alignment editor and analysis program for Windows 95/98/NT. Nuc. Acids Symp. Ser., **41**: 95-98.

HAMANO, T., K. IDEGUCHI & K. NAKATA, 2005. Larval drift and juvenile recruitment of amphidromous freshwater shrimps (Decapoda: Caridea) in the Nishida River, Western Japan. Aquacul. Sci., **53**: 439-446. [In Japanese.]

HAMANO, T., M. KAMADA & T. TANABE, 2000. Distributions of freshwater decapods crustaceans in Tokushima Prefecture, Japan, with notes on conservation methods for local populations. Bull. Tokushima Pref. Mus., **10**: 1-47. [In Japanese.]

HAYASHI, K., 2007. Caridean shrimps (Crustacea: Decapoda: Pleocyemata) from Japanese Waters. Part 1. Oplophoroidea, Nematocarcinoidea, Atyoidea, Stylodactyloidea, Pasiphaeoidea and Psalidopodoidea: 138-142, 154-155. (Seibutsukenkyusha, Tokyo). [In Japanese.]

HEBERT, P. D. N., S. RATNASINGHAM & J. R. DEWAARD, 2003. Barcoding animal life: cytochrome c oxidase subunit 1 divergences among closely related species. Proc. Roy. Soc., (B, Biol. Sciences) **270**: S96-S99.

HICKEY, A. J. R., S. D. LAVERY, D. A. HANNAN, C. S. BAKER & K. D. CLEMENTS, 2009. New Zealand triplefin fishes (family Tripterygiidae): contrasting population structure and mtDNA diversity within a marine species flock. Mol. Ecol., **18**: 680-696.

HURWOOD, D. A. & J. M. HUGHES, 2001. Nested clade analysis of the freshwater shrimp, *Caridina zebra* (Decapoda: Atyidae), from north-eastern Australia. Mol. Ecol., **10**: 113-125.

KAMITA, T., 1970. Studies on the fresh-water shrimps, prawns and crawfishes of Japan: 35-45, 64-74. (Sonoyamashoten, Matsue). [In Japanese.]

LIBRADO, P. & J. ROZAS, 2009. DnaSP v5: a software for comprehensive analysis of DNA polymorphism data. Bioinformatics, **25**: 1451-1452.

NAKAHARA, Y., A. HAGIWARA, Y. MIYA & K. HIRAYAMA, 2005. Larval rearing of three amphidromous shrimp species (Atyidae) under different feeding and salinity conditions. Aquacul. Sci., **53**: 305-310. [In Japanese.]

NEI, M., 1987. Molecular evolutionary genetics: 1-512. (Columbia Univ. Press, New York).

NIWA, N. & T. HAMANO, 1990. Population ecology of *Neocaridina denticulata* (De Haan, 1849) (Caridea, Decapoda) in the Sugow River, Japan. Res. Crus., **19**: 43-54. [In Japanese.]

NIWA, N., J. OHTOMI, A. OHTAKA & S. R. GELDER, 2005. The first record of the ectosymbiotic branchiobdellidan *Holtodrilus truncatus* (Annelida, Clitellata) and on the freshwater shrimp *Neocaridina denticulata* denticulata (Caridea, Atyidae) in Japan. Fisher. Sci., **71**: 685-687.

PAGE, T. J., S. C. CHOY & J. M. HUGHES, 2005. The taxonomic feedback loop: symbiosis of morphology and molecules. Biol. Lett., **1**: 139-142.

PAGE, T. J., B. D. COOK, T. V. RINTELEN, K. V. RINTELEN & J. M. HUGHES, 2008. Evolutionary relationships of atyid shrimps imply both ancient Caribbean radiations and common marine dispersals. Journ. North American Bent. Soc., **27**: 68-83.

PAGE, T. J. & J. M. HUGHES, 2007a. Phylogeographic structure in an Australian freshwater shrimp largely pre-dates the geological origins of its landscape. Heredity, **98**: 222-231.

— — & — —, 2007b. Radically different scales of phylogeographic structuring within cryptic species of freshwater shrimp (Atyidae: Caridina). Limnol. Oceanogr., **52**: 1055-1066.

PAGE, T. J., K. V. RINTELEN & J. M. HUGHES, 2007. Phylogenetic and biogeographic relationships of subterranean and surface genera of Australian Atyidae (Crustacea: Decapoda: Caridea) inferred with mitochondrial DNA. Invert. Systematics, **21**: 137-145.

POSADA, D. & K. CRANDALL, 1998. MODELTEST: testing the model of DNA substitution. Bioinformatics, **14**: 817-818.

SHIH, H. & Y. CAI, 2007. Two new species of the land-locked freshwater shrimps genus, *Neocaridina* Kubo, 1938 (Decapoda: Caridea: Atyidae), from Taiwan, with notes on speciation on the island. Zool. Stud., **46**: 680-694.

SHOKITA, S., 1979. The distribution and speciation of the inland water shrimps and prawns from the Ryukyu Islands — II. Bull. Coll. Sci., Univ. Ryukyus, **28**: 193-278. [In Japanese.]

— —, 1981. Life history of the family Atyidae (Decapoda, Caridea). Aquabiology, **12**: 15-23. [In Japanese.]

STEELE, C. A., J. BAUMSTEIGER & A. STORFER, 2009. Influence of life-history variation on the genetic structure of two sympatric salamander taxa. Mol. Ecol., **18**: 1629-1639.

TAMURA, K., J. DUDLEY, M. NEI & S. KUMAR, 2007. MEGA4: Molecular Evolutionary Genetics Analysis (MEGA) software version 4.0. Mol. Biol. Evol., **24**: 1596-1599.

YAMAHIRA, K., A. INOUE, S. OISHI & K. IDEGUCHI, 2007. Upstream and downstream variation in demography of the amphidromous shrimp *Caridina leucosticta*. Japanese Journ. Bent., **62**: 9-16. [In Japanese.]

First received 13 November 2009.
Final version accepted 2 March 2010.

# DIVERSITY, ABUNDANCE AND DISTRIBUTION OF RIVER SHRIMPS (DECAPODA, CARIDEA) IN THE LARGEST RIVER BASIN OF COSTA RICA, CENTRAL AMERICA

BY


LUIS RÓLIER-LARA[1,3]) and INGO S. WEHRTMANN[2,4])

[1]) Instituto Costarricense de Electricidad-Proyecto Hidroeléctrico El Diquís, Puntarenas, Costa Rica

[2]) Unidad de Investigación Pesquera y Acuicultura (UNIP) of the Centro de Investigación en Ciencias del Mar y Limnología (CIMAR); and Escuela de Biología, Museo de Zoología, Universidad de Costa Rica, 2060 San José, Costa Rica


## ABSTRACT


Our knowledge about the biodiversity of freshwater decapods is more limited compared to that of species inhabiting estuarine and marine ecosystems. Costa Rica is not an exception of this tendency, and so far three families (Palaemonidae, Atyidae, Pseudothelphusidae) have been reported to occur in its freshwater systems. The basin of the river Grande de Térraba, Pacific slope of Costa Rica, was selected to collect data on diversity, distribution and abundance of its shrimp fauna before the largest hydroelectric construction in Central America will be implemented in these Pacific lowlands. During 2005-2008, we collected 4247 specimens of freshwater shrimps, representing two families, four genera and 13 species. *Macrobrachium digueti*, *M. occidentale*, and *M. tenellum* comprised 90.1% of all palaemonid shrimps, and the first two species showed the widest distribution in the river basin. Palaemonid shrimps were abundant in the vicinity of estuarine waters. Among atyid shrimp, *A. margaritacea* was the most abundant and collected in all river sections. The study area harbors 59% of all caridean freshwater shrimp species so far reported from Costa Rica, and 20% of the 35 *Macrobrachium* spp. known to occur in the eastern hemisphere. We strongly recommend that the design of any hydroelectric project includes structures permitting the passage of the migrating shrimp to mitigate the negative effects of this type of construction on the associated freshwater fauna.


## INTRODUCTION

In recent decades, our knowledge regarding the taxonomic diversity of decapods has increased considerably. Most studies, however, concern decapods

---


[3]) e-mail: luisrolier@gmail.com

[4]) e-mail: ingowehrtmann@gmx.de

inhabiting marine-coastal ecosystems, and information about the diversity of freshwater decapods is much more limited (Chia & Ng, 2006; De Grave et al., 2008; Yeo et al., 2008). Currently there exist 3268 extant caridean shrimp species (De Grave et al., 2009), and 655 (20%) of these are freshwater species (De Grave et al., 2008). In the case of brachyuran crabs, the presence of 6559 species has been reported so far, and 1300 species (20%) can be found in freshwater systems (De Grave et al., 2009).

Research on freshwater decapods in Central America focuses mainly on taxonomic aspects (Smalley, 1963, 1964, 1970; Bott, 1967, 1968; Rodríguez, 1982). For Costa Rica, the presence of representatives of three families inhabiting freshwater systems has been documented (Holthuis, 1952; Chace & Hobbs, 1969; Hobbs & Harts, 1982; Rodríguez, 1982; Rodríguez & Hedström, 2000): shrimps of the families Palaemonidae and Atyidae, and crabs of the family Pseudothelphusidae. Bowles et al. (2000) mentioned the presence of 35 species of *Macrobrachium* (Palaemonidae) for the eastern hemisphere; 12 of these species have been reported from Costa Rica (Holthuis, 1952), as well as one additional endemic species from the offshore Isla del Coco, Pacific Costa Rica (Abele & Kim, 1984).

Atyid shrimps have a known geographic distribution including eastern Africa and both slopes of the Americas (Hobbs & Harts, 1982). In Costa Rica, this family is comprised by two genera: *Atya* (4 spp.) and *Potimirim* (2 spp.) (Chace & Hobbs, 1969; Hobbs & Harts, 1982). Additionally, *Micratya* and *Jonga* are present in Costa Rica with at least one species each (L. R. Lara, Y. Gutiérrez & I. S. Wehrtmann, unpubl. data).

Knowledge about species diversity and ecology of freshwater species is of special importance when evaluating the possible environmental impacts, which may cause the construction of hydroelectric plants. The Costa Rican Institute of Electricity (Instituto Costarricense de Electricidad, ICE) is considering the construction, in the southern Pacific lowlands, of one of the largest hydroelectric plants in Central America (ICE, 2009). Although decapods are one of the most prominent elements of the freshwater fauna, studies about freshwater decapods in Costa Rica are sparse: Álvarez et al. (1996) analyzed morphometric aspects, breeding season, and commercial size of *Macrobrachium americanum* Bate, 1868 from the Pacific slope, while Lara & Wehrtmann (2009) described the reproductive ecology of *M. carcinus* (Linnaeus, 1758) from the Caribbean slope. Here we present results about the taxonomic diversity, abundance and distribution of freshwater shrimps in the basin of the river Grande de Térraba, the area where the Costa Rican government is planning to construct the largest hydroelectric plant in Central America.


## MATERIAL AND METHODS

### Study area

The study was carried out in the basin of the river Grande de Térraba (Pacific slope of southern Costa Rica), with 5084 km$^2$ the largest river catchment in Costa Rica (EGIRH, 2005). A total of 84 different locations were sampled, ranging in altitude (expressed as meters above sea level) between 0 and 1075 m (fig. 1). The basin was divided into four zones according to altitude and environmental variables (table I).

Between July 2005 and July 2008, the freshwater shrimps were collected with an electro fishing equipment Samus 725-GN. Abundance represents the number of individuals collected during the duration of the sampling. Each shrimp was identified to species level (Palaemonidae: Holthuis, 1952; Hendrickx, 1995; Valencia & Campos, 2007; Atyidae: Hobbs & Harts, 1982; Chace & Hobbs, 1969).

Total abundances were expressed in percentage of the total number of shrimps collected during the sampling period. The abundance index represents the number of individuals divided by the sampling effort (number of visits in each sampling location). The abundance index was related to the altitude of the sampling site, and a Student t-test was applied to analyze the possible correlation between the two variables.


## RESULTS

Environmental variables. — Dissolved oxygen, water temperature and conductivity of our sampling sites ranged from 4.9 to 9.9 mg/l, 18.6 to 32.0°C, and 33.1 to 474.1 $\mu$S/cm, respectively (table I). Shrimps were encountered in areas with dissolved oxygen varying between 5.9 and 9.9 mg/l, water temperatures and conductivity ranging from 20.1 to 30.9°C and from 50.7 to 474.1 $\mu$S/cm, respectively.

Diversity. — Shrimps were found in 59 (70%) of the 84 sampling sites, ranging in altitude between 0 and 765 m. We collected a total of 4247 shrimps, representing two families, four genera and 13 species. Palaemonids were

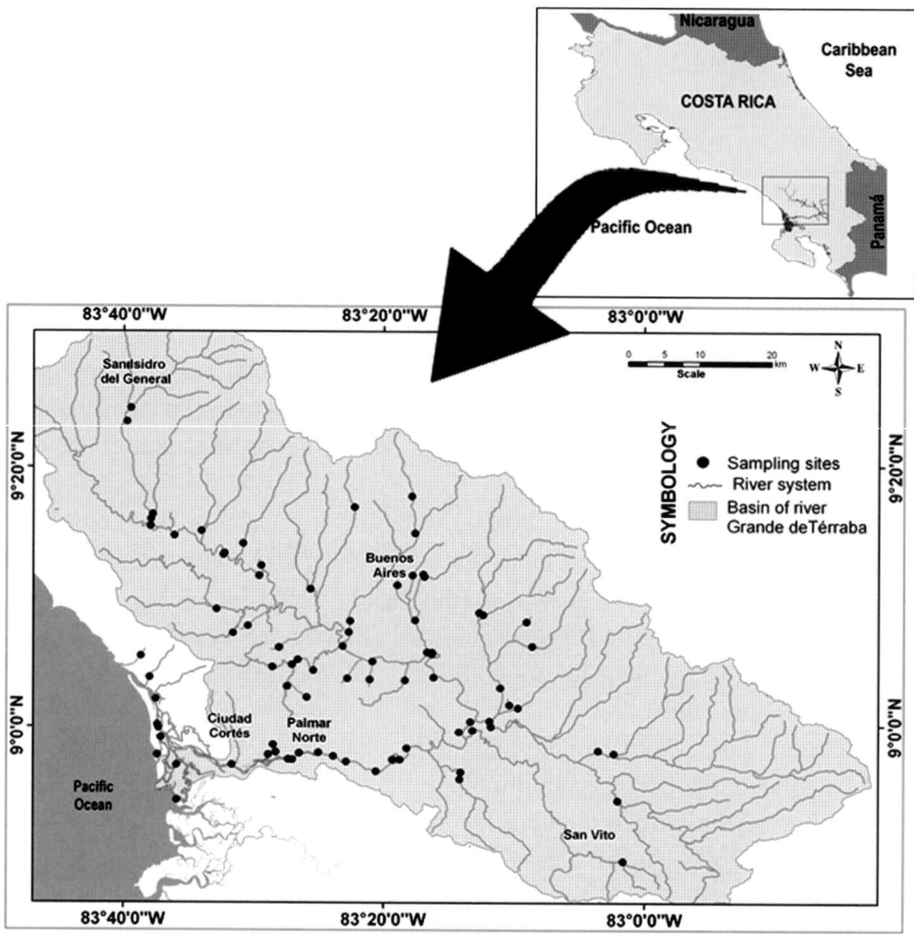


Fig. 1. Location of sampling sites (July 2005 to July 2008) in the basin of the river Grande de Térraba, Pacific slope of southern Costa Rica.

represented by members of the genera *Macrobrachium* and *Palaemon*, Atyidae by *Atya* and *Potimirim*. Most species belonged to the genus *Macrobrachium* (7 spp.).

Abundance. — *Macrobrachium* spp. were the predominant group, representing 96.6% of all shrimps collected. Representatives of the genus *Palaemon* comprised 1.9%, and shrimps of the genera *Atya* and *Potimirim* were even less abundant (0.7 and 0.9%, respectively). Among Palaemonidae, *M. digueti* (Bouvier, 1895), was the most abundant species, followed by *M. occidentale* Holthuis, 1950 and *M. tenellum* (Smith, 1871) (fig. 2A). These three species comprised 90.1% of all palaemonid shrimps collected. Shrimps of the

## TABLE I

The four sampling zones of the river Grande de Térraba basin, Pacific slope of southern Costa Rica, with its corresponding temperature, dissolved oxygen (DO) and conductivity values (± standard deviation; SD) during dry and rainy season, respectively. The number of sample sites per river zone is also indicated

| Zone of basin | Altitude (m) | Distance to estuarine system (km) | Average temperature (°C ± SD) maximum-minimum °C | | Average DO (mg/l ± SD) maximum-minimum | | Average conductivity ($\mu$S/cm ± SD) maximum-minimum | | Number of sites sampled |
|---|---|---|---|---|---|---|---|---|---|
| | | | Dry season | Rainy season | Dry season | Rainy season | Dry season | Rainy season | |
| Low | 0-100 | 0-65 | 28.1 (1.6), 31.1-25.1 | 25.8 (1.2), 28.9-23.8 | 7.8 (0.9), 9.9-5.9 | 7.6 (0.4), 8.4-5.4 | 128.6 (52.7), 342.0-50.7 | 121.5 (64.4), 325.9-55.8 | 26 |
| Middle-low | 101-300 | 65-112 | 26.7 (2.4), 32.0-22.8 | 25.0 (1.4), 25.9-22.3 | 7.9 (0.8), 9.5-5.2 | 7.8 (0.7), 9.9-5.6 | 134.3 (109.8), 474.1-58.6 | 117.3 (98.2), 344.3-33.1 | 22 |
| Middle-medium | 301-500 | 112-146 | 24.9 (2.2), 30.3-20.3 | 22.3 (1.2), 24.4-20.1 | 7.8 (0.7), 8.9-6.1 | 8.2 (0.5), 8.9-7.5 | 110.1 (50.4), 205.1-42.2 | 69.0 (18.9), 99.4-33.1 | 22 |
| Middle-high | 501-1075 | 146-170 | 23.1 (2.4), 27.8-18.6 | 21.1 (1.6), 23.4-19.0 | 8.2 (1.1), 9.9-4.9 | 7.8 (0.6), 9.4-7.2 | 94.4 (20.0), 148.4-87.0 | 100.8 (42.7), 144.6-35.7 | 14 |

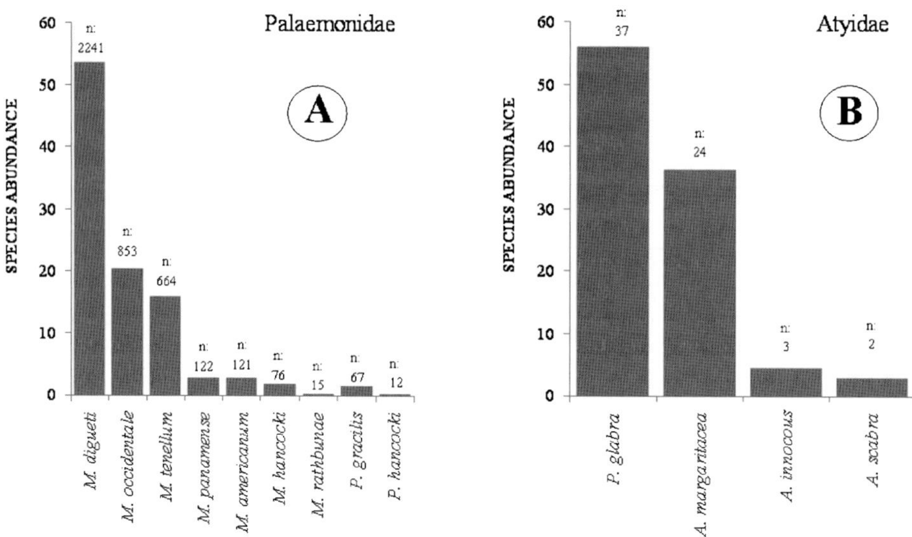



Fig. 2. Species abundance (expressed as percentages of total number of individuals collected) for representatives of Palaemonidae (A) and Atyidae (B), collected in the basin of the river Grande de Térraba, Pacific slope of southern Costa Rica.

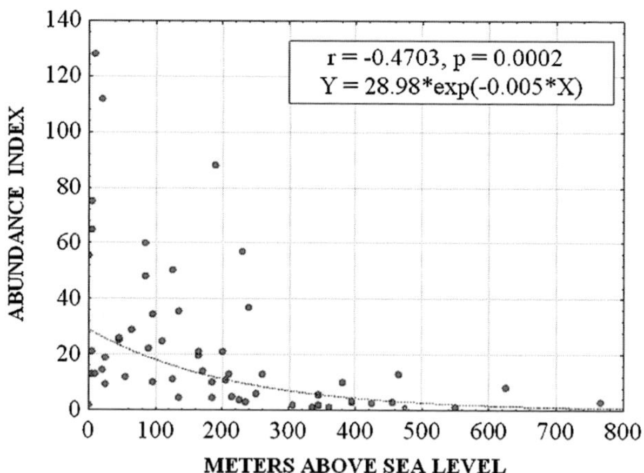



Fig. 3. Relation between shrimp abundance (all collected species) and altitude in the basin of the river Grande Térraba, Pacific slope of southern Costa Rica.

genus *Palaemon* were rare (fig. 2A). Among atyid shrimps, *P. glabra* (Kingsley, 1878) and *A. margaritacea* A. Milne-Edwards, 1864 were the predominant species, representing 92.4% of all atyid shrimp (fig. 2B).

Shrimp abundance was negatively correlated ($p < 0.001$) with altitude (fig. 3). The total shrimp abundance increased toward the estuarine system:

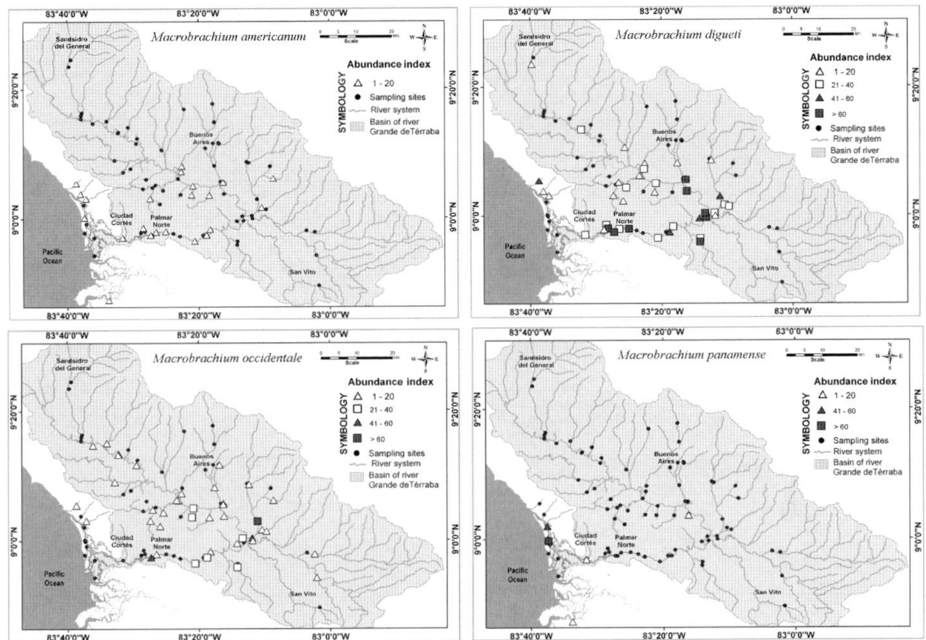

Fig. 4. Distribution and abundance (number of individuals per sampling time) of four *Macrobrachium* species in the basin of the river Grande de Térraba, Pacific slope of southern Costa Rica. The symbol for "sampling sites" indicates places where shrimps were absent.

56.4% of all shrimps were collected in the lower basin, and 37.9% in the midpart of the river basin. Lowest abundance was found in the middle-medium and middle-high zone with 4.1% and 1.6%, respectively.

Species diversity and distribution. — Highest species diversity (all 13 spp.) was detected in the lower river section (0-100 m) (table II). Only three species (*M. americanum*, *M. digueti* and *M. occidentale*) were found in all river sections (table II), and the highest abundances of these species were obtained in the lower and middle sections (table II, figs. 4 and 5). *Macrobrachium digueti* inhabited altitudes between 5 and 765 m, representing the highest altitude where we collected freshwater shrimp. *Macrobrachium occidentale* and *M. americanum* were found between 10-625 m and 0-550 m, respectively.

*Macrobrachium panamense* Rathbun, 1912 and *M. tenellum* were encountered mainly in the lower part of the river basin: 99.1% of these species were collected in altitudes between 0-25 m (figs. 4D and 5B), where they were highly abundant (up to 136 individuals per sampling station). Other species such as *P. gracilis* (Smith, 1871) and *P. hancocki* Holthuis, 1950 were found exclusively in the lower part (<10 m) of the river basin (table II, fig. 5C, D), inhabiting only the main river and the estuarine zone of the river mouth.

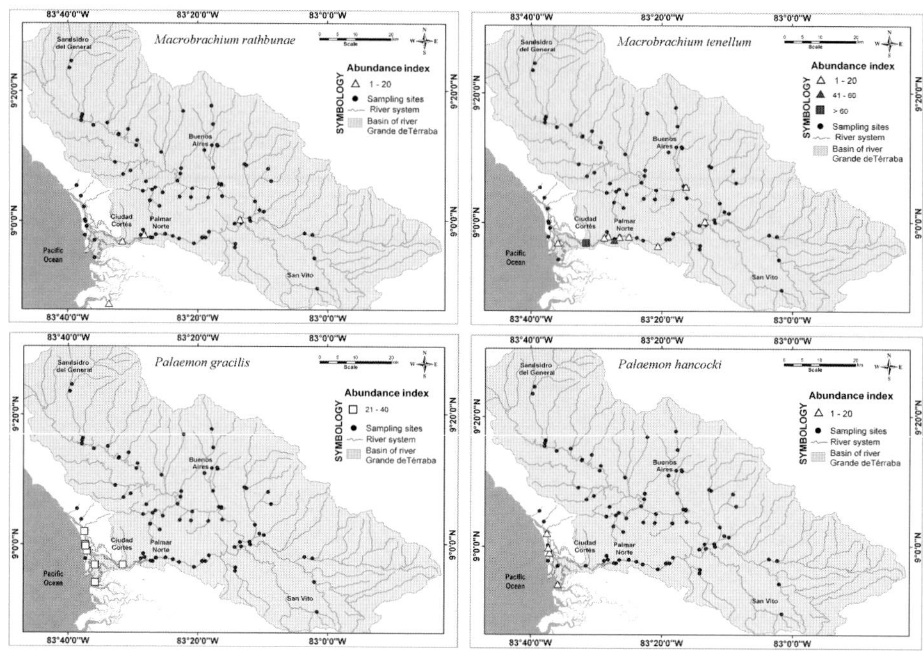

Fig. 5. Distribution and abundance (number of individuals per sampling time) of four palaemonids in the basin of the river Grande de Térraba, Pacific slope of southern Costa Rica. The
symbol for "sampling sites" indicates places where shrimps were absent.

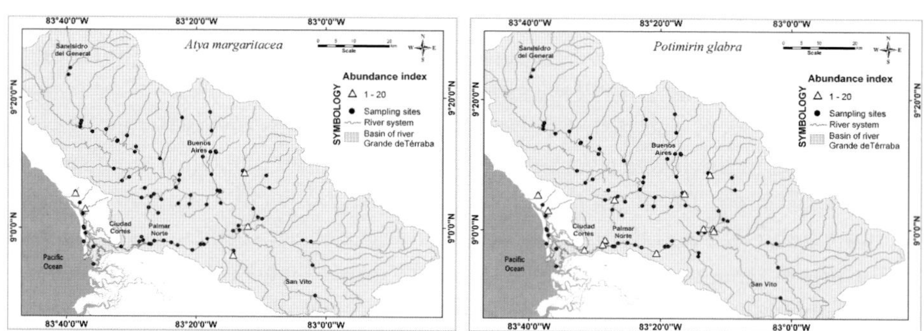

Fig. 6. Distribution and abundance (number of individuals per sampling time) of two freshwater
shrimp species in the basin of the river Grande de Térraba, Pacific slope of southern Costa Rica.
The symbol for "sampling sites" indicates places where shrimps were absent.

Atyids with the widest distribution along the river basin were *A. margaritacea* (between 0 and 465 m) and *P. glabra* (between 0 and 135 m) (fig. 6A, B);
both species were found in the main river and its tributaries. In contrast, *A. innocous* (Herbst, 1792) and *A. scabra* (Leach, 1815) were collected exclusively
in the tributaries of the lower river basin.

*Macrobrachium americanum, M. panamense, M. rathbunae* Holthuis, 1950, *M. tenellum* and *P. gracilis* were found in freshwater as well as in near-coastal environments, while *P. hancocki* was collected exclusively in the river mouth area. The following species were encountered only in freshwater habitats: *M. occidentale, M. digueti, M. hancocki* Holthuis, 1950, and all species of *Atya*.

Habitat preferences. — Both *M. hancocki* and *M. americanum* inhabited river sections with rapid as well as slow flowing waters. *Macrobrachium digueti* preferred sections with low stream velocities, and *M. occidentale* was collected mainly in waters with high stream velocities. *Macrobrachium rathbunae, M. tenellum, M. panamense*, and the two *Palaemon* species were collected mainly in slow flowing river sections, where they inhabited the emerged vegetation (Poaceae). The presence of atyid species was restricted to the close vicinity of water cascades.


## DISCUSSION

Diversity. — The basin of the river Grande de Térraba harbors an impressive diversity of shrimps, corroborating similar findings concerning the decapod fauna of the adjacent Térraba-Sierpe mangrove area (Echeverría-Sáenz et al., 2003). We collected seven *Macrobrachium* and two *Palaemon* species as well as four species of the family Atyidae (fig. 2). Of all *Macrobrachium* species reported so far from the Pacific slope of Costa Rica (Holthuis, 1952), only the endemic species *M. cocoense* Abele and Kim, 1984 from Isla del Coco was not collected by us. Moreover, the study area harbors 54% of the *Macrobrachium* species from Costa Rica, and 20% of the 35 *Macrobrachium* spp. known to occur in the eastern hemisphere (Bowles et al., 2000).

Vargas and Wehrtmann (2009) listed for the Pacific coast of Costa Rica a total of 23 palaemonid species comprising 13 genera. These authors mentioned two species of the genus *Palaemon*, which were collected by us, too. All atyid species obtained by us have been already reported for the Pacific slope of Costa Rica (Chace & Hobbs, 1969; Hobbs & Harts, 1982).

Abundance. — Shrimp abundance increased towards the river mouth area (fig. 3). This finding might be related to the life history of the *Macrobrachium* and atyid species: some of these species are completely adapted to freshwater, but many other of these species still depend upon saline waters for the successful development of their early life stages (see Rome et al., 2009). In general, ovigerous females of these amphidromous river shrimps migrate to the coast where they liberate their offspring. After completing the zoeal phase, massive upstream migrations of juvenile shrimps can be observed (Latin America and adjacent areas: see Bauer & Delahoussaye, 2008). According to our results (I. S. Wehrtmann & L. R. Lara, unpubl. data), all *Macrobrachium* and most probably also the atyid species collected in the study area are amphidromous shrimps. Therefore, it can be expected that the lower parts of the basin the river Grande de Térraba harbor a higher number of individuals than the higher parts. However, our results may also reflect a preference of these shrimps for low altitude areas. More studies are needed to clarify the habitat utilization of these freshwater decapods in tropical systems.

Distribution. — Highest species diversity was found in the lower river sections (0-100 m), which may be related to the fact that these species need salt water to complete successfully their life cycle. These results coincide with similar findings of freshwater shrimps in Venezuela (Gamba, 1982) and Mexico (Mejía-Ortíz et al., 2001). The spatial distribution of palaemonid freshwater shrimps may be grouped into three categories (Montoya, 2003): Group I represents species with a distribution concentrated in river sections near the sea; *M. rathbunae*, *M. tenellum*, *M. panamense*, *P. gracilis*, and *P. hancocki* can be considered as part of this category (table II). Group II contains species, which are independent of the proximity of marine habitats (Montoya, 2003). In this category fall all the remaining species collected by us. Group III is comprised by species whose geographic distribution is restricted to rivers far from the sea (Montoya, 2003), thus shrimps with a larval development independent from saline waters. To our best knowledge, the Costa Rican freshwater fauna does not contain shrimps other than amphidromous species, which explains the absence of representatives of Group III.

Information concerning the altitudinal distribution of freshwater shrimps in Latin America is extremely scarce. Gamba (1982) found most *Macrobrachium* species in northern Venezuela well below 100 m altitude (exception: *M. heterochirus* (Wiegmann, 1836), up to 470 m). The latter species was collected in Mexico between 105 and 535 m altitude (Mejía-Ortíz et al., 2001). Our data extend the altitudinal range of *Macrobrachium* species: three species (*M. americanum*, *M. digueti*, *M. occidentale*) were found well above 500 m altitude, with *M. digueti* collected in a maximum altitude of 765 m. These findings demonstrate the capacity of these three species to inhabit riverine environments located in relatively high altitudes, which may indicate a better adaptation to purely freshwater habitats. However, to verify this interpretation, more studies about the reproductive and larval ecology of these species are suggested to shed light on the larval dependence on saline waters for their successful development.

*Palaemon* spp. were less abundant compared to *Macrobrachium* spp. (fig. 2). This result is not surprising, considering that these two palaemonid species are typical inhabitants of mangrove areas (Echeverría-Sáenz et al., 2003), which were not extensively sampled by us (see fig. 1).

The *Atya* species were encountered preferably in well-oxygenated river sections with high stream velocities. They can be found under rocks in rapidly flowing stream segments, but also in vegetated areas, where they burrow shallow depressions under rocks or inhabit existing crevices to cover themselves (Hobbs & Harts, 1982). Therefore, the presence of atyid shrimps is more restricted to specific habitats with well-defined environmental conditions (well-oxygenated and fast-flowing river segments) than most of the other caridean shrimps living in the study area.

*Potimirim glabra* was by far the most abundant atyid shrimp collected during our study (fig. 2). The species prefers rocky bottom and moderately swift to swift currents (Smalley, 1963), and thus may co-occur with shrimps of the genus *Atya* (Chace & Hobbs, 1959). Our results confirm this observation: in several locations, we collected representatives of both genera. According to Smalley (1963), *P. glabra* occurs exclusively near the sea. Our results are in general agreement with this observation; however, we collected this species also in the middle-low portion of the river basin (table II) and up to 135 m, which seems to indicate that its distribution is not restricted to river segments directly influenced by marine waters.

Conclusion. — More than half of all caridean freshwater shrimp species so far reported from Costa Rica (59%) can be found in the basin of the river Grande de Térraba. To maintain this impressive species diversity, it is of utmost importance to preserve the present environment in the river basin. We are dealing with amphidromous shrimps which depend on the migration of the adults to the vicinity of estuarine habitats as well as migration of juveniles to upstream habitats; thus, any study aimed to evaluate possible impacts of human activities in the area needs to consider these life history features. In particular, the construction of dams and/or reservoirs is known to alter significantly the distribution and abundance of freshwater faunas by blocking migratory pathways (Holmquist et al., 1998; March et al., 2003). The studied river basin is dominated by migratory shrimps which are vulnerable to hydrologic modification; therefore, we strongly recommend that the design of any hydroelectric project in the study area includes structures permitting the passage of the migrating shrimp (e.g., shrimp ladders: March et al., 2003) to mitigate the negative effects of this type of construction on the associated freshwater fauna.

## TABLE II

Abundance (%) of different caridean shrimps in the four sections of the basin of the river Grande de Térraba, Pacific slope of southern Costa Rica. "Presence": number of total sampling sites where shrimp were collected; "Total": total number of species encountered in each of the four river sections

| Section of river basin | Sampling sites | Presence | *M. americanum* | *M. digueti* | *M. hancocki* | *M. occidentale* | *M. panamense* | *M. rathbunae* | *M. tenellum* | *P. gracilis* | *P. hancocki* | *A. innocous* | *A. margaritacea* | *A. scabra* | *P. glabra* | Total |
|---|---|---|---|---|---|---|---|---|---|---|---|---|---|---|---|---|
| Low | 26 | 21 | 54.2 | 41.6 | 46.2 | 22.2 | 99.6 | 100.0 | 99.9 | 100.0 | 100.0 | 100.0 | 28.8 | 100.0 | 73.9 | 13 |
| Middle low | 22 | 21 | 32.6 | 56.2 | 53.8 | 59.8 | 0.4 | 0.0 | 0.1 | 0.0 | 0.0 | 0.0 | 25.1 | 0.0 | 26.1 | 8 |
| Middle medium | 22 | 13 | 4.4 | 1.1 | 0.0 | 13.1 | 0.0 | 0.0 | 0.0 | 0.0 | 0.0 | 0.0 | 46.1 | 0.0 | 0.0 | 4 |
| Middle high | 14 | 3 | 8.8 | 1.0 | 0.0 | 4.9 | 0.0 | 0.0 | 0.0 | 0.0 | 0.0 | 0.0 | 0.0 | 0.0 | 0.0 | 3 |

## ACKNOWLEDGMENTS

We thank Rita Vargas (Universidad de Costa Rica) for their collaboration with the identification of shrimps. Luis Salazar, Leonel Delgado, Yonder Guzmán, Edgar Chinchilla, Carlos Canales and Diego Valerín helped us during the fieldwork. Jorge Picado (ICE) provided technical support. The Hydroelectric Project El Diquís-ICE (PHED) supported the present investigation. We are grateful to Paulo Bermúdez and M. Sofía Artavia from the Department of Geographic Information System, PHED, for preparing the distribution maps. Prime Catch Seafood GmbH, Germany, and the German Investment and Development Company (Deutsche Investitions- und Entwicklungsgesellschaft, DEG) provided financial support within the Public-Private-Partnership project E 2118, which is greatly appreciated.

## REFERENCES

ABELE, L. G. & WOM-KIM, 1984. Notes on the freshwater shrimp of Isla del Coco with the description of *Macrobrachium cocoense*, new species. Proceedings of the Biological Society Washington, **97**(4): 951-960.

ÁLVAREZ, M. D., J. CABRERA & Y. LOPEZ, 1996. Morfometría, época reproductiva y talla comercial de *Macrobrachium americanum* (Crustacea: Palaemonidae) en Guanacaste, Costa Rica. Revista de Biología Tropical, **44**: 127-132.

BAUER, R. T. & J. DELAHOUSSAYE, 2008. Life history migrations of the amphidromous river shrimp *Macrobrachium ohione* from a continental large river system. Journal of Crustacean Biology, **28**(4): 622-632.

BOTT, R., 1967. Fluß-Krabben aus dem westlichen Mittelamerika (Crust., Decap.). Senckenbergiana Biologica, **48**: 373-380.

— —, 1968. Fluß-Krabben aus dem östlichen Mittel-Amerika und von den Großen Antillen. Senckenbergiana Biologica, **49**: 39-49.

BOWLES, D. E., K. AZIZ & C. L. KNIGHT, 2000. *Macrobrachium* (Decapoda: Caridea: Palaemonidae) in the contiguous United States: a review of the species and an assessment of threats to their survival. Journal of Crustacean Biology, **20**: 158-171.

CHACE, F. A. & H. H. HOBBS, 1969. The freshwater and terrestrial decapod crustaceans of the West Indies with special reference to Dominica. United States National Museum Bulletin, **292**: 1-258.

CHIA, O. K. S. & P. K. L. NG, 2006. The freshwater crabs of Sulawesi, with descriptions of two new genera and four new species (Crustacea: Decapoda: Brachyura: Parathelphusidae). Raffles Bulletin of Zoology, **54**: 383-428.

DE GRAVE, S., Y. CAI & A. ANKER, 2008. Global diversity of shrimps (Crustacea: Decapoda: Caridea) in freshwater. Hydrobiologia, **595**: 287-293.

DE GRAVE, S., N. D. PENTCHEFF, S. T. AHYONG, T.-Y. CHAN, K. A. CRANDALL, P. C. DWORSCHAK, D. L. FELDER, R. M. FELDMANN, C. H. J. M. FRANSEN, L. Y. D. GOULDING, R. LEMAITRE, M. E. Y. LOW, J. W. MARTIN, P. K. L. NG, C. E. SCHWEITZER, S. H. TAN, D. TSHUDY & R. WETZER, 2009. A classification of living and fossil genera of decapod crustaceans. Raffles Bull. Zool., (Supplement) **21**: 1-114.

ECHEVERRÍ-SÁEENZ, S., R. VARGAS & I. S. WEHRTMANN, 2003. Diversity of decapods inhabiting the largest mangrove system of Pacific Costa Rica. Nauplius, **11**(2): 91-97.

EGIRH, 2005. Estrategia para la gestión integrada de los recursos hídricos en Costa Rica, diagnóstico: primera etapa del plan de manejo integral del recurso hídrico: una estrategia nacional para el MIRH. Proyecto BID ATN/WP-8467-CR: 1-114. (Ministerio de Ambiente y Energía Instituto Meteorológico Nacional, San José).

GAMBA, A. L., 1982. *Macrobrachium*: its presence in estuaries of the northern Venezuelan coast (Decapoda, Palaemonidae). Caribbean Journal of Science, **18**: 23-25.

HENDRICKX, M. E., 1995. Camarones. In: Guía FAO para la identificación de especies para los fines de la pesca, Pacífico Centro-Oriental. Vol. 1, Plantas e Invertebrados: 417-537. (FAO, Organización de las Naciones Unidas para la Agricultura y la Alimentación, Rome).

HOBBS, H. H. & C. W. HART, 1982. The shrimp genus *Atya* (Decapoda: Atyidae). Smithsonian Contribution to Zoology, **364**: 1-143.

HOLMQUIST, J. G., J. M. SCHMIDT-GENGENBACH & B. BUCHANAN-YOSHIOKA, 1998. High dams and marine-freshwater linkages: effects on native and introduced fauna in the Caribbean. Conservation Biology, **12**: 621-630.

HOLTHUIS, L. B., 1952. The subfamily Palaemoninae. A general revision of the Palaemonidae (Crustacea: Decapoda: Natantia) of the Americas. Allan Hancock Foundation Occasional Papers, **12**: 1-396.

ICE, 2009. Déjenos contarle, revista informativa del Proyecto Hidroeléctrico El Diquís: 1-15. (Septiembre 2009, 1 ed, año 1, Costa Rica).

LARA, L. R. & I. S. WEEHRTMANN, 2009. Reproductive biology of the freshwater shrimp *Macrobrachium carcinus* (L.) (Decapoda: Palaemonidae) from Costa Rica, Central America. Journal of Crustacean Biology, **29**(3): 343-349.

MARCH, J. G., J. P. BENSTEAD, C. M. PRINGLE & F. N. SCATENA, 2003. Damming tropical island streams: problems, solutions, alternatives. Bioscience, **53**: 1069-1078.

MEJÍA-ORTÍZ, L. M., F. ALVAREZ, R. ROMAN & J. A. VICCON-PALE, 2001. Fecundity and distribution of freshwater prawns of the genus *Macrobrachium* in the Huitzilapan river, Veracruz, Mexico. Crustaceana, **74**: 69-77.

MONTOYA, J. V., 2003. Freshwater shrimps of the genus *Macrobrachium* associated with roots of *Eichhornia crassipes* (Water Hyacinth) in the Orinoco Delta (Venezuela). Caribbean Journal of Science, **39**: 155-159.

RODRÍGUEZ, G., 1982. Les crabes d'eau douce d'Amerique. Famille des Pseudothelphusidae. Faune Tropicale, **22**: 1-224. (ORSTOM, Paris).

RODRÍGUEZ, G. & I. HEDSTRÖM, 2000. The freshwater crabs of the Barbilla National Park, Costa Rica (Crustacea: Brachyura: Pseudothelpusidae), with notes on the evolution of structures for spermatophore retention. Proceedings of the Biological Society Washington, **113**(2): 420-425.

ROME, N. E., S. L. CONNER & R. T. BAUER, 2009. Delivery of hatching larvae to estuaries by an amphidromous river shrimp: tests of hypotheses based on larval moulting and distribution. Freshwater Biology, **54**(9): 1924-1932. doi:10.1111/j.1365-2427.2009.02244.x

SMALLEY, A. E., 1963. The genus *Potimirim* in Central America (Crustacea, Atyidae). Revista de Biología Tropical **11**: 177-183.

— —, 1964. The river crabs of Costa Rica, and the subfamilies of the Pseudothelphusidae. Tulane Studies in Zoology, **12**: 5-13.

— —, 1970. A new genus of freshwater crabs from Guatemala, with a key to the middle American genera (Crustacea Decapoda, Pseudothelphusidae). American Midland Naturalist, **83**: 96-106.

VALENCIA, D. M. & M. R. CAMPOS, 2007. Freshwater prawns of the genus *Macrobrachium* Bate, 1868 (Crustacea: Decapoda: Palaemonidae) of Colombia. Zootaxa, **1456**: 1-44.

VARGAS, R. & I. S. WEHRTMANN, 2009. Decapods. In: I. S. WEHRTMANN & J. CORTÉS (eds.), Marine biodiversity of Costa Rica, Central America. Monographiae Biologicae, **86**: 209-228. (Springer and Business Media B.V., Berlin).

YEO, D. C. J., P. K. L. NG, N. CUMBERLIDGE, C. MAGALHÃES, S. R. DANIELS & M. R. CAMPOS, 2008. Global diversity of crabs (Crustacea: Decapoda: Brachyura) in freshwater. Hydrobiologia, **595**: 275-286.

First received 2 February 2010.
Final version accepted 13 March 2010.

# SEXUAL SELECTION IN CRAYFISH: A REVIEW

BY


FRANCESCA GHERARDI[1]) and LAURA AQUILONI[2])

Department of Evolutionary Biology "Leo Pardi", University of Florence, Via Romana 17, 50125 Florence, Italy


## ABSTRACT


Crayfish are used as model organisms in many research fields but this potential is not fully expressed in behavioral studies. This review paper attempts to organize the results of the abundant literature on crayfish reproduction within the framework of the sexual selection theory. Our aim is to stimulate further research in this promising field of study.


## INTRODUCTION

Crayfish are used as model organisms of study in a variety of research fields, from ecology to molecular evolution. However, at least in the field of behavior, this taxon appears to be an "almost perfect" model system (Hazlett, 2009). In fact, notwithstanding the about 640 species described so far (Crandall & Buhay, 2008), over 75% of studies have focused on 10 species only (Gherardi et al., 2010). As a consequence, important questions about the evolution of crayfish behavior are difficult to be addressed. This drawback is particularly evident when the aim is to understand the evolutionary mechanisms through which sexual selection operates.

Here, we attempt to organize the results of the abundant literature on the reproductive biology of crayfish within the framework of the sexual selection theory. Sexual selection is a term coined by Charles Darwin (1871) to indicate the process leading to the evolution of morphological and/or behavioral traits that increase the reproductive success of the individuals of each sex but that

---

[1]) e-mail: francesca.gherardi@unifi.it

[2]) e-mail: laura.aquiloni@unifi.it

    New frontiers in crustacean biology: 213-223

cannot be accounted for by natural selection. Females, who invest in few costly gametes, are highly choosy, whereas males, who produce large numbers of cheap sperm, are limited in their reproductive output only by the frequency of matings (Trivers, 1972). Because of this intersexual difference in investment, typically selection gives rise to 'reluctant' females on one hand and 'ardent' males with exaggerated morphological and/or behavioral traits on the other.


## INTRASEXUAL SELECTION

### Contest and scramble competition

Intrasexual competition occurs when two or more individuals are dependent on the same resource and when this resource is limited. There are two modes of competition: contest competition, in which one individual gains the exclusive access to a given resource by impeding its access to competitors, and scramble competition, in which each individual tries to maximize its share with that resource without directly interfering with the others (Barki, 2008). However, especially when the resource is a receptive female, these two modes of competition are extremes in a *continuum* and often may co-occur or occur in sequence during the mating season.

This is well illustrated in the case of *Orconectes rusticus* in southern Ontario (Berrill & Arsenault, 1982). The mating season of this species is short during spring and female receptivity is synchronous. During the first 10 days of mating activity when the operational sex ratio (OSR) is about 1:1, males and females wander and copulations is frequent. Aggressive interruptions of copulations by other males begin 8-9 days after, when OSR starts to rise due to the increased number of already mated females that sequester themselves in shelters to incubate the extruded eggs. When receptive females become extremely rare (after 1-2 additional weeks) feeding replace competition in males and copulations stop.

Intermale competition for the access of a female occurs in *Austropotamobius pallipes* (cf. Gherardi et al., 2006) and *Pacifastacus trowbridgii* (cf. Mason, 1970). It is particularly intense in laboratory groups of *A. pallipes* (cf. Gherardi et al., 2006), which results in a decreased number and duration of mating attempts if compared with a competition-free context. The ability to dominate over other males depends on a number of extrinsic (e.g., experience of wins/losses) and intrinsic factors (e.g., body size, weight, and chelar dimensions, neurochemical state) (Tricarico & Gherardi, 2007). An obvious determinant of dominance is size: in laboratory groups of *A. pallipes*, larger males even arrive to kill the smaller males in the presence of a receptive female (Woodlock & Reynolds, 1988); as a result of larger-male dominance, small males are typically involved in little copulations (Gherardi et al., 2006).

## Recognition of species and sex

The success in scramble competition relies on the ability to detect a reproductive conspecific of the other sex at a distance and to orient towards it: to identify the correct species, sex and reproductive status, crayfish typically rely on pheromones. Indeed, crayfish have an efficient system for broadcasting and detecting chemical stimuli (Moore & Bergman, 2005). This has been confirmed by studies investigating recognition between syntopic species, e.g., *Orconectes virilis* and *Orconectes propinquus* (cf. Tierney & Dunham, 1984); when this mechanism fails, mating may occur between individuals of two different species, resulting in either reproductive interference (between *A. pallipes* and *Astacus leptodactylus* or *Pacifastacus leniusculus* in England; Holdich et al., 1995) or hybridization (with the eventual genetic assimilation of one species, as in the case of the invasive *O. rusticus* replacing in this way *O. propinquus* in Michigan; Perry et al., 2001).

Pheromones are also involved in sex recognition. Receptive females of several species (*Cambarus robustus*, *O. propinquus*, *O. virilis*, and *Procambarus clarkii*) are known to emit urine-born sex pheromones. The exposure to water conditioned by females leads *P. leniusculus* males to exhibit the behavioral patterns typical of mating, i.e., seizure, mounting, and spermatophore deposition (Stebbing et al., 2003). Pheromones are primarily detected by the male antennules (Ameyaw-Akumfi & Hazlett, 1975). However, since antennule ablation do not completely annul sex discrimination (Corotto et al., 1999), other chemosensory organs may play a role, such as the smooth setae on the major chelae of reproductive males (Belanger et al., 2008).

Crayfish may also use vision for sex identification (Dunham & Oh, 1996), particularly in species, such as *P. clarkii*, with a more diurnal timing of activity. More recent studies pinpoint the bimodal nature of sex recognition: crayfish rely on both smell and sight but the relative importance of chemical and visual cues varies between sexes. Chemical recognition seems to be a male prerogative (Aquiloni et al., 2009), whereas females seem to make more extensive use of vision. However, visual stimuli are not sufficient to identify the sex of a conspecific but they should be combined with the male odor to suppress aggressiveness (Aquiloni et al., 2009).

## Sperm competition

Males of some crayfish species have evolved means to reduce the risks of decreased paternity due to sperm competition. Multiple paternity seems to be common in crayfish, as shown by genetic studies: the use of the microsatellite technique on 15 wild broods of *Orconectes placidus* proved that most of the females had mated with two to four males (Walker et al., 2002). However, even in multiple mated females, 80% and more of the brood descended from one male only (Walker et al., 2002).

When spermatophores are deposited externally, such as in the Astacidae, males either deposit their spermatophores on the top of the first (Woodlock & Reynolds, 1988) or feed on those deposited by the previous mates (Villanelli & Gherardi, 1998), removing over 70% of the previous male's spermatophores (Galeotti et al., 2007). The last-male prevalence in paternity is in any case large: about 85% of the spermatophores deposited onto *A. pallipes* females after the second mating derives from the second male, which also obtained the extra benefit of a proteic meal by feeding on the spermatophores of his rival (Galeotti et al., 2007).

In Cambaridae, on the contrary, spermatophores are inaccessible for the subsequent males because they are inserted into the *annulus ventralis*, a cuticular spermatheca originating from the posterior part of the seventh sternite. Sperm competition may be thus avoided by adjusting the length of copulation as a function of the female mating status (suggested for *O. rusticus*; Snedden, 1990), depositing a mating plug in the opening of the receptacle (suggested for *O. rusticus*; Crocker & Barr, 1968), removing the plug of a previous male with copulatory stylets (Berrill & Arsenault, 1984), or selecting virgin females (in *P. clarkii*, cf. Aquiloni & Gherardi, 2008a), identified through pheromones (in *Orconectes quinebaugensis*, cf. Durgin et al., 2008), and defending them.

## Alternative mating strategies

In various crustaceans, particularly in the species in which sexual selection through male contest competition is strong (Barki, 2008), males exhibit discontinuous variation in mating behavior and morphology, thus displaying alternative mating strategies. In Cambaridae, two morphotypes have been described. For instance, *P. clarkii* reproductive males (Form I) have rectangular and greatly inflated chelae, prominent copulatory hooks at the bases of the third and fourth pairs of walking legs and cornified gonopodia (Huner, 2002). Form I males are agonistically dominant over the non-reproductive males (Form II) (Guiasu & Dunham, 1998), but we do not know whether the latter have evolved an alternative mating tactic, as on the contrary found in the freshwater prawn *Macrobrachium rosenbergii* (e.g., Ráan & Sagi, 1985).


## INTERSEXUAL SELECTION

### Mating behavior

Observations on mating of crayfish under natural conditions are rare; on the contrary, mating was described in the laboratory in several crayfish species, including *Cambaroides japonicus* (cf. Kawai & Saito, 2001). In *A. pallipes*, it consists of three sequential phases, i.e., contact, turning of the female, and mounting by the male (Villanelli & Gherardi, 1998). Receptive females are always hard-shelled and, interestingly, seem to play an active role during mating: mate acceptance is denoted by their freezing during the contact phase, which determines the subsequent success of copulation. In *A. pallipes* (cf. Gherardi et al., 2006) and other species (e.g., *Cherax quadricarinatus*; cf. Barki & Karplus, 1999), females almost invariably initiate mating by approaching the male. Sometimes the females seem to resist mating attempts, which might provide information regarding the male's vigor and condition: in *O. rusticus*, the females may delay copulation by keeping the abdomen curled, precluding the access of the male's copulatory stylets (Berrill & Arsenault, 1984). Male–female fighting occurs in several species (*O. rusticus*, cf. Berrill & Arsenault, 1984; *P. trowbridgii*, cf. Mason, 1970). However, resistance might be sometimes risky: *A. pallipes* females may be even killed by the males before or after mating, at least in the laboratory setting (Woodlock & Reynolds, 1988). A reversal of sex roles in mating, with females initiating copulation-like behaviors directed towards other females or males, has been also observed in laboratory groups of *P. clarkii* when sex ratio was kept skewed towards females (1 male and 2 females) (Kasuya et al., 1996).


### Mate choice by females

Females of a wide array of species were found to select mates with large body size: *Astacus astacus* (cf. Furrer, 2004), *A. pallipes* (cf. Villanelli & Gherardi, 1998; Gherardi et al., 2006), *O. rusticus* (cf. Berrill & Arsenault, 1984), and *P. clarkii* (cf. Aquiloni & Gherardi, 2008a). This preference might have evolved because large males are relatively more fertile with respect to smaller individuals (but in *A. pallipes* the extent of ejaculates decreases with the increased male size as the effect of senescence; Rubolini et al., 2006, 2007). Being dominant in intrasexual competition, they also offer the females with high-quality vital resources, such as breeding burrows. Indirect benefits, yet to be investigated, may be those hypothesized by the "good genes" model (i.e., large size might be the expression of high quality genes and females, mating with a large male, will transmit this quality to their offspring; Hunt et al., 2005) or by the "runaway selection" model (i.e., a slight ancestral preference for large males might have led to both an increased frequency of hereditary "large-body genes" in males and the female preference for large males; Weatherhead & Robertson, 1979). Indeed, *P. clarkii* females invest relatively less when they are forced to copulate with small males (Aquiloni & Gherardi, 2008b) but seem unable to annul the time and energy expended to take care of seemingly low-fitness clutches: the size of the spawned eggs (reflecting their content in yolk) is significantly reduced although their number remains unchanged. Larger eggs produce larger and heavier offspring with greater competitive ability and higher chance of survival.

Large chelae serve as powerful weapons during intraspecific fights, being obvious determinants of wins (in: *C. robustus*, cf. Guiasu & Dunham, 1997; *O. propinquus,* cf. Stein, 1976; *O. rusticus*, cf. Schroeder & Huber, 2001; *P. clarkii*, cf. Gherardi et al., 1999). Chelae are sexually dimorphic: in *O. propinquus*, for instance, the allometric increase of cheliped weight with body size is much faster in males than in females, so that males have usually longer, wider, and heavier chelae than similarly-sized females (Stein, 1976). A similar pattern has been reported for *O. rusticus* (cf. Schroeder & Huber, 2001) and *P. clarkii* (cf. Gherardi et al., 1999).

In *O. rusticus*, males with larger chelipeds, but of similar cephalothorax length, are better able to displace rivals and to secure females, preventing their escape, and/or more easily handle larger, more fecund females (Snedden, 1990). Large- and small-clawed males also differ in the duration of copulation: pairs involving males with large chelipeds remain in copula longer than pairs involving small clawed males (Snedden, 1990).

Similarly, in the Australian slender crayfish, *Cherax dispar*, males with large chelae are more successful during copulation than those with small chelae (Stein, 1976). But larger chelae are associated with decreased performance under the risk of predators, which suggests that their function of signaling quality may be ensured via a handicap (Wilson et al., 2009). It is thus expected that the male chelae are subject to intersexual selection and that their bigger size has evolved because of the females' preference (Villanelli & Gherardi, 1998).

The above statement has been however questioned in *P. clarkii*: females do not show any preference when simultaneously offered with equally-sized males with large and small chelae (Aquiloni & Gherardi, 2008a). Similarly, *A. pallipes* females adopt two co-occurring strategies of maternal allocation depending on male traits, laying fewer but larger eggs for relatively small-sized and large-clawed males and the reverse for relatively large-sized and small-clawed males (Galeotti et al., 2008). Chelar asymmetry, due to the loss of one of the two chelae, has also no apparent effect on mate choice, at least in *P. clarkii* (cf. Aquiloni & Gherardi, 2008a), whereas in *A. pallipes* symmetry may decrease the ability of males to win fights, to secure females for copulation (Rubolini et al., 2006), and to remove the spermatophores of the previous mate when they are the second mates (Galeotti et al., 2008).

Some male ornaments, such as the soft, uncalcified red patch on the outer surface of the cheliped propodi in mature males of the redclaw crayfish *C. quadricarinatus*, have been suggested to serve as choice criteria, being a honest signal of male quality as a reflection of the physiological health of a crayfish. In fact, red color derives from carotenoids that crayfish cannot synthesize but obtain from the diet and is resistant to cheating (Karplus et al., 2003).

Apparently complex is the proximate mechanisms used by the females to recognize high-quality mates: visual and chemical stimuli combined are used by *P. clarkii* females to choose a large male, whereas the sight alone and the odor alone of an individual of the other sex, independently of its size, elicit aggression (Aquiloni & Gherardi, 2008c). That is, the sight of a mate of a larger size is not *per se* an index of the "best" partner but it must be confirmed by chemical stimuli that, in turn, inform about the species, the sex, and the reproductive condition of the potential mate.

Finally, also the importance that the male hierarchical status plays in sexual selection is controversial. *P. clarkii* females make a choice between equally-sized dominant or subordinate individuals only after having eavesdropped on them fighting (Aquiloni et al., 2008). By eavesdropping, *P. clarkii* females seem to make low-cost, direct comparisons between the two potential mates, obtain information about the quality of the signalers, and can then use this information to guide their future decisions. Again, this ability seems to rely on the combination of visual and chemical stimuli; interestingly, the winner is selected only after the female has eavesdropped on that individual male and not on a generic one, thus denoting the intervention of a refined form of individual recognition (Aquiloni & Gherardi, 2010).


## Mate choice by males

Due to their polygynous habit and the long time needed to produce sperm, males may be limited in their sperm supply: in *A. pallipes*, the ejaculate size decreases with consecutive matings (Rubolini et al., 2007) and at the end of the mating season, *vasa deferentia* weight was 55% lower than in the pre-mating season (Woodlock & Reynolds, 1988). Male sperm limitation might induce the females to choose non-sperm-depleted males (but no evidence has been collected so far; Aquiloni & Gherardi, 2008a) and the males to evolve forms of mate selection. Male choosiness might also be determined by the long copulation time of some species, large investment in sperm production, and restricted mating periods (less than one month in *A. pallipes*; Villanelli & Gherardi, 1998). Indeed, *P. clarkii* males select their mate using size as a criterion, the larger and more fecund females being preferred (Aquiloni & Gherardi, 2008a). Pleopodal egg number, in fact, increases with female body size (in *A. pallipes*, cf. Rubolini et al., 2006; *C. japonicus*, cf. Nakata & Goshima, 2004).

Finally, *A. pallipes* males may adjust the volume of their ejaculate to the size of the females (Rubolini et al., 2006): males individually paired in the laboratory with receptive females of different size allocated more sperm to larger females, i.e., the mates that provide to them the greatest fertilization returns.


## CONCLUDING REMARKS

Along with their taxonomic diversity, ease of collection, relatively large size, and conspicuousness in the ecological community, crayfish are particularly suitable for sexual selection studies because of the potential by researchers to separate male and female effects during reproduction. An intensification of studies in this field and an increase in the taxonomic breadth of crayfish as model organisms will certainly provide valuable insight into some currently debated issues, such as sexual conflict and parental allocation (Chapman et al., 2003).

Gherardi & Aquiloni, SEXUAL SELECTION IN CRAYFISH 221

## ACKNOWLEDGEMENTS

F.G. acknowledges the organizers of the TCS Summer Meeting in Tokyo for their kind invitation. Particular thanks are directed to Prof. Akira Asakura, Secretary General of the Organizing Committee, and to the chairs of the symposium on the Reproductive Behavior of Decapod Crustaceans, Prof. Keiji Wada and Prof. Satoshi Wada.

# REFERENCES

AMEYAW-AKUMFI, C. & B. A. HAZLETT, 1975. Sex recognition in the crayfish, *Procambarus clarkii*. Science, **190**: 1225-1226.

AQUILONI, L., M. BUŘIČ & F. GHERARDI, 2008. Crayfish females eavesdrop on fighting males before choosing the dominant mate. Current Biology, **18**: 462-463.

AQUILONI, L. & F. GHERARDI, 2008a. Mutual mate choice in crayfish: large body size is selected by both sexes, virginity by males only. Journal of Zoology, **274**: 171-179.

—— & ——, 2008b. Evidence of cryptic mate choice in crayfish. Biology Letters, **4**: 163-165.

—— & ——, 2008c. Mate assessment by size in the red swamp crayfish *Procambarus clarkii*: effects of chemical *versus* visual cues. Freshwater Biology, **53**: 461-469.

—— & ——, 2010. Crayfish females eavesdrop on fighting males and use smell and sight to recognize the identity of the winner. Animal Behaviour, **79**: 265-269.

AQUILONI, L., A. MASSOLO & F. GHERARDI, 2009. Sex identification in female crayfish is bimodal. Naturwissenschaften, **96**: 103-110.

BARKI, A., 2008. Mating behaviour. In: E. MENTE (ed.), Reproductive biology of crustaceans. Case studies of decapod crustaceans: 223-265. (Science Publishers, Enfield).

BARKI, A. & I. KARPLUS, 1999. Mating behavior and a behavioral assay for female receptivity in the red-claw crayfish *Cherax quadricarinatus*. Journal of Crustacean Biology, **19**: 493-497.

BELANGER, R., X. REN, K. MCDOWELL, S. CHANG, P. MOORE & B. ZIELINSKI, 2008. Sensory setae on the major chelae of male crayfish, *Orconectes rusticus* (Decapoda, Astacidae) — impact of reproductive state on function and distribution. Journal of Crustacean Biology, **28**: 27-36.

BERRILL, M. & M. ARSENAULT, 1982. Spring breeding of a northern temperate crayfish *Orconectes rusticus*. Canadian Journal of Zoology, **60**: 2641-2645.

—— & ——, 1984. The breeding behaviour of a northern temperate orconectid crayfish, *Orconectes rusticus*. Animal Behaviour, **32**: 333-339.

CHAPMAN, T., G. ARNQVIST, J. BANGHAM & L. ROWE, 2003. Sexual conflict. Trends in Ecology and Evolution, **18**: 41-47.

CRANDALL, K. A. & J. E. BUHAY, 2008. Global diversity of crayfish (Astacidae, Cambaridae, and Parastacidae — Decapoda) in freshwater. Hydrobiologia, **595**: 295-301.

COROTTO, F. S., D. M. BONENBERGER, J. M. BOUNKEO & C. C. DUKAS, 1999. Antennule ablation, sex discrimination, and mating behavior in the crayfish *Procambarus clarkii*. Journal of Crustacean Biology, **19**: 708-712.

CROCKER, D. W. & D. W. BARR, 1968. Handbook of the crayfish of Ontario: 1-158. (University of Toronto Press, Toronto).

DARWIN, C., 1871. The descent of man, and selection in relation to sex: 1-395. (John Murray, London).

DUNHAM, D. W. & J. W. OH, 1992. Chemical sex-discrimination in the crayfish *Procambarus clarkii* — role of antennules. Journal of Chemical Ecology, **18**: 2363-2372.

DURGIN, W. S., K. E. MARTIN, H. R. WATKINS & L. M. MATHEWS, 2008. Distance communication of sexual status in the crayfish *Orconectes quinebaugensis*: female sexual history mediates male and female behavior. Journal of Chemical Ecology, **34**: 702-707.

FURRER, S. C., 2004. Untersuchungen des Partnerwahlverhaltens beim Edel- und Galizierkrebs sowie der Life History beim Steinkrebs. (Ph.D. Thesis Dissertation, University of Zürich, Zürich).

GALEOTTI, P., F. PUPIN, R. SACCHI, P. A. NARDI & M. FASOLA, 2007. Effects of female mating status on copulation behaviour and sperm expenditure in the freshwater crayfish *Austropotamobius italicus*. Behavioral Ecology and Sociobiology, **61**: 711-718.

GALEOTTI, P., D. RUBOLINI, F. PUPIN, R. SACCHI & M. FASOLA, 2008. Sperm removal and ejaculate size correlate with chelae asymmetry in a freshwater crayfish species. Behavioral Ecology and Sociobiology, **62**: 1739-1745.

GHERARDI, F., S. BARBARESI & A. RADDI, 1999. The agonistic behaviour in the red swamp crayfish, *Procambarus clarkii*: functions of the chelae. Freshwater Crayfish, **12**: 233-243.

GHERARDI, F., B. RENAI, P. GALEOTTI & D. RUBOLINI, 2006. Nonrandom mating, mate choice, and male–male competition in the crayfish *Austropotamobius italicus*, a threatened species. Archiv für Hydrobiologie, **165**: 557-576.

GHERARDI, F., C. SOUTY-GROSSET, G. VOGT, J. DIÉGUEZ-URIBEONDO & K. CRANDALL, 2010. Infraorder Astacidea Latreille, 1802 P.P.: The freshwater crayfish, Chapter 67. In: F. SCHRAM (ed.), Treatise on zoology — Crustacea, vol. **9A**: 269-423. (Brill, Leiden).

GUIASU, R. C. & D. W. DUNHAM, 1998. Inter-form agonistic contests in male crayfishes, *Cambarus robustus* (Decapoda, Cambaridae). Invertebrate Biology, **117**: 144-154.

HAZLETT, B. A., 2009. Crayfish: an almost perfect model system for the study of behaviour. (Atti del Convegno CCDM, Museo Regionale di Scienze Naturali di Torino, in press).

HOLDICH, D. M., J. P. READER, W. D. ROGERS & M. HARLIOĞLU, 1995. Interactions between three species of crayfish (*Austropotamobius pallipes*, *Astacus leptodactylus* and *Pacifastacus leniusculus*). Freshwater Crayfish, **10**: 46-56.

HUNER, J. V., 2002. *Procambarus*. In: D. M. HOLDICH (ed.), Biology of freshwater crayfish: 541-584. (Blackwell Science, Oxford).

HUNT, J., R. BROOKS & M. D. JENNIONS, 2005. Female mate choice as a condition-dependent life-history trait. American Naturalist, **166**: 79-92.

KARPLUS, I., A. SAGI, I. KHALAILA & A. BARKI, 2003. The soft red-patch of the Australian freshwater crayfish *Cherax quadricarinatus* (von Martens): a review and prospects for future research. Journal of Zoology, **259**: 375-379.

KASUYA, E., S. TSURUMAKI & M. KANIE, 1996. Reversal of sex roles in the copulatory behavior of the imported crayfish *Procambarus clarkii*. Journal of Crustacean Biology, **16**: 469-471.

KAWAI, T. & K. SAITO, 2001. Observations on the mating behavior and season, with no form alternation, of the Japanese crayfish, *Cambaroides japonicus* (Decapoda, Cambaridae), in Lake Komadome, Japan. Journal of Crustacean Biology, **2**: 885-890.

MASON, J. C., 1970. Copulatory behavior of the crayfish *Pacifastacus trowbridgii* (Stimpson). Canadian Journal of Zoology, **48**: 969-976.

MOORE, P. A. & D. A. BERGMAN, 2005. The smell of success and failure: the role of intrinsic and extrinsic chemical signals on the social behavior of crayfish. Integrative and Comparative Biology, **45**: 650-657.

NAKATA, K. & S. GOSHIMA, 2004. Fecundity of the Japanese crayfish, *Cambaroides japonicus*: ovary formation, egg number and egg size. Aquaculture, **242**: 335-343.

PERRY, W. L., J. E. FEDER & D. M. LODGE, 2001. Implications of hybridization between introduced and resident *Orconectes* crayfish. Conservation Biology, **15**: 1656-1666.

RÁAN, Z. & A. SAGI, 1985. Alternative mating strategies in males of the freshwater prawn *Macrobrachium rosenbergii* (de Man). Biological Bulletin, **169**: 592-601.

RUBOLINI, D., P. GALEOTTI, G. FERRARI, M. SPAIRANI, F. BERNINI & M. FASOLA, 2006. Sperm allocation in relation to male traits, female size, and copulation behaviour in a freshwater crayfish species. Behavioral Ecology and Sociobiology, **60**: 212-219.

RUBOLINI, D., P. GALEOTTI, F. PUPIN, R. SACCHI, A. P. NARDI & M. FASOLA, 2007. Repeated matings and sperm depletion in the freshwater crayfish *Austropotamobius italicus*. Freshwater Biology, **52**: 1898-1906.

SCHROEDER, L. & R. HUBER, 2001. Fighting strategies in small and large individuals of the crayfish, *Orconectes rusticus*. Behaviour, **138**: 1437-1449.

SNEDDEN, W. A., 1990. Determinants of male mating success in the temperate crayfish *Orconectes rusticus*: chela size and sperm competition. Behaviour, **115**: 100-113.

STEBBING, P. D., M. G. BENTLEY & G. J. WATSON, 2003. Mating behaviour and evidence for a female released courtship pheromone in the signal crayfish *Pacifastacus leniusculus*. Journal of Chemical Ecology, **29**: 465-475.

STEIN, R. A., 1976. Sexual dimorphism in crayfish chelae: functional significance linked to reproductive activities. Canadian Journal of Zoology, **54**: 220-227.

TIERNEY, A. J. & D. W. DUNHAM, 1984. Behavioral mechanisms of reproductive isolation in the crayfishes of the genus *Orconectes*. The American Midland Naturalist, **111**: 304-310.

TRICARICO, E. & F. GHERARDI, 2007. Biogenic amines influence aggressiveness in crayfish but not their force or hierarchical rank. Animal Behaviour, **74**: 1715-1724.

TRIVERS, R. L., 1972. Parental investment and sexual selection. In: B. CHAMPBELL (ed.), Sexual selection and the descent of man: 136-179. (Aldine, Chicago).

VILLANELLI, F. & F. GHERARDI, 1998. Breeding in the crayfish, *Austropotamobius pallipes*: mating patterns, mate choice and intermale competition. Freshwater Biology, **40**: 305-315.

WALKER, D., B. A. PORTER & J. C. AVISE, 2002. Genetic parentage assessment in the crayfish *Orconectes placidus*, a high-fecundity invertebrate with extended maternal brood care. Molecular Ecology, **11**: 2115-2122.

WEATHERHEAD, P. J. & R. J. ROBERTSON, 1979. Offspring quality and the polygyny threshold: "the sexy son hypothesis". American Naturalist, **113**: 201-208.

WILSON, R. S., R. S. JAMES, C. BYWATER & F. SEEBACHER, 2009. Costs and benefits of increased weapon size differ between sexes of the slender crayfish, *Cherax dispar*. Journal of Experimental Biology, **212**: 853-858.

WOODLOCK, B. & J. D. REYNOLDS, 1988. Laboratory breeding studies of freshwater crayfish, *Austropotamobius pallipes* (Lereboullet). Freshwater Biology, **19**: 71-78.


First received 11 November 2009.
Final version accepted 19 December 2009.

# MORPHOLOGY AND ELECTROPHYSIOLOGY OF CRAYFISH ANTENNULES

BY

HAROLD MONTECLARO, KAZUHIKO ANRAKU[1]) and TATSURO MATSUOKA
Faculty of Fisheries, Kagoshima University, Shimoarata 4-50-20, Kagoshima 890-0056, Japan

ABSTRACT

Flicking behavior and morphological features reveal contrasts between the lateral and medial flagella in crayfish antennules suggesting that both flagella may have functional differences. In this study, we report the major setae present on the biramous antennules, which fork into the lateral and medial flagella, of American crayfish *Procambarus clarkii*. Chemoreceptor setae referred to as aesthetascs were present only on the lateral flagellum. Non-aesthetasc sensilla were observed in both lateral and medial flagella. We examined the sensitivity of both flagella to mechanical stimuli by isolating the antennular flagellum from the test animal. Techniques in recording neural activities from an isolated antennular flagellum are also described.

## INTRODUCTION

Aquatic organisms are equipped with sensory organs that receive many cues from the aquatic environment. In decapod crustaceans, the surface of the body and appendages bear cuticular hair organs referred to as setae that function as receptors of chemical and mechanical stimuli (Tautz et al., 1981; Tierney et al., 1986; Breithaupt & Tautz, 1990; Mellon, 1997; Douglass & Wilkens, 1998). Investigations of these setae using behavioural and electrophysiological techniques provided information on their functions as well as the mechanisms involved in processing information in the ambient environment. In this study, we identified the major setae present on the antennules of American crayfish *Procambarus clarkii*. We also studied the sensitivity of antennules to hydrodynamic stimuli by isolating the antennules from the animal body to limit the effects of interference from chordotonal

---

[1]) e-mail: anraku@fish.kagoshima-u.ac.jp

        New frontiers in crustacean biology: 225-233

organs and avoid the increase in spike number which can be generated by the movement of the test animal. We present morphological and behavioural differences between the lateral and medial flagella which suggest functional differences between the two antennular flagella.

## MATERIAL AND METHODS

### Test animals

Adult American crayfish *Procambarus clarkii* (Girard, 1852), 35 to 40 mm in carapace length, were obtained from a local supplier in Kagoshima, Japan. Animals were kept in aerated tanks with water at 25°C until used for experiments. Holding tanks were provided with *Elodea* and animals were fed twice a week with crayfish pellets (Japan Pet Drugs Co., Ltd., Tokyo, Japan) and were kept under a light/dark regime of 12 h:12 h (L:D).

### SEM preparation

Antennules from newly-molted crayfish were prepared for microscopy. Eight lateral and medial flagella were fixed in 2.5% glutaraldehyde in PBS solution (0.1 mM, pH 7.2). After 24 h, fixed samples were rinsed in PBS, dehydrated in ethanol series, dried using t-BuOH freeze dryer (VFD-21S, Vacuum Device Co., Ltd., Ibaragi, Japan), and gold-coated (Magnetron spatter MSP-10, Vacuum Device Co., Ltd., Ibaragi, Japan). Scanning electron microscopy (SEM; Hitachi S-4100, Tokyo, Japan) observations were performed all over the surface of the medial and lateral flagella of crayfish antennules.

### Electrical recording

Neural activity was recorded on an isolated crayfish antennular flagellum which was cut at the base where the lateral and medial flagella fork. The carapace in at least three segments at the proximal end was removed and the nerves exposed. Fig. 1 illustrates the recording set-up for the isolated flagellum. The isolated flagellum was suspended in a plexiglass container ($10 \times 10 \times 10$ cm, L × W × H) containing chilled and oxygenated van Harreveld saline solution (van Harreveld, 1936). The pH of the saline solution was 7.8 and its temperature ranged from 15-20°C. Exposed nerves were sucked into a small (40 to 70 $\mu$m inner diameter) glass electrode, initially filled with van Harreveld saline solution, using a threaded suction pump (A-M Systems, Inc.,

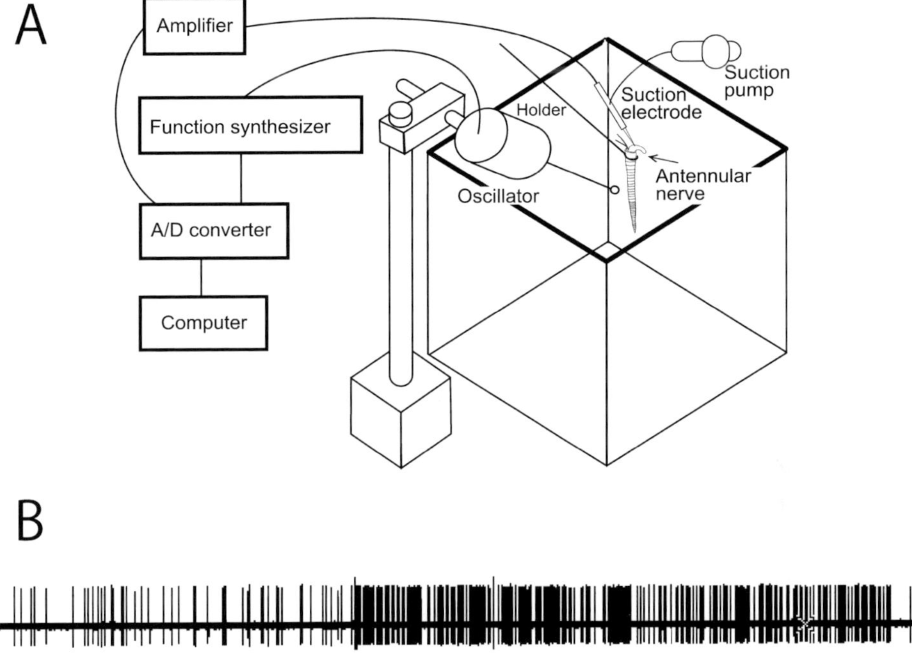



Fig. 1. A, experimental set-up for recording neural responses of an isolated crayfish antennular flagellum to sinusoidal stimulus; B, neural discharge in crayfish antennular flagellum increased as a response to sinusoidal stimulation. Response to sinusoidal stimulus showed phased-lock property. Bar indicates duration of stimulus.

Carlsborg, Washington, U.S.A.). Electrical responses were amplified (Nihon Kohden, Tokyo, Japan) and digitized with a PowerLab 4/20 A/D converter (ADInstruments, Milford, Massachusetts, U.S.A.). A speaker was used to listen to the spike activity.

Six crayfish antennules were stimulated following the procedures of Watanabe & Anraku (2007). A sinusoidal dipole source was generated by a small ($\phi = 4.4$ mm) plastic sphere that was attached to an oscillator (Akashi Corporation, Kanagawa, Japan) by a stainless steel shaft (15 cm long, 1 mm diameter). The shaft was mounted perpendicular to the flagellum, with the sphere placed 4 mm from the distal half of the dorsal section of the antennule. The oscillator was driven by a function synthesizer (NF Electronics, Yokohama, Japan) controlled by a computer (NEC Valuestar NX VE26/4). Sinusoidal signals were digitized and synchronized with neural activity recordings. Data were monitored and stored in an Apple Macintosh computer.

## RESULTS

### Morphological observations

Based on the setal classification system of Watling (1989) and Cate & Derby (2001), the antennular setae fall into eleven morphological types (fig. 2) which, from comparison with past works, are listed in table I. The aesthetascs (fig. 2A), characterized by their blunt tip, were observed on the ventral side of the distal half of the lateral flagellum. The most distal segments contained 3-5 aesthetascs occurring in two rows while the segments near the middle of the flagellum had a single row with 1-3 aesthetascs. Flanking the aesthetascs were setae associated with them- the larger guard hairs and the smaller companion cells (fig. 2A). Among the setal types, the procumbent plumose hairs (fig. 2B), which were present at the proximal half on both lateral and medial flagella, were the largest and most numerous. These setae originated from one segment and bridged to the next distal segment. They lie flat along the flagellar surface with a length that ranged from 130-420 $\mu$m. A few standing feathered setae (fig. 2C) were observed mostly on the ventral portion of the proximal half in both flagella.

Another major setal type present in crayfish antennules were the smooth setae. These setae possessed a single annulation, but differences in shaft length, base diameter, distribution along the flagellar surface, and shape allow them to be grouped into four. Long simple setae (fig. 2D) and medium simple setae (fig. 2E) ranged in length from 110-330 $\mu$m and 60-90 $\mu$m, respectively. These setae were mainly present on the distal half in both flagella. The long setae were sparsely distributed on the dorsal side of the lateral flagellum; in the medial flagellum, they were present on the dorsal and ventro-lateral sides. On the proximal half in both flagella, the short simple setae (fig. 2F) and smooth conate setae (fig. 2G) were mainly present. Both setal types possessed thin walls and had wide base diameters. Two other smooth setae were observed in the lateral flagellum: a smooth seta with a few non-rigid setules (fig. 2G) and another simple seta with denticulations at the tip of the setal shaft (fig. 2H).

---

Fig. 2. Setal types present on the biramous antennules of American crayfish *Procambarus clarkii*. A, aesthetascs (Ae), guard cell (Gd) and companion cell (Co) present on the ventral side of the lateral flagellum's distal half; B, procumbent plumose setae on the proximal side of both flagella; C, standing plumose seta; D, long simple seta found in both flagella; E, medium simple seta found in both flagella; F, short simple seta found in both flagella; G, smooth conate seta present in both flagella; H, setuled seta with a few non-rigid setules on the shaft (arrow); I, denticulated seta with projections at the tip of the shaft.

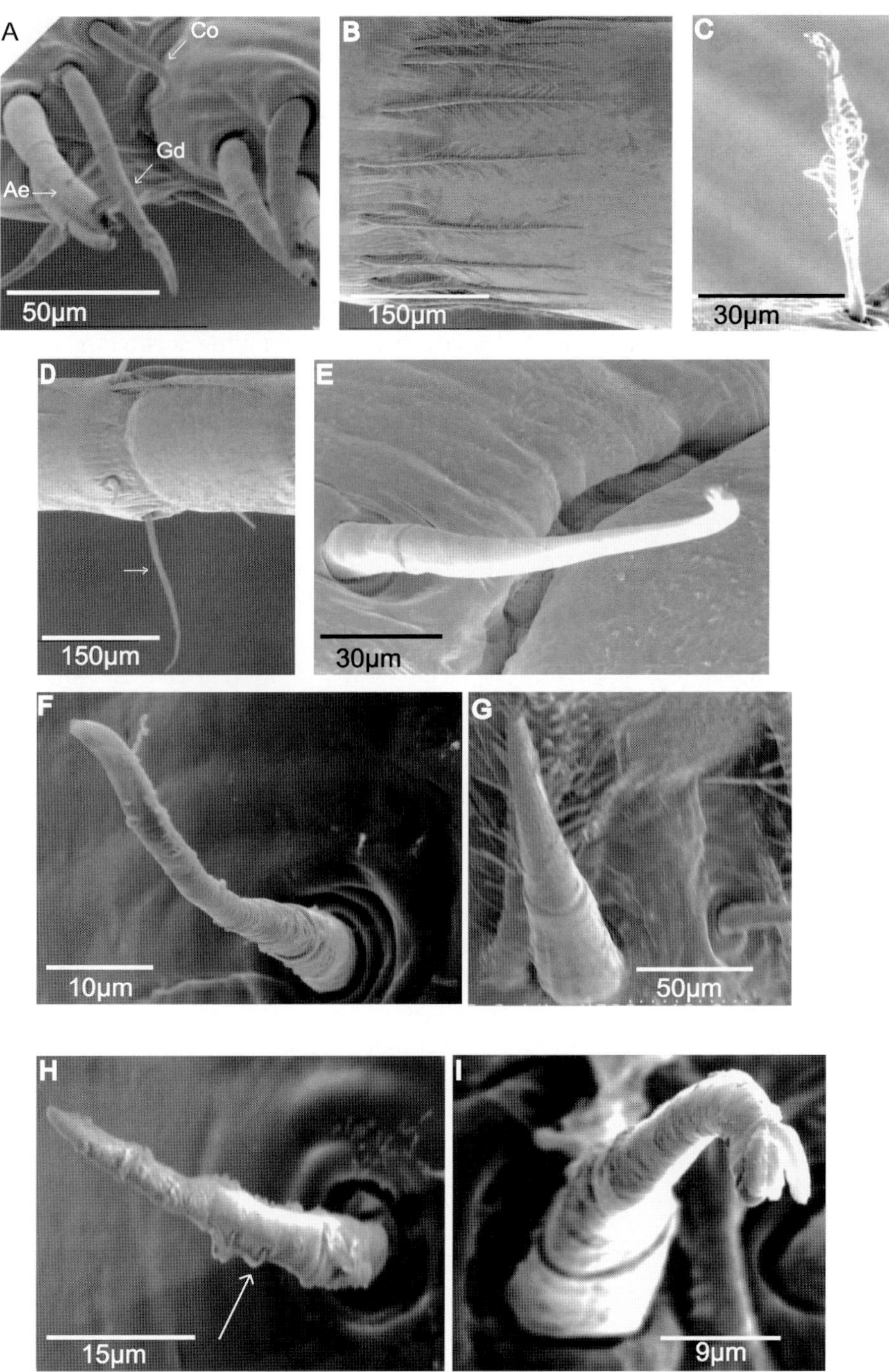

TABLE I

Morphometerics and distribution of antennular setal types in *Procambarus clarkii* ($n = 30$) and equivalent names as reported in previous works

| Setal type | Length[a] ($\mu$m) | Base diameter[a] ($\mu$m) | Distribution | Equivalent names | Type of study in previous works[b] | References |
|---|---|---|---|---|---|---|
| Aesthetasc-associated | | | | | | |
| 1.  Aesthetasc | $81 \pm 2$ | $17 \pm 1$ | Lateral | | Morphological and functional | Tierney et al., 1986; Sandeman & Sandeman, 1996; Horner et al., 2008 |
| 2.  Guard | $89 \pm 4$ | $10 \pm 0$ | Lateral | L-type, large guard hair | Morphological | Sandeman & Luff, 1974; Sandeman & Sandeman, 1996 |
| 3.  Companion | $48 \pm 2$ | $7 \pm 0$ | Lateral | Acuminate, small guard hair | Morphological | Sandeman & Luff, 1974; Tierney et al., 1986; Sandeman & Sandeman, 1996 |
| Non-aesthetasc | | | | | | |
| 4.  Simple long | $200 \pm 14$ | $11 \pm 1$ | Both | L-type, guard hair | Morphological | Sandeman & Luff, 1974; Chichibu et al., 1978a; Sandeman & |
| 5.  Simple medium | $79 \pm 2$ | $9 \pm 0$ | Both | S-type, smooth conical, smooth hair, beaked | Morphological and functional | Chichibu et al., 1978a, b; Mellon & Humphrey, 2007; Humphrey & Mellon, 2007 |
| 6.  Simple short | $42 \pm 1$ | $10 \pm 0$ | Both | | None | |
| 7.  Smooth conate | $120 \pm 8$ | $16 \pm 1$ | Both | Smooth conical hair | Morphological and functional | Tautz et al., 1981*; Bender et al., 1984* |
| 8.  Procumbent plumose | $292 \pm 17$ | $15 \pm 1$ | Both | Feathered hair, procumbent feathered hair | Morphological and functional | Tautz et al., 1981*; Bender etal., 1984*; Sandeman, 1989; Sandeman & Sandeman, 1996 |
| 9.  Standing plumose[b] | $144 \pm 7$ | $10 \pm 1$ | Both | Standing feathered hair | Morphological and functional | Mellon & Lagay, 2008 |
| 10.  Smooth denticulated | 20 | 15 | Lateral | | None | |
| 11.  Smooth setuled | 36 | 8 | Lateral | | None | |

[a]mean $\pm$ s.e.m.; [b] $n = 15$; *Studied morphologically similar setae on the crayfish second antennae.

The last two types did not occur frequently and were not seen in all flagella examined.

## Response to mechanical stimulation

Neural activities in the isolated antennular flagellum could be recorded satisfactorily. Recording remained viable from 30 min to 5 h as long as it was placed in oxygenated van Harreveld saline solution at 15-20°C. Immersion in tap water was also possible, although recording time was drastically reduced. Sinusoidal stimulation generated an increase in the number of neural activities from both antennular flagella (fig. 1B). In each recorded fiber, spikes generated were typically uniform in height and generally void of noise. Both the lateral and medial flagella generated response to sinusoidal stimulation that ranged from 10 to 200 Hz frequency, though the medial flagellum showed more sensitivity especially at higher frequencies. Response to sinusoidal stimulation was tonic and showed phase-locked property. In most cases, spike height decreased over time and frequency of ongoing spike activity was reduced.

## Discussion

A number of difficulties hinder the electrophysiological study of setae especially those present on appendages of crayfish. To study the functional properties of setae on crayfish antennules, we isolated the flagella to avoid disturbances generated by even the slightest movement of the test animal. In contrast to lobster, where cannulation is possible to keep the excised antennule perfused with oxygenated saline (Derby, 1995), arteries in crayfish antennules are too small for such operation. However, placing the antennule in pre-oxygenated and chilled van Harreveld saline solution was usually satisfactory even without cannulation.

Among the setae present on crayfish antennules, the aesthetascs have been largely studied. Their occurrence on the lateral flagellum probably established the lateral flagellum as an olfactory organ in crayfish (Dunham & Oh, 1992; Mellon, 1997; Giri & Dunham, 1999). Behavioural observations also showed that the lateral flagellum is involved in crayfish antennular flicking, a behaviour that is believed to enhance detection of odor (Schmitt & Ache, 1979). Aesthetascs have been demonstrated to play an important role in intraspecific communication; crayfish lacking these setae failed to establish dominance hierarchies and continued to engage in fighting among other crayfish (Horner et al., 2008).

The medial flagellum, while devoid of aesthetasc sensilla, may have other sensory functions. Although it was shown to respond to chemical cues as well as the lateral flagellum (Mellon, 1997), ablation studies suggest that the medial flagellum was not as efficient in detecting long distance sources of food odor (Giri & Dunham, 1999). The non-aesthetasc setae present on the medial flagellum must have other functions. For example, results from our tests indicated that the medial flagellum is more sensitive to hydrodynamic stimuli than the lateral flagella (data not shown). This suggests that this flagellum probably functions as a sensory organ to detect the presence and movements of other aquatic animals, including conspecifics, predators and prey (Breithaupt & Tautz, 1990). Sensory hairs present on the thorax, chelae, pereiopods and tailfan (Mellon, 1963; Douglass & Wilkens, 1998; Belanger et al., 2008) can similarly receive such hydrodynamic stimuli. Further study (Monteclaro et al., in prep.) will determine the response properties of the crayfish antennules to hydrodynamic stimuli to provide insights on the role of both flagella in the sensory activities in crayfish. Specific setae that are responsible for this property will be identified.

## REFERENCES

BELANGER, R., X. REN, K. MCDOWELL, S. CHANG, P. MOORE & B. ZIELINSKI, 2008. Sensory setae on the major chelae of male crayfish, *Orconectes rusticus* (Decapoda: Astacidae) — impact of reproductive state on function and distribution. J. Crust. Biol., **28**: 27-36.

BENDER, M., W. GNATZY & J. TAUTZ, 1984. The antennal feathered hairs in the crayfish: a non-innervated stimulus transmitting system. J. Comp. Physiol., (A) **154**: 45-47.

BREITHAUPT, T. & J. TAUTZ, 1990. The sensitivity of crayfish mechanoreceptors to hydrodynamic and acoustic stimuli. In: K. WIESE, W. D. KRENZ, J. TAUTZ, H. REICHERT & B. MULLONEY (eds.), Frontiers in crustacean neurobiology: 114-120. (Birkhauser, Basel).

CATE, H. S. & C. D. DERBY, 2001. Morphology and distribution of setae on the antennules of the Caribbean spiny lobster *Panulirus argus* reveal new types of bimodal chemo-mechanosensilla. Cell Tissue Res., **304**: 439-454.

CHICHIBU, S., S. TAKAMIZAWA & M. TSUKADA, 1978a. Impulse response patterns of short tactile hairs to mechanical stimulations. Acta Med. Kinki Univ., **3**: 155-165.

CHICHIBU, S., T. WADA, H. KOMIYA & K. SUZUKI, 1978b. Structure of mechanoreceptive hairs on the crayfish first antenna. Acta Med. Kinki Univ., **3**: 27-39.

DERBY, C. D., 1995. Single unit electrophysiological recordings from crustacean chemoreceptor neurons. In: A. I. SPIELMAN & J. G. BRAND (eds.), Experimental cell biology of taste and olfaction. (CRC Press, Boca Raton).

DOUGLASS, J. K. & L. A. WILKENS, 1998. Directional selectivities of near-field filiform hair mechanoreceptors on the crayfish tailfan (Crustacea: Decapoda). J. Comp. Physiol., (A) **183**: 23-34.

DUNHAM, D. W. & J. W. OH, 1992. Chemical sex-discrimination in the crayfish *Procambarus clarkii* — role of antennules. J. Chem. Ecol., **18**: 2363-2372.

GIRI, T. & D. W. DUNHAM, 1999. Use of the inner antennule ramus in the localisation of distant food odours by *Procambarus clarkii* (Girard, 1852) (Decapoda, Cambaridae). Crustaceana, **72**: 123-127.

HORNER, A. J., M. SCHMIDT, D. H. EDWARDS & C. D. DERBY, 2008. Role of the olfactory pathway in agonistic behavior of crayfish, *Procambarus clarkii*. Invert. Neurosci., **8**: 11-18.

HUMPHREY, J. A. C. & D. J. MELLON, 2007. Analytical and numerical investigation of the flow past the lateral antennular flagellum of the crayfish *Procambarus clarkii*. J. Exp. Biol., **210**: 2969-2978.

MELLON, D. J., 1963. Electrical responses from dually innervated tactile receptors on the thorax of the crayfish. J. Exp. Biol., **40**: 137-148.

— —, 1997. Physiological characterization of antennular flicking reflexes in the crayfish. J. Comp. Physiol., (A) **180**: 553-565.

MELLON, D. J. & K. CHRISTISON-LAGAY, 2008. A mechanism for neuronal coincidence revealed in the crayfish antennule. Proc. Natl. Acad. Sci., **105**: 14626-14631.

MELLON, D. J. & J. A. C. HUMPHREY, 2007. Directional asymmetry in responses of local interneurones in the crayfish deutocerebrum to hydrodynamic stimulation of the lateral antennular flagellum. J. Exp. Biol., **210**: 2961-2968.

SANDEMAN, D. C. & S. E. LUFF, 1974. Regeneration of the antennules in the Australian freshwater crayfish, *Cherax destructor*. J. Neurobiol., **5**: 475-488.

SANDEMAN, R. E. & D. C. SANDEMAN, 1996. Pre- and postembryonic development, growth and turnover of olfactory receptor neurones in crayfish antennules. J. Exp. Biol., **199**: 2409-2418.

SCHMITT, B. C. & B. W. ACHE, 1979. Olfaction: responses of a decapod crustacean are enhanced by flicking. Science, **205**: 204-206.

TAUTZ, J., W. M. MASTERS, B. AICHER & H. MARKL, 1981. A new type of water vibration receptor on the crayfish antenna. I. Sensory physiology. J. Comp. Physiol., (A) **144**: 533-541.

TIERNEY, A. J., C. S. THOMPSON & D. W. DUNHAM, 1986. Fine structure of aesthetasc chemoreceptors in the crayfish *Orconectes propinquus*. Can. J. Zool., **64**: 392-399.

VAN HARREVELD, A., 1936. A physiological solution for fresh-water crustaceans. Proc. Soc. Exp. Biol. Med., **34**: 428-432.

WATANABE, K. & K. ANRAKU, 2007. The activities of afferent lateral line nerves induced by swimming motions. Nippon Suisan Gakkaishi, **73**: 1096-1102. [In Japanese with English abstract.]

WATLING, L., 1989. A classification system for crustacean setae based on the homology concept. In: B. E. FELGENHAUER, L. WATLING & A. B. THISTLE (eds.), Functional morphology of feeding and grooming in Crustacea: 15-26. (A.A. Balkema, Rotterdam).

First received 13 November 2009.
Final version accepted 19 December 2009.

# EFFECT OF PH AND WATER TEMPERATURE ON THE DEVELOPMENT OF THE JAPANESE CRAYFISH, *CAMBAROIDES JAPONICUS* EGGS IN VITRO

BY


HAJIME MATSUBARA[1,3]), AYAKA CHIBA[1]), YOSHIFUMI HORIE[1]),
DAISUKE IWATA[1]), MASAHARU SHIMIZU[1]), TAKAHIRO KINOSHITA[1])
and KAZUYOSHI NAKATA[2])

[1]) Laboratory of Aquatic Genome Science, Department of Aquatic Biology, 196 Yasaka, Abashiri, Hokkaido 099-2493, Japan

[2]) River Restoration Research Team, Water Environment Research Group, Public Works Research Institute, 1-6 Minami-Hara, Tsukuba 305-8516, Japan


## ABSTRACT


We incubated the endangered Japanese crayfish (*Cambaroides japonicus*) eggs in vitro at the different water conditions, i.e., pH (from 3 to 12 using 20°C water) and water temperature (from 2 to 20°C using pH 7 water) in order to understand the suitable egg rearing conditions. Crayfish eggs sampled just after spawning (15 May, 2009) were not able to survive until hatch out. This result strongly suggested that early stage of eggs effectively develop when attached to the pleopods of the mother. In contrast, crayfish eggs that were before hatching (31 July, 2009) were able to survive in a wide range of water conditions. However, they were weak for acid condition. These findings suggested that acid conditions (i.e., acid rain) caused by human activities might be spurred the drastic decline of Japanese crayfish.


## INTRODUCTION

The Japanese crayfish, *Cambaroides japonicus*, is the only native crayfish species in Japan (Miyake, 1982). However, local populations of the Japanese crayfish have drastically declined, prompting the Japanese Fisheries Agency in 1998 and the Environment Agency in 2000 to declare it as an endangered species, and necessitating the collection of information for its cultivation and conservation.

---


[3]) Corresponding author; e-mail: h3matsub@bioindustry.nodai.ac.jp

*Cambaroides japonicus* spawn in May, and carries eggs and hatchlings for a long period (from two to three months) until the juveniles become independent (Kawai et al., 1994). Therefore, the survival ratio of eggs and hatchlings depends on also their mother. However, the potential negative effects of the mother, for example, egg loss, death of the mother and a decline in water quality sometimes occur in wild. So far, Nakata et al. (2004) developed a simple, easy method with a microplate to artificially incubate eggs for their cultivation to prevent these effects. However, the suitable water qualities (pH and water temperature) for artificially cultivating their eggs were still unknown. Therefore, we examined the tolerance to the pH and water temperature of *C. japonicus* eggs in vitro.

## MATERIAL AND METHODS

We collected ovigerous female *C. japonicus* which had over 100 eggs by hand from a river in eastern Hokkaido on 15 May (just after spawning: female numbers $=$ 11) and 31 July (just before hatching: female numbers $=$ 4) 2009. The number of eggs of a female is from 30 to 70 (Kawai et al., 1994). In order to conserve the population, only three to five eggs/1 treatment (i.e., 19 treatments used 57 or 95 eggs/a *C. japonicus*) were gently and randomly collected from the pleopods by using forceps and then the females which had over 10 eggs were released in the river. The egg stage of just before hatching were decided to observe embryo in eggs or not. Using these eggs, we incubated the eggs according to Nakata et al. (2004). Nakata et al. (2004) clarified that a suitable water temperature for the eggs to develop would be approximately 15°C. Therefore, we set pre-boiled water at pH 3, 4, 5, 6, 7, 8, 9, 10, 11 and 12 using hydrochloric acid or sodium hydroxide. The methods and conditions of this experiment were the same as the experiments using sterile water at 15°C. In contrast, using pH 7 water, we investigated the detailed suitable water temperature at 2, 4, 6, 8, 10, 12, 14, 16, 18 and 20°C. The methods and conditions of this experiment were the same as the experiments using sterile water at pH 7. The water temperature was controlled in an incubator (Cool. Incubator; As one, Japan). The water volumes were 1 ml in 48-well microplates. With the collecting eggs, one egg was placed in each well of the microplates and the plates were capped. An egg was judged to have hatched when the stage-1 juvenile according to the criteria of Scholtz & Kawai (2002) was observed. An egg was judged to have been dead when it appeared bright orange or was infected by fungus (Nakata et al., 2004). An egg was reared until dead or hatch. The photoperiod was 24-h dark. The water was gently changed every week using sterilized micropipet.

Data are presented as mean $\pm$ SEM. The data were analyzed using a one-way ANOVA followed by Fisher's PLSD post-hoc test. Significance was accepted at $P < 0.05$.

## RESULTS

The *C. japonicus* eggs sampled just after spawning for the temperature and pH experiments were no other hatching and appeared bright orange during 3 months incubation (data not shown). In contrast, the experiments that used the eggs just before hatching, the hatching rate of pH 3 was 0, pH 4 was $37.5 \pm 23.9$, pH 5 was $75 \pm 14.4$, pH 6 was $75 \pm 14.4$, pH 7 was $75 \pm 14.4$, pH 8 was $87.5 \pm 12.5$, pH 9 was $37.5 \pm 23.9$, pH 10 was $75 \pm 25$, pH 11 was $12.5 \pm 12.5$ and pH 12 was $37.5 \pm 23.9\%$, respectively (fig. 1A). In addition, the water temperature experiments using eggs just before hatching, the hatching rate of 2°C was $75 \pm 25$, 4°C was $87.5 \pm 12.5$, 6°C was 100, 8°C was 100, 10°C was 100, 12°C was 100, 14°C was 100, 16°C was $75 \pm 14.4$, 18°C was $62.5 \pm 23.9$ and 20°C was $0\%$, respectively (fig. 1B). All of eggs just before hatching were hatched within 1 week after incubation. The rest eggs were appeared bright orange during 1 month incubation.

## DISCUSSION

The *C. japonicus* eggs sampled just after spawning for the temperature and pH experiments were no other hatching. In our previous study, we succeeded to get the larvae using 1 month after spawn eggs (Nakata et al., 2004). Therefore, we hypothesized that the early stage eggs might be necessary to be bred by mother. However, further study is necessary to solve this problem.

On the other hand, the experiments that used the eggs just before hatching, high levels of hatching were observed. They hatched within 1 month after incubation. In pH experiment, the suitable pH to incubate the eggs was pH 5-pH 8 and pH 10, especially pH 8. However, strong alkaline condition (pH 11 and 12) was showed low levels of hatching. And strong acid condition (pH 3) showed the eggs failed to hatch. From this result, the eggs might be weak for acid condition. Recently, acid conditions of river water owe to acid rain etc. make trouble with wild living resources all over the world. The pH in *C.*

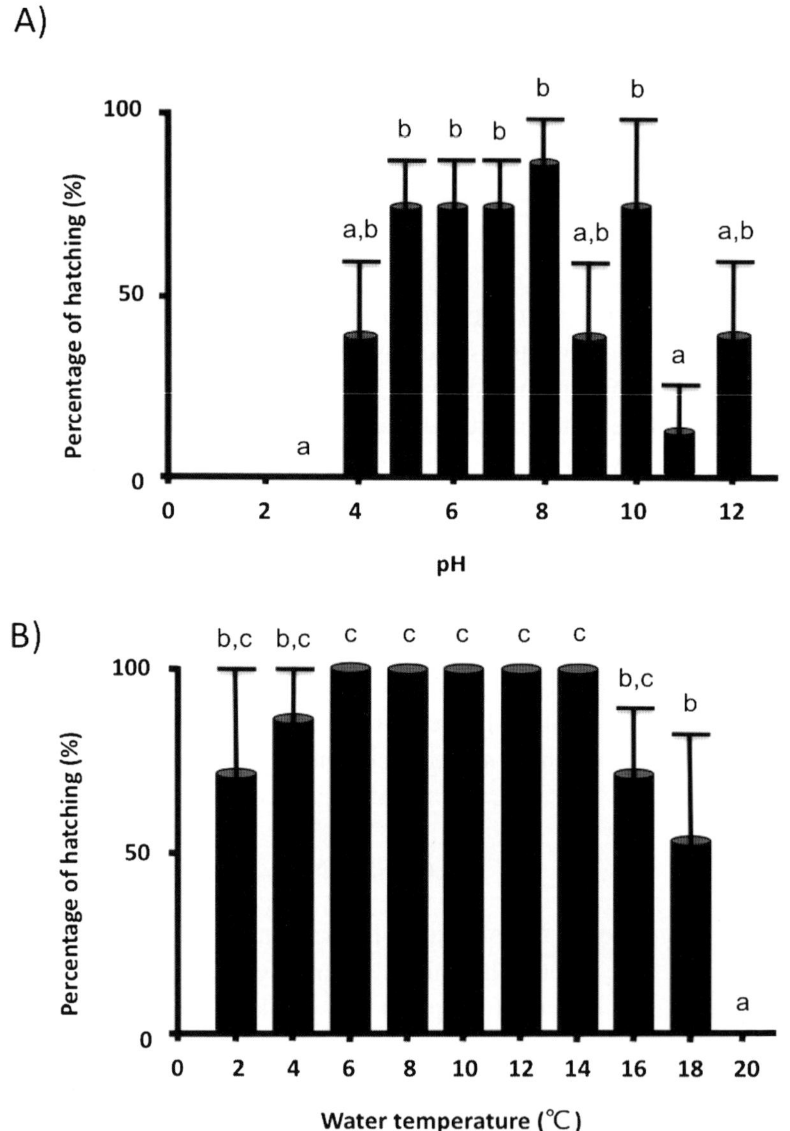



Fig. 1. The experiment of artificially cultivating *C. japonicus* eggs in vitro. A, shows the pH experiment. The hatching rate of pH 3 was 0, pH 4 was 37.5 ± 23.9, pH 5 was 75 ± 14.4, pH 6 was 75 ± 14.4, pH 7 was 75 ± 14.4, pH 8 was 87.5 ± 12.5, pH 9 was 37.5 ± 23.9, pH 10 was 75 ± 25, pH 11 was 12.5 ± 12.5, pH 12 was 37.5 ± 23.9%, respectively; B, shows the water temperature experiment. The hatching rate of 2°C was 75 ± 25, 4°C was 87.5 ± 12.5, 6°C was 100, 8°C was 100, 10°C was 100, 12°C was 100, 14°C was 100, 16°C was 75 ± 14.4, 18°C was 62.5 ± 23.9 and 20°C was 0%, respectively. Letters above bars indicate statistically differ.

*japonicus* living brooks, are actually pH 7-8 all year around. Therefore, the coming oxidization in brooks might also be spurred the decrease of juvenile and adult *C. japonicus*. The present result is a new insight into the factors causing decrease of endangered aquatic specimens, however further detailed study are necessary.

In water temperature experiment that used the eggs of just before hatching, all eggs hatched in the 6, 8, 10, 12 and 14°C. Therefore, the suitable temperature to incubate the eggs was 6-14°C. However, all eggs were dead in the 20°C treatment. From these results, the water temperature of 20°C may be too high for hatching *C. japonicus* eggs. Actually, adult *C. japonicus* begins to be affected by thermal shock above 20°C and dies above 25°C (Nakata et al., 2002). In natural conditions, this species grows well at 14-18°C (Kawai et al., 1994). In addition, our previous study showed that water temperature is suitable for crayfish eggs to develop at 15°C for *C. japonicus* (Nakata et al., 2004). However, the present result indicated that the experimental groups at low water temperature (6, 8 and 10°C) could hatch well. This difference might have occurred in relation to the conditions of the sampling station. Actually, the water temperature in the sampling station of Nakata et al. (2004) was 15.1°C, although those of present study was 11.6°C. Even though we used the same eggs in this experiments, the hatching rate of water temperature experiment (6-14°C) at pH 7 were higher than those of pH experiment (pH 7) at 15°C. Therefore, we considered that the hatching rate of *C. japonicus* eggs might most critically be depended on the living water temperature, naturally enough, it made all the difference between local populations.

We conclude that the suitable pH and temperature to incubate the eggs were pH 5-8 and 6-14°C, respectively. In addition, we estimate the eggs of *C. japonicus* could not hatch acidified water and their optimal hatching temperature is different with each local population. These means that to conserve local population of *C. japonicus* is firstly necessary to understand the characteristics of each living station and to protect their habitat.


## ACKNOWLEDGEMENTS

The authors thank Yoshiaki Watanabe (Abashiri city) for his encouragement of the Japanese crayfish sampling. The authors thank Dr. Chris Norman (Japan Scientific Texts) for critical reading of the manuscript. This research was supported by grants from the Tokyo University of Agriculture for Young Scientists.

## REFERENCES

KAWAI, T., T. HAMANO & S. MATSUURA, 1994. Molting season and reproductive cycle of the Japanese crayfish, *Cambaroides japonicus* in a stream and a small lake in Hokkaido. Suisan Zoshoku, **42**: 465-470. [In Japanese with English abstract.]

MIYAKE, S., 1982. Japanese crustacean decapods and stomatopods in color I: 1-264. (Hoiku-sya, Osaka). [In Japanese.]

NAKATA, K., T. HAMANO, K. HAYASHI & T. KAWAI, 2002. Lethal limits of high temperature for two crayfishes, the native species *Cambaroides japonicus* and the alien species *Pacifastacus leniusculus* in Japan. Fisheries Science, **68**: 763-767.

NAKATA, K., H. MATSUBARA & S. GOSHIMA, 2004. Artificial incubation of Japanese crayfish (*Cambaroides japonicus*) eggs by using a simple, easy method with a microplate. Aquaculture, **230**: 273-279.

SCHOLTZ, G. & T. KAWAI, 2002. Aspects of embryonic and postembryonic development of the Japanese crayfish *Cambaroides japonicus* (Crustacea, Decapoda) including a hypothesis on the evolution of maternal care in the Astacida. Acta Zoologica, **83**: 203-212.

First received 1 December 2009.
Final version accepted 17 February 2010.

# AN ASSESSMENT OF THE DISTRIBUTION, BIOLOGY, THREATENING PROCESSES AND CONSERVATION STATUS OF THE FRESHWATER CRAYFISH, GENUS *EUASTACUS* (DECAPODA, PARASTACIDAE), IN CONTINENTAL AUSTRALIA. I. BIOLOGICAL BACKGROUND AND CURRENT STATUS

BY


JAMES M. FURSE[1]) and JASON COUGHRAN[2])

The Environmental Futures Centre, Griffith School of Environment, Gold Coast Campus, Griffith University, Queensland 4222, Australia



## ABSTRACT

Of the 10 Australian crayfish genera, *Euastacus* is the largest with 49 species described, including some of the largest, and some of the rarest, species in the World. Many species are slow growing, late maturing, and the lifespan of some exceeds 30 years. The distribution of the genus encompasses most of the east coast of Continental Australia, a latitudinal range of approximately 23° (a distance of over 2500 km). *Euastacus* also has the widest altitudinal range of all Australian genera, occurring from sea level to over 1500 m above sea level. Due to these broad geographical and altitudinal ranges, *Euastacus* inhabit most of the climatic zones in Australia, and are found in a wide range of habitats including large temperate lowland rivers, high altitude rainforest gullies and Alpine streams. We review the distribution, habitat, population data, and biology for these crayfish, and their current conservation status where available. Sixteen species were listed on the IUCN Red List in 1996, and there have been increasing concerns regarding the conservation status of the genus since then. The current review was undertaken in order to allow, 1) assessment of the conservation status of all species against current IUCN Red List Criteria, and, 2) a discussion of research, conservation and management imperatives for the genus, presented in two accompanying papers.


## INTRODUCTION

### The freshwater crayfish of Australia

Australia features the Worlds' second most diverse freshwater crayfish fauna with more than 130 species known from 10 genera, and in terms of

---


[1]) Corresponding author; e-mail: j.furse@griffith.edu.au

[2]) e-mail: j.coughran@griffith.edu.au

species richness, Australia is second only to North America (more than 380 species from 12 genera) (Crandall & Buhay, 2008). The Australian freshwater crayfish fauna features many of the Worlds' largest species (from the genera *Astacopsis*, *Euastacus* and *Cherax*), the largest of which, *Astacopsis gouldi* Clark, 1936, can reach ~1 m in overall length and ~4.5 kg in weight (Holdich, 2002). In contrast, at 25 mm overall length *Tenuibranchiurus glypticus* Riek, 1951 is one of the Worlds' smallest species (Riek, 1951).

## Genus *Euastacus*

*Euastacus* is the largest of the 10 Australian genera with 49 species described (Coughran, 2008; McCormack & Coughran, 2008) and comprises 37% of the total Australian freshwater crayfish species. Sixteen of these species (one third of the genus) have only been described in the last 12 years (Morgan, 1997; Coughran, 2002, 2005; Coughran & Leckie, 2007; McCormack & Coughran, 2008), with further probable new species collected and awaiting description (Coughran, unpubl.). The very large and remote geographical areas of the Australian Continent that have not yet been surveyed, but feature suitable *Euastacus* habitat, may yield a number of additional new species in the future. Some *Euastacus* species are very large, with 21 species exceeding 50 mm Occipital-Carapace Length (OCL, Morgan, 1986), including the Worlds' second largest species *Euastacus armatus* (von Martens, 1866) which can reportedly reach half a metre overall length and weigh more than 3 kg (Horwitz, 1990; Geddes et al., 1993). However, while the genus is recognised as containing several of the largest crayfish in the World, many *Euastacus* species are quite small, such as *Euastacus jagabar* Coughran, 2005, reaching a maximum OCL of 30 mm. A number of the larger species are highly aggressive (e.g., *Euastacus sulcatus* Riek, 1951) and/or very powerful (e.g., *Euastacus valentulus* Riek, 1951) and a great deal of care is required when handling these species as the chelae can inflict serious injuries. *Euastacus* are often referred to as the "Spiny Crayfish" of Australia as some species feature impressive arrays of spines on the thorax, abdomen and chelae (some of which are needle sharp). However some species lack any appreciable spination (Morgan, 1986, 1988, 1997; Coughran, 2002, 2005, 2008; McCormack et al., 2010) and can easily be confused with the "Smooth Crayfish" from genus *Cherax*. Many species are brightly coloured in vivid blues or various shades of red. Other species are less colourful overall (i.e., brown, green or dark black) but feature spines that are tipped in bright fluorescent colours, or have striking red, yellow, orange, green, blue or lilac undersides (see Merrick, 1993;

Coughran, 2006b; Davie, 2007; AABIO, 2008; McCormack, 2008). Given their often large size and vivid colouration *Euastacus* are conspicuous, and in some places species are well known to landowners and bushwalkers due to their tendency to the leave the water (Morey, 1998) and/or move considerable distances overland (Clark, 1937; Riek, 1951; Morgan, 1988; Furse & Wild, 2002b; Furse et al., 2004).

## Distribution and habitats occupied

The distribution of *Euastacus* is fragmented and covers most of the east coast of mainland Australia, a latitudinal range of about 23° or over 2500 km: the 3$^{rd}$ largest latitudinal range of all the Worlds' freshwater crayfish genera (*Cherax* >3500 km, *Procambarus* ~3000 km) (Taylor, 2002). The majority of species are distributed proximal to the east and southeast coasts, a notable exception is *E. armatus* which extends into the semi-arid zone. Species are not known to exist on any offshore islands. The distributions of many species are associated with the Great Dividing Range (in the east), the only major mountain range on the Continent (fig. 1): unusually, several species occur on both sides of this north-south Continental divide (McCormack et al., 2010). *Euastacus* has the widest altitudinal range of all Australian genera, occurring from sea-level to over 1500 m above sea level (a.s.l.) which is well above the snowline in winter (and near to the highest point in Australia, Mt Kosciuszko 2228 m a.s.l.) (Morgan, 1986, 1997). Some species have particular and in some cases narrow altitudinal ranges (e.g., *Euastacus maidae* (Riek, 1956)), and many are only found at higher altitudes which restricts them to isolated mountain tops, consistent with a relic distribution (e.g., *Euastacus monteithorum* Morgan, 1989). The genus separates into two distinct ecological groups, lowland and highland, and in general 200 m a.s.l. is the altitudinal minimum that defines the highland group (Morgan, 1997). Generally, the minimum altitude of *Euastacus* increases with decreasing latitude (Horwitz, 1990; Morgan, 1991), such that the lowland group only extends to around the middle of the latitudinal extent of the genus, and only the highland group extends north of ~28°S (Coughran & Furse, unpubl.).

Due to their broad latitudinal and altitudinal range *Euastacus* inhabit all climatic zones in Australia (semi-arid, temperate, subtropical, and tropical: the only exception is the arid zone). They also occupy a wide variety of habitat types ranging from large lowland rivers, through high altitude heathland and Alpine bogs, to high altitude tropical rainforest streams and damp gullies (Morgan, 1986, 1991, 1997; Coughran, 2007), and even cattle pastures at

Fig. 1. Distribution of the freshwater crayfish genus, *Euastacus*. Although most of the 49 species of *Euastacus* have restricted distributions, the genus itself is widely distributed across most of the eastern mainland of Australia. The Great Dividing Range is indicated by the unbroken line running approximately north-south along the eastern seaboard.

high altitude in the case of *Euastacus maccai* McCormack & Coughran, 2008. While species are found in a wide range of habitats, relatively few are habitat generalists, and many have very specific habitat requirements with a large number of species associated with particular vegetation types and habitat characteristics. In addition, many species are associated with a vegetative canopy over stream-habitats, and cool water temperatures typically <21°C (Riek, 1951, 1969; Morgan, 1986, 1988; Horwitz, 1990; Morgan, 1991, 1997; Furse & Wild, 2002a; Coughran, 2005, 2007; Coughran & Leckie, 2007; McCormack et al., 2010).

## The current state of knowledge

Despite the large number of species, relatively few have received any specific biological research, i.e., *Euastacus bispinosus* Clark, 1936 (Hoey, 1990; Honan & Mitchell, 1995b, 1995a, 1995c; Honan, 1998; Johnston et al., 2008), *E. armatus* (Sandeman & Wilkens, 1982; Geddes et al., 1993; Gilligan et al., 2007; Ryan et al., 2008), *Euastacus spinifer* (Heller, 1865) (Merrick, 1997; Turvey & Merrick, 1997a, 1997e, 1997d, 1997b, 1997c; Growns & Marsden, 1998), *E. sulcatus* (Furse & Wild, 2002a, 2004; Furse et al., 2004; Wild & Furse, 2004; Furse et al., 2006), *Euastacus gumar* Morgan, 1997 (Coughran, in press-b), *Euastacus mirangudjin* Coughran, 2002 (Coughran, in press-a), *Euastacus hystricosus* Riek, 1951 (Smith et al., 1998), *Euastacus urospinosus* (Riek, 1956) (Borsboom, 1998), and *Euastacus kershawi* (Smith, 1912) (Morey, 1998). Researchers working on the genus are currently forced to draw heavily (and cautiously) on this work, however all of these studies indicate a consistent pattern in biological traits. The phylogeny and biogeography of the genus (Shull et al., 2005) and evolution of the Queensland species (Ponniah & Hughes, 1998, 2004, 2006) have been studied using molecular genetic techniques and will provide an important basis for prioritising and directing future research, conservation and management efforts.

The conservation status of *Euastacus* and the threats facing them were first highlighted, and then updated, by Horwitz (1990, 1995), and this catalysed much of the abovementioned research and, more recently, attention on the conservation status and management of a number of other species (Merrick, 1995, 1997; Coughran, 2007; Gilligan et al., 2007; O'Brien, 2007; McCormack et al., 2010).

## Biology and ecology

The consensus in studies to date is that *Euastacus* are slow-growing (growth increments of a few mm OCL yr$^{-1}$), long-lived and can take decades to reach the very large sizes that are recorded: >35 years in some species (Honan & Mitchell, 1995b; Turvey & Merrick, 1997b, 1997c; Morey, 1998; Furse & Wild, 2004). Similarly, reproductive studies establish that *Euastacus* are late-maturing with females only becoming reproductive at >5-8 yrs in most species (Honan & Mitchell, 1995c; Turvey & Merrick, 1997e; Borsboom, 1998; Wild & Furse, 2004). In addition, this genus has a slow reproductive cycle, many species are "winter brooders" (mating in late summer/autumn and brooding eggs over winter), long brooding periods of 6-10 months are typical, and some species only breed biennially (Clark, 1937; Barker, 1992; Honan & Mitchell, 1995c; Turvey & Merrick, 1997e; Borsboom, 1998; Honan, 1998; Morey, 1998; Wild & Furse, 2004; Coughran, 2006a, in press-a, in press-b). Some species are highly fecund, others rather less so, and pleopodal egg fecundity varies considerably both between species (20-1500 eggs per female), and within species (i.e., small *versus* large females) (Clark, 1937; Borsboom, 1998; Morey, 1998; Coughran, 2006a, in press-a, in press-b; McCormack et al., 2010).

## Abundance and population densities

Some species are extremely rare (*E. maidae*, cf. Coughran & Furse, 2010), but others can be quite abundant or may be locally very abundant indeed (*E. sulcatus*, cf. Furse et al., 2006; Furse & Wild, unpubl.). While density estimates are scarce, species can occur in moderate to high densities of 0.32 to 5.0 crayfish m$^{-2}$ (*E. bispinosus* and *E. gumar* respectively, Honan & Mitchell, 1995a; Coughran, in press-b). For many species (e.g., *Euastacus bindal* Morgan, 1989) nothing has been documented about their status in the wild since their original description.

## Legislative protection and conservation status

Few species are listed under any Australian State conservation legislation, however many species are afforded incidental protection when occurring within the boundaries of National Parks and other protected areas. With the exception of Queensland where all *Euastacus* are "no take" species, total bans, closed seasons and/or recreational catch limits are specified by various State fisheries regulations. The conservation status of the genus was last assessed for the International Union for Conservation of Nature (IUCN) Red List of Threatened Species in 1996, at which time 16 species were listed in a threat category (table I). The conservation status of the other 33 species in the genus remains unclear.

TABLE I

Current IUCN Red List status of the genus *Euastacus* (as assessed in the 1996 IUCN Red List), status current at time of writing in 2009 (Source: IUCN 2009)

| Species | IUCN Status |
| --- | --- |
| *armatus* | VU |
| *bindal* | EN |
| *bispinosus* | VU |
| *crassus* | EN |
| *diversus* | EN |
| *eungella* | VU |
| *fleckeri* | VU |
| *hystricosus* | VU |
| *jagara* | EN |
| *maidae* | EN |
| *monteithorum* | EN |
| *neodiversus* | VU |
| *robertsi* | EN |
| *setosus* | VU |
| *urospinosus* | EN |
| *yigara* | EN |

To summarise, *Euastacus* is the largest Australian genus with additional species yet to be described, it contains many of the Worlds' most charismatic species: these crayfish typically have specific habitat requirements, are slow-growing, long-lived and slow to reproduce. The overall distribution of the genus essentially encompasses the entire spectrum of Australian climatic zones and habitat types, and extends to almost the highest point in Australia, thus potentially exposing these crayfish to a wide range of threats. While a number of species are quite well studied, overall our understanding of the genus is poor, and although it has been evident for some time that this genus is of considerable conservation concern (i.e., Horwitz, 1990) little has been done to address the various identified threats. Given the high proportion of species assessed as belonging in IUCN threat categories in 1996, and the uncertain status of the rest of the genus (including many species described since that time), the conservation status of these crayfish clearly is in need of urgent review.

In the accompanying papers (Furse & Coughran, 2011a: 253-263; 2011b: 265-274, this volume), we assess the conservation status of all species in the genus against Internationally recognised criteria (the IUCN Red List), and provide a case study of a species from each IUCN category in order to highlight the key characteristics that render such a large proportion of this genus endangered. We also present and discuss research imperatives, and management and conservation recommendations for this highly threatened genus of freshwater crayfish.

## ACKNOWLEDGEMENTS

Generous assistance with travel costs was extended us by the organising committee and sponsors, which enabled us to present this work at the "Conservation Biology of Freshwater Crayfishes" Symposium at The Crustacean Society Summer Meeting in Ikebukuro, Tokyo (September 2009). Particular thanks are due to Tadashi Kawai, Yakichi Kobayashi and Hirotomo Kumaki for their considerable assistance in facilitating our attendance, and warm hospitality. Many thanks to Alistair Richardson, Todd Walsh and Robert McCormack who kindly provided crayfish images used in our conference presentation. This research project was supported by the Environmental Futures Centre, and the Griffith School of Environment, Griffith University, Gold Coast, Australia.

## REFERENCES

AABIO, 2008. Online freshwater crayfish species list & photographs. (Australian Aquatic Biological Pty Ltd., Karuah, NSW, Australia). Available online via: http://www.aabio.com.au/list_crayfish.html

BARKER, J., 1992. The spiny freshwater crayfish monitoring program, **1990**: 1-31. (Inland Fisheries Management Branch, Fisheries Management Division, Department of Conservation and Environment. Victoria).

BORSBOOM, A., 1998. Aspects of the biology and ecology of the Australian freshwater crayfish, *Euastacus urospinosus* (Decapoda: Parastacidae). Proc. Linn. Soc. N.S.W., **119**: 87-100.

CLARK, E., 1937. The life history of the Gippsland crayfish. The Australian Museum Magazine, **6**: 186-192.

COUGHRAN, J., 2002. A new species of the freshwater crayfish genus *Euastacus* (Decapoda: Parastacidae) from Northeastern New South Wales: Australia. Rec. Aust. Mus., **54**: 25-30.

— —, 2005. New crayfishes (Decapoda: Parastacidae: *Euastacus*) from northeastern New South Wales, Australia. Rec. Aust. Mus., **57**(3): 361-374.

— —, 2006a. Biology of the freshwater crayfishes of Northeastern New South Wales, Australia. (Ph.D Thesis, School of Environmental Science and Management, Southern Cross University, Lismore, New South Wales).

— —, 2006b. Field guide to the freshwater crayfishes of Northeastern New South Wales (Compact Disk Format). (Natureview Publishing, Bangalow, ISBN: 0 9581149 3 5).

— —, 2007. Distribution, habitat and conservation status of the freshwater crayfishes, *Euastacus dalagarbe, E. girurmulayn, E. guruhgi, E. jagabar* and *E. mirangudjin*. Australian Zoologist, **34**(2): 222-227.

— —, 2008. Distinct groups in the genus *Euastacus*? Freshwater Crayfish, **16**: 125-132.

— —, in press-a. Aspects of the biology and ecology of the Orange-Bellied Crayfish, *Euastacus mirangudjin* Coughran 2002, from northeastern New South Wales. Australian Zoologist, in press.

— —, in press-b. Biology of the Blood Crayfish, *Euastacus gumar* Morgan 1997, a small freshwater crayfish from the Richmond Range, northeastern New South Wales. Australian Zoologist, in press.

COUGHRAN, J. & J. M. FURSE, 2010. An assessment of genus *Euastacus* (49 species) versus IUCN Red List criteria. Report prepared for the global species conservation assessment of crayfishes for the IUCN Red List of Threatened Species: 1-170. (The International Association of Astacology, Auburn, Alabama, USA. ISBN: 978-0-9805452-1-0).

COUGHRAN, J. & S. LECKIE, 2007. *Euastacus pilosus* n. sp., a new crayfish from the highland forests of northern New South Wales, Australia. Fishes of Sahul, **21**(1): 309-316.

CRANDALL, K. A. & J. E. BUHAY, 2008. Global diversity of crayfish (Astacidae, Cambaridae, and Parastacidae–Decapoda) in freshwater. Hydrobiologia, **595**(1): 295-301.

DAVIE, P., 2007. Crustaceans. In: I. GALLOWAY (ed.), Wildlife of Greater Brisbane (2$^{nd}$ ed.): 69-83. (Queensland Museum, Brisbane, ISBN: 978-0-9775943-1-3).

FURSE, J. M. & J. COUGHRAN, 2011a. An assessment of the distribution, biology, threatening processes and conservation status of the freshwater crayfish, genus *Euastacus* (Decapoda: Parastacidae), in continental Australia. II. Threats, conservation assessments and key findings. In: A. ASAKURA (ed.), New frontiers in crustacean biology: Proceedings of the TCS Summer Meeting, Tokyo, 20-24 September 2009. Crustaceana Monographs, **15**: 253-263.

— — & — —, 2011b. An assessment of the distribution, biology, threatening processes and conservation status of the freshwater crayfish, genus *Euastacus* (Decapoda: Parastacidae), in continental Australia. III. Case studies and recommendations. In: A. ASAKURA (ed.), New frontiers in crustacean biology: Proceedings of the TCS Summer Meeting, Tokyo, 20-24 September 2009. Crustaceana Monographs, **15**: 265-274.

FURSE, J. M. & C. H. WILD, 2002a. Prediction of crayfish density from environmental factors for *Euastacus sulcatus* (Crustacea: Decapoda: Parastacidae). Freshwater Crayfish, **13**: 316-329.

— — & — —, 2002b. Terrestrial activities of *Euastacus sulcatus*, the Lamington Spiny Crayfish (Decapoda: Parastacidae). Freshwater Crayfish, **13**: 604 (note).

— — & — —, 2004. Laboratory moult increment, frequency, and growth in *Euastacus sulcatus*, the Lamington Spiny Crayfish. Freshwater Crayfish, **14**: 205-211.

FURSE, J. M., C. H. WILD, S. SIROTTII & H. PETHYBRIDGE, 2006. The daily activity patterns of *Euastacus sulcatus* (Decapoda: Parastacidae) in Southeast Queensland. Freshwater Crayfish, **15**: 139-147.

FURSE, J. M., C. H. WILD & N. N. VILLAMAR, 2004. In-stream and terrestrial movements of *Euastacus sulcatus* in the Gold Coast hinterland: developing and testing a method of accessing freshwater crayfish movements. Freshwater Crayfish, **14**: 213-220.

GEDDES, M. C., R. J. MUSGROVE & N. J. H. CAMPBELL, 1993. The feasibility of re-establishing the River Murray Crayfish, *Euastacus armatus*, in the Lower Murray River. Freshwater Crayfish, **9**: 368-379.

GILLIGAN, D., R. ROLLS, J. MERRICK, M. LINTERMANS, P. DUNCAN & J. KOHEN, 2007. Scoping the knowledge requirements for Murray crayfish (*Euastacus armatus*). New South Wales Department of Primary Industries, Narrandera, New South Wales, Australia. Fisheries Final Report Series, **89**: 1-107.

GROWNS, I. & T. MARSDEN, 1998. Altitude separation and pollution tolerance in the freshwater crayfish *Euastacus spinifer* and *E. australasiensis* (Decapoda: Parastacidae) in coastal flowing streams of the Blue Mountains, New South Wales. Proc. Linn. Soc. N.S.W., **120**: 139-145.

HOEY, J. A., 1990. The biology of the freshwater crayfish, *Euastacus bispinosus* Clark (Decapoda: Parastacidae), and its management in the Lower Glenelg River drainage. (M.Sc. Thesis, Deakin University, Warrnambool).

HOLDICH, D. M., 2002. General biology: background and functional morphology. In: D. M. HOLDICH (ed.), Biology of freshwater crayfish: 3-29. (Blackwell Science, Oxford, ISBN: 0-632-05431-X).

HONAN, J. A., 1998. Egg and juvenile development of the Australian freshwater crayfish, *Euastacus bispinosus* Clark (Decapoda: Parastacidae). Proc. Linn. Soc. N.S.W., **119**: 37-54.

HONAN, J. A. & B. D. MITCHELL, 1995a. Catch characteristics of the large freshwater crayfish, *Euastacus bispinosus* Clark (Decapoda: Parastacidae), and implications for management. Freshwater Crayfish, **10**: 57-69.

— — & — —, 1995b. Growth of the large freshwater crayfish *Euastacus bispinosus* Clark (Decapoda: Parastacidae). Freshwater Crayfish, **10**: 118-131.

— — & — —, 1995c. Reproduction of *Euastacus bispinosus* Clark (Decapoda: Parastacidae), and trends in the reproductive characteristics of freshwater crayfish. Mar. Freshwater Res., **46**: 485-499.

HORWITZ, P., 1990. The conservation status of Australian freshwater crayfish with a provisional list of threatened species, habitats and potentially threatening processes. Australian National Parks and Wildlife Service, Canberra, Australia, Report series, **14**: 1-121.

— —, 1995. The conservation status of Australian freshwater crayfish: review and update. Freshwater Crayfish, **10**: 70-80.

IUCN, 2009. IUCN Red list of threatened species (version 2009.2). (The International Union for Conservation of Nature (Red List Unit), Cambridge, England). Available online via: http://www.iucnredlist.org/

JOHNSTON, K., B. J. ROBSON & C. M. AUSTIN, 2008. Population structure and life history characteristics of *Euastacus bispinosus* and *Cherax destructor* (Parastacidae) in the Grampians National Park, Australia. Freshwater Crayfish, **16**: 165-173.

MCCORMACK, R. B., 2008. The freshwater crayfishes of New South Wales Australia: 138 pages. (Australian Aquatic Biological Pty Ltd., Karuah, NSW, Australia. ISBN: 0 978-0-9805144-1-4).

MCCORMACK, R. B. & J. COUGHRAN, 2008. *Euastacus maccai*, a new freshwater crayfish from New South Wales. Fishes of Sahul, **22**(4): 471-476.

MCCORMACK, R. B., J. COUGHRAN, J. M. FURSE & P. VAN-DER-WERF, 2010. Conservation of imperiled crayfish — *Euastacus jagara* (Decapoda: Parastacidae), a highland crayfish from the Main Range, South-Eastern Queensland, Australia. J. Crustacean Biol., **30**(3): 531-535.

MERRICK, J. R., 1993. Freshwater crayfishes of New South Wales: 1-127. (The Linnean Society of New South Wales, Milsons Point, NSW, ISBN: 0 9590535 1 4).

— —, 1995. Diversity, distribution and conservation of freshwater crayfishes in the eastern highlands of New South Wales. Proc. Linn. Soc. N.S.W., **151**: 247-258.

— —, 1997. Conservation and field management of the freshwater crayfish, *Euastacus spinifer* (Decapoda: Parastacidae), in the Sydney region, Australia. Proc. Linn. Soc. N.S.W., **118**: 217-255.

MOREY, J. L., 1998. Growth, catch rates and notes on the biology of the Gippsland Spiny Freshwater Crayfish, *Euastacus kershawi* (Decapoda: Parastacidae), in West Gippsland, Victoria. Proc. Linn. Soc. N.S.W., **119**: 55-69.

MORGAN, G. J., 1986. Freshwater crayfish of the genus *Euastacus* Clark (Decapoda, Parastacidae) from Victoria. Memoirs of the Museum of Victoria, **47**(1): 1-57.

— —, 1988. Freshwater crayfish of the genus *Euastacus* Clark (Decapoda: Parastacidae) from Queensland. Memoirs of the Museum of Victoria, **49**(1): 1-49.

— —, 1991. The spiny freshwater crayfish of Queensland. Queensland Naturalist, **31**(1-2): 29-36.

— —, 1997. Freshwater crayfish of the genus *Euastacus* Clark (Decapoda: Parastacidae) from New South Wales, with a key to all species of the genus. Rec. Aust. Mus., (Supplement) **23**: 110.

O'BRIEN, M. B., 2007. Freshwater and terrestrial crayfish (Decapoda, Parastacidae) of Victoria, status, conservation, threatening processes and bibliography. The Victorian Naturalist, **14**(4): 210-229.

PONNIAH, M. & J. M. HUGHES, 1998. Evolution of Queensland spiny mountain crayfish of the genus *Euastacus* Clark (Decapoda: Parastacidae): preliminary 16S mtDNA phylogeny. Proc. Linn. Soc. N.S.W., **119**: 9-19.

— — & — —, 2004. The evolution of Queensland spiny mountain crayfish of the genus *Euastacus*. I. Testing vicariance and dispersal with intraspecific mitochondrial DNA. Evolution, **58**(5): 1073-1085.

— — & — —, 2006. The evolution of Queensland spiny mountain crayfish of the genus *Euastacus*. II. Investigating simultaneous vicariance with intraspecific genetic data. Mar. Freshwater Res., **57**(3): 349-362.

RIEK, E. F., 1951. The freshwater crayfish (family Parastacidae) of Queensland, with an appendix describing other Australian species. Rec. Aust. Mus., **22**: 368-388.

— —, 1969. The Australian freshwater crayfish (Crustacea: Decapoda: Parastacidae), with descriptions of new species. Aust. J. Zool., **17**: 855-918.

RYAN, K. A., B. C. EBNER & R. H. NORRIS, 2008. Radio-tracking interval effects on the accuracy of diel scale crayfish movement variables. Freshwater Crayfish, **16**: 87-92.

SANDEMAN, D. C. & L. A. WILKENS, 1982. Sound production by abdominal stridulation in the Australian Murray River Crayfish, *Euastacus armatus*. J. Exp. Biol., **99**: 469-472.

SHULL, H. C., M. PÉRES-LOSADA, D. BLAIR, K. SEWELL, E. A. SINCLAIR, S. LAWLER, M. PONNIAH & K. A. CRANDALL, 2005. Phylogeny and biogeography of the freshwater crayfish *Euastacus* (Decapoda: Parastacidae) based on nuclear and mitochondrial DNA. Mol. Phylogenet. Evol., **37**: 249-263.

SMITH, G., A. BORSBOOM, R. LLOYD, N. LEES & J. KEHL, 1998. Habitat changes, growth and abundance of juvenile Giant Spiny Crayfish, *Euastacus hystricosus* (Decapoda: Parastacidae), in the Conondale Ranges, South-east Queensland. Proc. Linn. Soc. N.S.W., **119**: 71-86.

TAYLOR, C. A., 2002. Taxonomy and conservation of native crayfish stocks. In: D. M. HOLDICH (ed.), Biology of freshwater crayfish: 236-257. (Blackwell Science, Oxford, ISBN: 0-632-05431-X).

TURVEY, P. & J. R. MERRICK, 1997a. Diet and feeding in the freshwater crayfish, *Euastacus spinifer* (Decapoda: Parastacidae), from the Sydney region, Australia. Proc. Linn. Soc. N.S.W., **118**: 175-185.

— — & — —, 1997b. Growth with age in the freshwater crayfish, *Euastacus spinifer* (Decapoda: Parastacidae), from the Sydney region, Australia. Proc. Linn. Soc. N.S.W., **118**: 205-215.

—— & ——, 1997c. Moult increments and frequency in the freshwater crayfish, *Euastacus spinifer* (Decapoda: Parastacidae), from the Sydney region, Australia. Proc. Linn. Soc. N.S.W., **118**: 188-204.

—— & ——, 1997d. Population structure of the freshwater crayfish *Euastacus spinifer*, (Decapoda: Parastacidae), from the Sydney region, Australia. Proc. Linn. Soc. N.S.W., **118**: 157-174.

—— & ——, 1997e. Reproductive biology of the freshwater crayfish, *Euastacus spinifer* (Decapoda: Parastacidae), from the Sydney region, Australia. Proc. Linn. Soc. N.S.W., **118**: 131-155.

WILD, C. H. & J. M. FURSE, 2004. The relationship between *Euastacus sulcatus* and Temnocephalan spp. (Platyhelminthes) in the Gold Coast hinterland, Queensland. Freshwater Crayfish, **14**: 236-245.

First received 1 December 2009.
Final version accepted 30 January 2010.

# AN ASSESSMENT OF THE DISTRIBUTION, BIOLOGY, THREATENING PROCESSES AND CONSERVATION STATUS OF THE FRESHWATER CRAYFISH, GENUS *EUASTACUS* (DECAPODA, PARASTACIDAE), IN CONTINENTAL AUSTRALIA. II. THREATS, CONSERVATION ASSESSMENTS AND KEY FINDINGS

BY


JAMES M. FURSE[1]) and JASON COUGHRAN[2])

The Environmental Futures Centre, Griffith School of Environment, Gold Coast Campus, Griffith University, Queensland 4222, Australia



## ABSTRACT

In our preceding paper we established that *Euastacus* is the largest of the Australian crayfish genera and includes some of the largest, and some of the rarest, species in the World. Species generally share a common suite of biological traits of slow growth, late maturation, and long lifespans (>30 years in some species). The distribution of the genus extends along most of the east coast of continental Australia, and from sea level to over 1500 m above sea level. Consequently, *Euastacus* inhabit most climatic zones in Australia, and are found in a wide range of habitats. In this paper we calculate distributions of the individual species, review threats, and assess the conservation status of all species against current IUCN Red List Criteria. Species' distributions range from highly restricted (2.5 km$^2$) to widespread (>150 000 km$^2$). Threats include: land use practices, pollution, recreational fishing, exotic species, and the known and anticipated effects of climate change. On these bases *Euastacus* separates into six conservation groups, with 80% (39 species) evaluated as belonging in IUCN threat categories, the majority of these Endangered or Critically Endangered: a bleak assessment.


## INTRODUCTION

It has long been recognised that *Euastacus* are of considerable conservation concern (e.g., Horwitz, 1990a, 1995; Merrick, 1997). The last assessment of the genus *versus* IUCN Red List criteria (in 1996) placed 16 species in threat

---


[1]) Corresponding author; e-mail: j.furse@griffith.edu.au

[2]) e-mail: j.coughran@griffith.edu.au

categories (IUCN, 2009). At the time of writing the conservation status of the other 33 known species is unclear, and a reassessment of the status of the genus is warranted (see Furse & Coughran, 2011a: 241-252, this volume). The aim of this project was to conduct the first systematic assessment of the conservation status of all species in genus *Euastacus* against Internationally recognised criteria (the IUCN Red List).

## METHOD

For each of the 49 species the literature was reviewed to identify and evaluate known and potential threats, and determine geographic locations and extent of species' distributions. Extent of species' distributions were calculated using topographical maps, remote imaging, and a Geographical Information System to give the Extent Of Occurrence (EOO) in $km^2$ for each species. By IUCN definition the EOO is "the area contained within the shortest continuous imaginary boundary which can be drawn to encompass all the known, inferred or projected sites of present occurrence of a taxon, excluding cases of vagrancy" (IUCN, 2006).

It was not possible to calculate Area Of Occupancy (AOO) or the "area of suitable habitat occupied by the taxon" (IUCN, 2006), as such data is not available for any *Euastacus*. Each species was then assessed versus current IUCN criteria (Version 6.1, 2006), and placed into one of six categories: Data Deficient (DD), Least Concern (LC), Near Threatened (NT), Vulnerable (VU), Endangered (EN), or Critically Endangered (CR). Definitions of these categories are provided in the IUCN Guidelines current at the time of our review (IUCN, 2006: 7).

## RESULTS & DISCUSSION

### Identified threats

Habitat destruction, fragmentation and modification threaten many species of *Euastacus* (Horwitz, 1990a, 1995; Merrick, 1995). Agriculture, forestry and urbanisation all involve clearing of native vegetation, which for most species appears to be an essential habitat component. Morgan (1986, 1988, 1997) documented examples of many species that were absent from sites that had been cleared of native vegetation, and Furse & Wild (2002) quantified a relationship between density of *Euastacus sulcatus* Riek, 1951 and vegetative canopy cover over streams. Construction of weirs and reservoirs also impacts on stream flow, temperature and oxygen levels, and thereby impacts on riverine *Euastacus* that require flowing, cool, well oxygenated conditions (Horwitz, 1990a). There appears to be at least one case of localised extinction due to this: *Euastacus bispinosus* Clark, 1936 disappeared when Rocklands Reservoir was built in Victoria (Horwitz, 1990a). Water quality and pollution has long been identified as a threat for the genus generally (Horwitz, 1990a, 1995; McKinnon, 1995; Merrick, 1995), and these threats include sedimentation, increased temperature, decreased oxygen levels, eutrophication and pesticide runoff. Given their extremely restricted ranges, many species are susceptible to small-scale, localised impacts and disturbance events ("local disasters", Merrick, 1995) such as forest management and agricultural practices, water harvesting, bushfires, and "accidents" such as chemical, pesticide or petroleum spills.

Climate change poses both direct and indirect threats to almost all species in the genus. Regional modelling of climate change impacts invariably predict increasing temperatures across the distribution of the genus (Howden, 2003; Hughes, 2003; Pittock, 2003; Westoby & Burgman, 2006), and this direct threat is of particular concern for many of the species that appear to require cool conditions. It is established that most *Euastacus* are already restricted to the extreme upper reaches of the catchments they inhabit (including isolated mountain tops), and evidence from Queensland suggests this is due to a retreat to cooler areas of habitat in response to the warming of the continent during the Pliocene (Ponniah & Hughes, 2004, 2006). If these species are indeed restricted by their thermal tolerance, then any further increase in temperature through climate change poses a significant threat, and could cause extinctions across this genus. Modelling in most areas where *Euastacus* occur also predicts altered hydrological regimes, including decreased rainfall, runoff and reduced soil moisture (Chiew & McMahon, 2002; Hughes, 2003). This is clearly a threat for riverine species that require flowing conditions, but is perhaps a greater concern for those species that occur in ephemeral habitats (see Coughran, 2007, 2008) and rely on sub-surface water and moist forest soils. It is also predicted that these climate change impacts will lead to other habitat changes, such as shifts in forest type from rainforest to sclerophyllous forest (Hilbert et al., 2001; Hughes, 2003). Since many *Euastacus* appear to be dependent on rainforest habitats or other specific vegetation types and habitats (e.g., high altitude heathland and Alpine bogs), any climate change induced alterations in vegetation communities and their distributions are clearly of concern (Hughes, 2003; Westoby & Burgman, 2006).

Climate change is also expected to result in an increased frequency of severe weather events (severe droughts, storms, floods, and tropical cyclones) (Hughes, 2003; IPCC, 2007; Specht, 2008), and these may impact on the quality of habitat and/or more directly on the crayfish themselves (McCormack et al., 2010). For example, a recent localised severe storm event resulted in a mass kill of *Euastacus valentulus* Riek, 1951 — several hundred crayfish had apparently been directly overwhelmed by the large volume of storm induced flow and had been buried in the alluvium, up to 50 m away from the stream channel (Furse, unpubl.). Mass crayfish kills and emersion events have recently been documented elsewhere in response to severe weather events, particularly floods (Parkyn & Collier, 2004; Lewis & Morris, 2008), therefore any climate change induced increase in severe flooding events is clearly a serious threat to endangered crayfish (Meyer et al., 2007).

Climate change will likely also lead to a range of other disturbance events that could contribute to declines in *Euastacus* or their habitat, such as increased incidence of widespread bushfires, which may in turn lead to "black water events", and mass emersions and mortalities as have been documented in *Euastacus armatus* (von Martens, 1866) in such conditions (McKinnon, 1995).

Over-exploitation is recognised as a threat for several species, and this extends to large species that are targeted by recreational fishing and smaller species that are targeted by 'collectors' (Coughran, 2007). It is also apparent that *Euastacus* are illegally traded for aquarium use, both domestically and internationally. Because of their striking colouration, size and armature, these crayfishes are highly prized as aquarium specimens, and illegal exploitation to supply this demand may present a considerable threat, particularly for the exceptionally rare species. All available information on the biology of this genus has revealed that they are extremely slow-growing with life history characteristics that are unsuited to even moderate levels of exploitation (e.g., Hoey, 1990; Honan & Mitchell, 1995a, 1995b; Turvey & Merrick, 1997; Furse & Wild, 2004; Wild & Furse, 2004; Coughran, 2006). Illegal poaching from protected areas has been reported for many species, and breaches of fishing regulations for recreational catch species also appears to be common (see Coughran, 2007; Coughran & Furse, 2010).

Exotic species are prevalent throughout the range of many *Euastacus*, and several of these have been known to impact either on crayfish directly, or on crayfish habitat (Green & Osbourne, 1981; Horwitz, 1990a, 1995; Merrick, 1995; O'Brien, 2007; Rowe et al., 2008).

These animals include cats (*Felis catus*), foxes (*Vulpes vulpes*), pigs (*Sus scrofa*), goats (*Capra hircus*), trouts (*Salmo trutta*, and *Oncorhynchus mykiss*), European carp (*Cyprinus carpio*), redfin perch (*Perca fluviatilis*), and toxic cane toads (*Bufo marinus*). Although there are no foreign crayfish in Australia, a few native species have been widely translocated outside of their natural range (Horwitz, 1990b) and two species are now becoming established on the eastern seaboard of Australia, *Cherax destructor* Clark, 1936 and *Cherax quadricarinatus* (von-Martens, 1868) (Coughran & Leckie, 2007; Coughran et al., 2009). Both of these species display a far superior reproductive biology to *Euastacus*, and thus have the potential to rapidly out-compete them. Many of the above threats compound by the fact that a number of species are distributed in close proximity to, or in some cases, partially within rapidly growing major population centres (ABS, 2009; Furse & Wild, unpubl.).

While there are not currently any alien crayfish species in Australia, any illegal importation or introductions of alien crayfish would pose a very serious threat to the native Australian crayfish fauna. Continued vigilance by authorities at Australian borders will be essential to minimise the risk of alien crayfish being introduced onto the Continent.

### Geographic range

The EOO of individual species ranges from 2.5 to >150 000 km². However, it is of note that apart from a few widespread species, most *Euastacus* have highly restricted ranges. Only 12 species have an EOO >5000 km². The remainder have restricted EOOs of <5000 km², and 12 of these species have an EOO <10 km². Ten species are known only from a single locality. While it was not possible to estimate AOO, it is noted that the overall EOO for many of these species contains a considerable amount of unsuitable habitat (in particular cleared land, unsuitable vegetation types, and eutrophic lowland habitats), and the AOO will be much smaller than the EOO. The altitudinal ranges of *Euastacus* reflect two clear groups, as identified by Morgan (1997). Some species occur at both high and low altitudes, in some cases at or near sea level. The majority, however, are restricted to highland areas >200 m above sea level. Most highland species extend across drainage divides, inhabiting the headwater reaches of several catchments. The distributions of these species are therefore fragmented by mountain ridge barriers and areas of unsuitable, lowland habitat (Morgan, 1997; Ponniah & Hughes, 2006).

There is genetic evidence to suggest that this fragmentation and isolation has led to appreciable speciation in the genus (Baker et al., 2004; Ponniah

& Hughes, 2004, 2006). Coughran & Furse (2010) also speculated that some fragmented populations may in fact be distinct taxa, and suggested this should be clarified through further analysis of population genetics.

## Conservation assessment

Five IUCN criteria are available and each is based on different indicators, but given the lack of quantified, historical data for this genus generally, we were largely restricted to employing Criterion B (based on Geographic Range). This criterion is well suited to a conservation assessment of this genus, since most species have highly restricted and fragmented distributions, and face a range of easily identified threats. The resulting assessment was clear-cut, in that no species were near to any of the boundaries within Criterion B (table I).

## Summary of revised conservation classifications

Eighty percent of all *Euastacus* were evaluated as belonging in a threat category (table II). More than two-thirds of the genus satisfies IUCN criteria for listing in the Endangered (EN) or Critically Endangered (CR) categories. This constitutes a very bleak assessment of the conservation status of this large and charismatic genus.

In summary, it is clear the conservation status of *Euastacus* is of concern and in our following paper (Furse & Coughran, 2011b: 265-274, this volume) we present a series of case studies where a species from each IUCN category is discussed. These case studies serve to highlight the characteristics that render so many species of *Euastacus* endangered. We also present research imperatives, and management and conservation recommendations for this highly threatened genus of freshwater crayfish.

## ACKNOWLEDGEMENTS

Generous assistance with travel costs was extended us by the organising committee and sponsors, which enabled us to present this work at the "Conservation Biology of Freshwater Crayfishes" Symposium at The Crustacean Society Summer Meeting in Ikebukuro, Tokyo (September 2009). Particular thanks are due to Tadashi Kawai, Yakichi Kobayashi and Hirotomo Kumaki for their considerable assistance in facilitating our attendance, and warm hospitality. Many thanks to Alistair Richardson, Todd Walsh and Robert McCormack who kindly provided crayfish images used in our conference presentation. This

TABLE I

Summary of conservation factors for the 49 species of *Euastacus*, and their resulting conservation status under current IUCN criteria. Key threats are coded as follows: CC, climate change; E, exotic species; F, fishing or collection pressure; H, habitat modification; and PP, pesticides and pollution. The code L is applied in cases where either the EOO, or the amount of suitable habitat within the EOO, is extremely restricted, thereby rendering the species as potentially susceptible to a range of small-scale, localised impacts such as bushfire, forestry management, habitat destruction or over-collection. Although in some cases it is less clear if an LC species is facing similar threats, potential threats are indicated in parentheses

| Species | Threats | EOO (km$^2$) | Fragmentation | Classification | Status |
|---|---|---|---|---|---|
| *armatus* | E, F, H, PP | >150 000 | – | – | DD |
| *australasiensis* | (E, F, H, PP) | <20 000 | – | – | LC |
| *balanensis* | CC, E, F | <5000 | yes | B1 + 2(a), (b) iii | EN |
| *bidawalus* | CC, E, F | <5000 | yes | B1 + 2(a), (b) iii | EN |
| *bindal* | CC, E, F, L | <10 | single locality | B1 + 2(a), (b) iii | CR |
| *bispinosus* | CC, E, F, H | ~10 000 | – | B1 (b) iii, iv (c) iv | VU |
| *brachythorax* | CC, E, L | ~2000 | yes | B1 + 2(a), (b) iii | EN |
| *clarkae* | CC, E, L | <10 | single locality | B1 + 2(a), (b) iii | CR |
| *claytoni* | CC, E, L | ~3000 | yes | B1 + 2(a), (b) iii | EN |
| *crassus* | CC, E, H, L | <5000 | yes | B1(a), (b) iii | EN |
| *dalagarbe* | CC, E, L | ~50 | yes | B1 + 2(a), (b) iii | CR |
| *dangadi* | (E, F) | <5000 | – | – | LC |
| *dharawalus* | CC, E, F, H, L | <10 | single locality | A2 + 3(d), (e), B1(a), (b) iii, v, (c) iv | CR |
| *diversus* | CC, E, H | <500 | yes | B1 + 2(a), (b) iii | EN |
| *eungella* | CC, E, F | <100 | yes | B1 + 2(a), (b) iii | CR |
| *fleckeri* | CC, E, F, H | <1000 | yes | B1 + 2(a), (b) iii | EN |
| *gamilaroi* | CC, E, L | <10 | single locality | B1 + 2(a), (b) iii | CR |
| *girurmulayn* | CC, E, L | <10 | yes | B1 + 2(a), (b) iii | CR |
| *gumar* | CC, E, L | <1000 | yes | B1 + 2(a), (b) iii | EN |
| *guruhgi* | CC, E, L | <10 | yes | B1 + 2(a), (b) iii | CR |
| *guwinus* | CC, E, L | <10 | single locality | B1 + 2(a), (b) iii | CR |
| *hirsutus* | CC, E, L | ~1200 | yes | B1(a), (b) iii | EN |
| *hystricosus* | CC, E, F | ~3000 | yes | B1 + 2(a), (b) iii | EN |
| *jagabar* | CC, E, L | <10 | single locality | B1 + 2(a), (b) iii | CR |
| *jagara* | CC, E, L | ~25 | yes | B1 + 2(a), (b) iii | CR |
| *kershawi* | CC, E, F | ~20 000 | – | – | LC |
| *maccai* | CC | ~2500 | yes | B1(a), (b) iii | EN |
| *maidae* | CC, E, H, L | <10 | single locality | B1 + 2(a), (b) iii | CR |
| *mirangudjin* | CC, E, L | ~45 | yes | B1 + 2(a), (b) iii | CR |
| *monteithorum* | CC, E, L | <10 | single locality | B1 + 2(a), (b) iii | CR |
| *neodiversus* | CC, E, F, H | ~1500 | yes | B1(a), (b) iii | EN |
| *neohirsutus* | (E, F) | <5000 | – | – | LC |
| *pilosus* | CC, E, L | ~1000 | yes | B1 + 2(a), (b) iii | EN |
| *polysetosus* | CC, E, L | ~750 | yes | B1(a), (b) iii | EN |
| *reductus* | (E, F) | <5000 | – | – | LC |
| *rieki* | CC, E, H | <5000 | yes | B1(a), (b) iii | EN |

## TABLE I
### (Continued)

| Species | Threats | EOO (km$^2$) | Fragmentation | Classification | Status |
|---|---|---|---|---|---|
| *robertsi* | CC, E, L | <100 | yes | B1 + 2(a), (b) iii | CR |
| *setosus* | CC, E, L | <10 | single locality | B1 + 2(a), (b) iii | CR |
| *simplex* | CC, E | <20 000 | yes | B1(a) + (b) iii | VU |
| *spinichelatus* | CC, E, L | <1000 | yes | B1(a), (b) iii | EN |
| *spinifer* | (F, H) | ~55 000 | – | – | LC |
| *sulcatus* | CC, E, F, H, L | ~8000 | yes | B1 + 2(a), (b) iii | VU |
| *suttoni* | CC, E, F | <20 000 | yes | B1(a) + (b) iii | VU |
| *urospinosus* | CC, E, F, L | ~200 | yes | B1 + 2(a), (b) iii | EN |
| *valentulus* | CC, E, F | <20 000 | – | – | LC |
| *woiwuru* | CC, E, F | <20 000 | uncertain | – | NT |
| *yanga* | CC, E, F | ~20 000 | – | – | LC |
| *yarraensis* | CC, E, F | <20 000 | yes | B1(a), (b) iii | VU |
| *yigara* | CC, E, L | <10 | single locality | B1 + 2(a), (b) iii | CR |

## TABLE II
Summary of findings for the genus *Euastacus*, assessed against
current IUCN conservation criteria

| IUCN category | Number of species | % of total species |
|---|---|---|
| Data Deficient | 1 | 2% |
| Least Concern | 8 | 16% |
| Near Threatened | 1 | 2% |
| Vulnerable | 5 | 10% |
| Endangered | 17 | 35% |
| Critically Endangered | 17 | 35% |

research project was supported by the Environmental Futures Centre, and the Griffith School of Environment, Griffith University, Gold Coast, Australia.

## REFERENCES

ABS, 2009. Regional Population Growth, Australia, 2007-08. (Report Number 3218.0, The Australian Bureau of Statistics, Canberra, ACT, Australia). Available online via: http://www.abs.gov.au/AUSSTATS/abs@.nsf/DetailsPage/3218.02007-08?OpenDocument

BAKER, A. M., J. M. HUGHES, J. C. DEAN & S. E. BUNN, 2004. Mitochondrial DNA reveals phylogenetic structuring and cryptic diversity in Australian freshwater macroinvertebrate assemblages. Mar. Freshwater Res., **55**(6): 629-640.

CHIEW, F. H. S. & T. A. MCMAHON, 2002. Modelling the impacts of climate change on Australian streamflow. Hydro. Process., **16**(6): 1235-1245.

COUGHRAN, J., 2006. Biology of the freshwater crayfishes of Northeastern New South Wales, Australia. (Ph.D. Thesis, School of Environmental Science and Management, Southern Cross University, Lismore).

— —, 2007. Distribution, habitat and conservation status of the freshwater crayfishes, *Euastacus dalagarbe*, *E. girurmulayn*, *E. guruhgi*, *E. jagabar* and *E. mirangudjin*. Australian Zoologist, **34**(2): 222-227.

— —, 2008. Distinct groups in the genus *Euastacus*? Freshwater Crayfish, **16**: 125-132.

COUGHRAN, J. & J. M. FURSE, 2010. An assessment of genus *Euastacus* (49 species) versus IUCN Red List criteria. Report prepared for the global species conservation assessment of crayfishes for the IUCN Red List of Threatened Species: 1-170. (The International Association of Astacology, Auburn, Alabama, USA. ISBN: 978-0-9805452-1-0).

COUGHRAN, J. & S. LECKIE, 2007. Invasion of a New South Wales stream by the Tropical Crayfish, *Cherax quadricarinatus* (von Martens). In: D. LUNNEY, P. EBY, P. HUTCH-INGS & S. BURGIN (eds.), Pest or guest: the zoology of overabundance: 40-46. (Royal Zoological Society of New South Wales, Mosman, NSW, ISBN: 978-0-9803272-1-2).

COUGHRAN, J., R. B. MCCORMACK & G. DALY, 2009. Translocation of the Yabby *Cherax destructor* into eastern drainages of New South Wales, Australia. Australian Zoologist, **35**(1): 100-103.

FURSE, J. M. & J. COUGHRAN, 2011a. An assessment of the distribution, biology, threatening processes and conservation status of the freshwater crayfish, genus *Euastacus* (Decapoda: Parastacidae), in continental Australia. I. Biological background and current status. In: A. ASAKURA (ed.), New frontiers in crustacean biology: Proceedings of the TCS Summer Meeting, Tokyo, 20-24 September 2009. Crustaceana Monographs, **15**: 241-252.

— — & — —, 2011b. An assessment of the distribution, biology, threatening processes and conservation status of the freshwater crayfish, genus *Euastacus* (Decapoda: Parastacidae), in continental Australia. III. Case studies and recommendations. In: A. ASAKURA (ed.), New frontiers in crustacean biology: Proceedings of the TCS Summer Meeting, Tokyo, 20-24 September 2009. Crustaceana Monographs, **15**: 265-274.

FURSE, J. M. & C. H. WILD, 2002. Prediction of crayfish density from environmental factors for *Euastacus sulcatus* (Crustacea: Decapoda: Parastacidae). Freshwater Crayfish, **13**: 316-329.

— — & — —, 2004. Laboratory moult increment, frequency, and growth in *Euastacus sulcatus*, the Lamington Spiny Crayfish. Freshwater Crayfish, **14**: 205-211.

GREEN, K. & W. S. OSBOURNE, 1981. The diet of foxes, *Vulpes vulpes* (L.), in relation to abundance of prey above the winter snowline in New South Wales. Aust. Wildlife Res., **8**(2): 349-360.

HILBERT, D. W., B. OSTENDORF & M. S. HOPKINS, 2001. Sensitivity of tropical forests to climate change in the humid tropics of north Queensland. Austral Ecol., **26**(6): 590-603.

HOEY, J. A., 1990. The biology of the freshwater crayfish, *Euastacus bispinosus* Clark (Decapoda: Parastacidae), and its management in the Lower Glenelg River drainage. (M.Sc. Thesis, Deakin University, Warrnambool).

HONAN, J. A. & B. D. MITCHELL, 1995a. Growth of the large freshwater crayfish *Euastacus bispinosus* Clark (Decapoda: Parastacidae). Freshwater Crayfish, **10**: 118-131.

— — & — —, 1995b. Reproduction of *Euastacus bispinosus* Clark (Decapoda: Parastacidae), and trends in the reproductive characteristics of freshwater crayfish. Mar. Freshwater Res., **46**: 485-499.

HORWITZ, P., 1990a. The conservation status of Australian freshwater crayfish with a provisional list of threatened species, habitats and potentially threatening processes. Australian National Parks and Wildlife Service, Canberra, Australia, Report series, **14**: 1-121.

— —, 1990b. The translocation of freshwater crayfish in Australia: potential impact, the need for control and global relevance. Biol. Conserv., **54**: 291-305.

— —, 1995. The conservation status of Australian freshwater crayfish: review and update. Freshwater Crayfish, **10**: 70-80.

HOWDEN, M., 2003. Climate trends and climate change scenarios. In: M. HOWDEN, L. HUGHES, M. DUNLOP, I. ZETHOVEN, D. HILBERT & C. CHILCOTT (eds.), Climate change impacts on biodiversity in Australia. Outcomes of a workshop sponsored by the Biological Diversity Advisory Committee, 1-2 October 2002: 8-13. (Commonwealth of Australia, Canberra, ACT, ISBN: 0 9580845 6 4).

HUGHES, L., 2003. Climate change and Australia: trends, projections and impacts. Austral Ecol., **28**(4): 423-443.

IPCC, 2007. Climate Change 2007 — Synthesis report. An assessment of the Intergovernmental Panel on Climate Change: 1-52. (Fourth Assessment Report, Intergovernmental Panel on Climate Change, Geneva).

IUCN, 2006. Guidelines for using the IUCN red list categories and criteria (version 6.1): 1-60. (Prepared by the Standards and Petitions Working Group for the IUCN SSC Biodiversity Assessments Sub-Committee in July 2006).

IUCN, 2009. IUCN Red list of threatened species (version 2009.2). (The International Union for Conservation of Nature (Red List Unit), Cambridge, England). Available online via: http://www.iucnredlist.org/

LEWIS, A. & F. MORRIS, 2008. Report on a major stranding of crayfish at Meldon, River Wansbeck, UK. Crayfish News, **30**(4): 7-8.

MCCORMACK, R. B., J. COUGHRAN, J. M. FURSE & P. VAN-DER-WERF, 2010. Conservation of imperiled crayfish — *Euastacus jagara* (Decapoda: Parastacidae), a highland crayfish from the Main Range, South-Eastern Queensland, Australia. J. Crustacean Biol., **30**(3): 531-535.

MCKINNON, L. J., 1995. Emersion of Murray crayfish, *Euastacus armatus* (Decapoda: Parastacidae), from The Murray River due to post-flood water quality. Proceedings of the Royal Society of Victoria, **107**(1): 31-38.

MERRICK, J. R., 1995. Diversity, distribution and conservation of freshwater crayfishes in the eastern highlands of New South Wales. Proc. Linn. Soc. N.S.W., **151**: 247-258.

— —, 1997. Conservation and field management of the freshwater crayfish, *Euastacus spinifer* (Decapoda: Parastacidae), in the Sydney region, Australia. Proc. Linn. Soc. N.S.W., **118**: 217-255.

MEYER, K. M., K. GIMPEL & R. BRANDL, 2007. Viability analysis of endangered crayfish populations. J. Zool., **273**(4): 364-371.

MORGAN, G. J., 1986. Freshwater crayfish of the genus *Euastacus* Clark (Decapoda, Parastacidae) from Victoria. Memoirs of the Museum of Victoria, **47**(1): 1-57.

— —, 1988. Freshwater crayfish of the genus *Euastacus* Clark (Decapoda: Parastacidae) from Queensland. Memoirs of the Museum of Victoria, **49**(1): 1-49.

— —, 1997. Freshwater crayfish of the genus *Euastacus* Clark (Decapoda: Parastacidae) from New South Wales, with a key to all species of the genus. Rec. Aust. Mus., (Supplement) **23**: 110.

O'BRIEN, M. B., 2007. Freshwater and terrestrial crayfish (Decapoda, Parastacidae) of Victoria, status, conservation, threatening processes and bibliography. The Victorian Naturalist, **14**(4): 210-229.

PARKYN, S. M. & K. J. COLLIER, 2004. Interaction of press and pulse disturbances on crayfish populations: flood impacts in pasture and forest streams. Hydrobiologia, **527**(1): 113-124.

PITTOCK, B. (ed.), 2003. Climate change: an Australian guide to the science and potential impacts: 1-239. (Australian Greenhouse Office, Canberra).

PONNIAH, M. & J. M. HUGHES, 2004. The evolution of Queensland spiny mountain crayfish of the genus *Euastacus*. I. Testing vicariance and dispersal with intraspecific mitochondrial DNA. Evolution, **58**(5): 1073-1085.

— — & — —, 2006. The evolution of Queensland spiny mountain crayfish of the genus *Euastacus*. II. Investigating simultaneous vicariance with intraspecific genetic data. Mar. Freshwater Res., **57**(3): 349-362.

ROWE, D. K., A. MOORE, A. GIORGETTI, C. MACLEAN, P. GRACE, S. WADHWA & J. COOKE, 2008. Review of the impacts of gambusia, redfin perch, tench, roach, yellowfin goby and streaked goby in Australia: 1-245. (Report prepared for the Australian Government Department of the Environment, Water, Heritage and the Arts. Canberra).

SPECHT, A., 2008. Extreme natural events and effects on tourism: central eastern coast of Australia: 1-66. (The Cooperative Centre for Sustainable Tourism Pty Ltd, Griffith University, Gold Coast).

TURVEY, P. & J. R. MERRICK, 1997. Growth with age in the freshwater crayfish, *Euastacus spinifer* (Decapoda: Parastacidae), from the Sydney region, Australia. Proc. Linn. Soc. N.S.W., **118**: 205-215.

WESTOBY, M. & M. BURGMAN, 2006. Climate change as a threatening process. Austral Ecol., **31**(5): 549-550.

WILD, C. H. & J. M. FURSE, 2004. The relationship between *Euastacus sulcatus* and Temnocephalan spp. (Platyhelminthes) in the Gold Coast hinterland, Queensland. Freshwater Crayfish, **14**: 236-245.

First received 1 December 2009.
Final version accepted 30 January 2010.

# AN ASSESSMENT OF THE DISTRIBUTION, BIOLOGY, THREATENING PROCESSES AND CONSERVATION STATUS OF THE FRESHWATER CRAYFISH, GENUS *EUASTACUS* (DECAPODA, PARASTACIDAE), IN CONTINENTAL AUSTRALIA. III. CASE STUDIES AND RECOMMENDATIONS

BY


JAMES M. FURSE[1]) and JASON COUGHRAN[2])

The Environmental Futures Centre, Griffith School of Environment, Gold Coast Campus,
Griffith University, Queensland 4222, Australia



## ABSTRACT

In our preceding paper 80% of *Euastacus* species were assessed as belonging in IUCN threat categories, the majority of these Endangered or Critically Endangered. As an essential part of our review and conservation assessment of this genus, we provide a species case study from each of the IUCN categories, and discuss research imperatives and conservation and management considerations for the genus in general. In addition to continuing basic taxonomic and biological research, on-going population monitoring should be implemented, and research urgently needs to be initiated into the susceptibility of most species to increasing temperature, increasing dryness and/or reduced flow, land use practices, and exotic species (particularly the toxic cane toad, *Bufo marinus*). For those species with fragmented distributions extending across a number of separate drainages, population genetics studies are required to clarify if the isolated populations are in fact distinct taxa. Conservation prospects of this genus would be enhanced by recognising threatened species under relevant State Conservation legislation, rectifying the chronic shortage of funding for basic, inexpensive and critical research, and adopting the well documented management recommendations. Given their wide distributional parameters, and their presence in essentially the entire spectrum of Australian climatic zones and habitats, *Euastacus* species will likely be the first to be broadly impacted by the combined effects of the various threats identified: a "sentinel" genus. We suggest *Euastacus* could serve as early warning indicators for other fauna in Australia, and perhaps other regions of the World.



---

[1]) Corresponding author; e-mail: j.furse@griffith.edu.au
[2]) e-mail: j.coughran@griffith.edu.au




New frontiers in crustacean biology: 265-274

## INTRODUCTION

In the accompanying papers (Furse & Coughran, 2011a: 241-252; 2011b: 253-263, this volume) we established that the 49 known *Euastacus* species share consistent biological traits (e.g., slow-growing and very long-lived), and that most species are either currently being impacted, or are generally threatened by a common suite of anthropogenic threats. A few species are quite common, but many are exceptionally rare. Many species have very specific habitat requirements, with restricted and fragmented distributions typical: 37 species have Extent of Occurrences (EOO) <5000 km², with much of their EOOs containing unsuitable habitat. Few species are listed under any State Conservation legislation, and protection is mostly incidental through their occurrence within National Parks. Assessment *versus* IUCN Red List criteria indicated that 80% (39 species) in the genus are threatened, with 34 of those either Endangered or Critically Endangered.

Although it is clearly beyond the scope of the present paper to describe the situation for each of the 49 species in the genus, we have included summaries for selected species here to serve as typical case studies for each of the IUCN conservation status categories of this genus (see Coughran & Furse, 2010 for a complete review of each species).

## CASE STUDIES

### Murray River spiny crayfish, *Euastacus armatus* (von Martens, 1866) (DD)

Murray River spiny crayfish have been comparatively well studied, largely on the initiative of various government management agencies, although curiously much of this information remains unpublished. This lowland species is by far the most widespread in the genus, with an EOO >150 000 km² that extends across three States and one Territory. However, the species is considered to be in decline, particularly in the lower reaches of the Murray River system, and a range reduction of 8% has been estimated by Gilligan et al. (2007). The species inhabits flowing, riverine environments, and various practices (e.g., river regulation, de-snagging and channelization) are thought to have impacted on this habitat throughout much of its range. Agricultural pesticides, exotic fish species and over-exploitation are also thought to have contributed to the decline of Murray River spiny crayfish. The species is large, slow-growing and late-maturing, with a slow reproductive cycle over the cooler months (Gilligan et al., 2007). These biological traits offer little resilience to the rapid pace of anthropogenic impacts, and it is conceivable that the species may continue to decline in the future. Although the species satisfies criteria for listing at State level, and is listed under conservation legislation in the Australian Capital Territory, Victoria and South Australia (and indirectly listed in New South Wales (NSW)), across its overall distribution the species does not satisfy the IUCN criteria for listing as Threatened. While this is one of the better understood species in the genus, and is facing numerous threats, there is a paucity of appropriate data on its contemporary distribution and abundance. Specifically, there appears to be no data supporting a >30% decline in any of the parameters listed for Criterion A, which is currently the most applicable criterion for assessing the species given its widespread geographical distribution. More information is required to fully assess this species *versus* IUCN criteria, and it is acknowledged here that results of future research will probably indicate that a threatened classification is warranted for the Murray River spiny crayfish. Importantly, the change of IUCN category from an earlier listing of VU to DD reflects a change in IUCN assessment criteria, and should not be interpreted as an improvement in the conservation status of the species. Conversely, *E. armatus* may well be experiencing continued declines and facing threat of extinction. Its listing as DD highlights that there are specific research gaps that need to be addressed before its status can be determined. This research needs to be conducted in a coherent manner across its entire range (i.e., in all the States and Territories it occurs in) with involvement from all stakeholders.

### Sydney giant spiny crayfish, *Euastacus spinifer* (Heller, 1865) (LC)

The Sydney giant spiny crayfish has been very well studied and while the research has been published (Merrick, 1997; Turvey & Merrick, 1997a, 1997e, 1997d, 1997b, 1997c), the findings have not as yet been incorporated into any management measures. Like *E. armatus*, this species is very long-lived and displays a slow-growing, late-maturing life cycle, with a slow reproductive cycle during the cooler months. The species has an EOO of ~55 000 km², although the distribution extends across several different coastal drainage systems and hence there may be some level of fragmentation within that range. Although there are threats identified for the species (e.g., exploitation, urbanisation and modification of habitat, exotic species and pollution) (Merrick, 1997), these are rather localised within its relatively broad overall distribution. The species is thus evaluated as of Least Concern under IUCN criteria. A number of specific management recommendations for this species have been developed and well documented (Merrick, 1997), and adoption and implementation of these should be a priority for resource management agencies.

## Woiwuru spiny crayfish, *Euastacus woiwuru* Morgan, 1986 (NT)

The Woiwuru spiny crayfish has a restricted distribution, with an EOO <20 000 km$^2$. The distribution extends across several different drainages, and may thus be fragmented. However, although the species usually only occurs at highland sites it is also occasionally found at lower altitudes (Morgan, 1986), so it is not clear if lowland habitats act as barriers to dispersal. The level of fragmentation within this species' distribution requires investigation as evidence of appreciable fragmentation would warrant its listing in a threat category under IUCN criteria given the threats it faces. The species is most common where native vegetation remains intact, although much has been cleared within the species' range (Morgan, 1986). Exotic species such as redfin perch (*Perca fluviatilis*) and European carp (*Cyprinus carpio*) occur within the species' distribution and may also threaten *E. woiwuru*. In addition to increased incidence of severe weather events, climate change modelling for the region has predicted both warmer and drier conditions, with reduced soil moisture and surface runoff, therefore climate change represents a threat to the species across its range. In summary, the Woiwuru spiny crayfish is close to satisfying criteria for listing as threatened, but further information is required, particularly regarding population genetics. In the interim, it should be listed as Near Threatened.

## Mountain crayfish or Lamington spiny crayfish, *Euastacus sulcatus* Riek, 1951 (VU)

The Mountain crayfish has an overall EOO ~8000 km$^2$, but within that EOO it is restricted to highland (typically >300 m above sea level), rainforest or wet sclerophyll habitats. Mountain ridges and unsuitable lowland habitats therefore fragment the species into 10 distinct populations in the upper reaches of different coastal drainages, some of which are smaller than 100 km$^2$ in extent. As such, if the species were assessed at the population level, some populations would be assessed as having a status of EN or CR, and analysis of population genetics may indicate this is warranted. However, in the present assessment we have assessed each species according to its overall range, and a status of VU applies at that level. The Mountain crayfish is one of the better understood species, it is slow-growing, late-maturing and long-lived, with a slow reproductive cycle over the cooler months (Morgan, 1988; Furse, 1999; Furse & Wild, 2004; Wild & Furse, 2004; Coughran, 2006; Furse & Wild, unpubl.). It is threatened by illegal exploitation, habitat degradation (e.g., loss of riparian vegetation), altered hydrology and temperatures due to climate change, pesticides and pollution from urban development and agriculture (Furse & Wild, 2002, 2004; Furse et al., 2004; Furse & Coughran, unpubl.). Exotic toxic cane toads (*Bufo marinus*) are established within the species' range, and research is urgently required into potential impacts of these amphibians. Some populations occur in areas subjected to considerable land-use disturbance and development pressure, and are situated very close to a major population centre: the fastest growing in Australia (ABS, 2009; Furse & Wild, unpubl.).

## Blood crayfish, *Euastacus gumar* Morgan, 1997 (EN)

The Blood crayfish has a highly restricted and fragmented, highland distribution. The overall EOO is <1000 km$^2$, but consists of disjunct fragments that are each much smaller than 100 km$^2$. The species is restricted to isolated pockets of rainforest habitat, where it inhabits small headwater streams, moist gullies and rainforest soaks. It does not require flowing water but appears to survive by burrowing into sub-surface moisture (Coughran, in press). Climate change modelling for the region predicts drier and warmer conditions and these ephemeral habitats can be expected to fluctuate in extent and quality with a changing climate. Climate change may also influence forest type, which is a particular concern for forest-specific crayfish such as *E. gumar*. This species is slow-growing and late-maturing, with a very low fecundity (Coughran, 2000, 2006, in press), biological characteristics that are not conducive to adapting to change. Exotic cane toads are also established within the range of this species. If assessed at the population level, a status of CR would be appropriate, but across its overall range the species is assessed as EN.

## Dharawal spiny crayfish or Fitzroy Falls spiny crayfish, *Euastacus dharawalus* Morgan, 1997 (CR)

The Dharawal spiny crayfish is known only from one locality, a small section of stream length (<2 km) above Fitzroy Falls, NSW. A water storage reservoir has been built upstream, and this could potentially impact on the species if not managed in accordance with environmental flow requirements. Exotic salmonids have also been reported from the habitat, and a translocated crayfish, *Cherax destructor* Clark, 1936 has also established. This invasive crayfish has far superior biological traits to those of the genus *Euastacus*, and recent surveys suggest that it is rapidly displacing *E. dharawalus* from the site (Coughran et al., 2009). The site is also targeted by recreational fishers, and there are currently no guidelines or education materials to alert the recreational fishing community to the conservation concerns for the imperilled *E. dharawalus*. The Dharawal spiny crayfish may be facing imminent extinction, and is assessed as CR. Urgent research is required into various aspects of its biology, ecology and life history, thermal tolerance, resilience to exotic species, and population parameters. For species such as *E. dharawalus*, the potential for establishing a conservation-based captive breeding program should be considered.

Next steps

Historically, the genus *Euastacus* has been very poorly studied and understood. The past two decades has seen a considerable increase in research into the taxonomy, biology and ecology of this group. Although many species remain poorly known there is a growing body of knowledge that highlights their precarious status, and their limited resilience to a wide range of threats. It is important that this biological, ecological and taxonomic research continues, but in light of the increasing awareness of their imperilled status it is imperative that more conservation-specific research questions are addressed. In particular, research needs to be initiated into their thermal tolerance; susceptibility to increased dryness and reduced flow; susceptibility to exotic species, such as the cane toad, *Bufo marinus*; susceptibility to land use management practices; and population genetics, particularly for those species that extend across different drainage systems with fragmented distributions, which in fact may be distinct taxa.

The IUCN Red List conservation categories effectively prioritise research needs, in that species in higher threat categories are of the most immediate concern. However, as many of these *Euastacus* are exceptionally rare, it would be inappropriate, and in some cases unfeasible, to obtain data on relevant research questions for such species (Horwitz, 1995). In some cases, therefore, suitable proxy species for research need to be identified. For example, research into thermal tolerance for highland *Euastacus* may need to be undertaken on suitably common candidate species with appropriate altitudinal ranges (e.g., *E. sulcatus*, or *Euastacus suttoni* Clark, 1941) and interpreted for the genus generally. In addition to specific research questions, it is imperative that basic field surveys and population monitoring efforts continue, and that such efforts are assigned appropriate funding.

There has been nothing documented on the status of some species in the wild since their original discovery, up to 20 years in some cases (e.g., *Euastacus bindal* Morgan, 1989, in far north Queensland), a situation that is clearly inadequate given the highly imperilled status of this group. In addition, it is worth noting that many species of high conservation concern in our assessment are comparatively small species with low fecundity (e.g., clutch size of 20-150 eggs in *E. gumar*, Coughran, in press). These smaller species have also been largely overlooked in research efforts, and this should therefore be addressed in assigning future research priorities, so that important life history traits such as fecundity and growth rates can be better factored into future conservation assessments or initiatives for these species.

All of the species evaluated as belonging in threat categories in this review should be nominated for listing under relevant State conservation legislation. Furthermore, other species that are not threatened across their entire range may nonetheless satisfy criteria for listing at the State level. This process will likely require assessment and nomination on a case-by-case basis, but is of fundamental importance in the conservation of these animals. It is also essential that fishery and forestry management agencies, and other stakeholders, are aware of the distribution, habitat and conservation status of these species, such that they can be factored into any management plans and regulations, and that the existing and specific management recommendations are adopted.

## *Euastacus*: a sentinel genus?

Considering, a) the biology and very specific habitat requirements of this genus, b) its wide distribution (both latitudinal and altitudinal), and, c) its presence in most Australian climatic zones and habitat types, we consider this genus will likely be the first to show evidence of population and/or distributional contractions due to the various known and anticipated threats. In particular this genus will likely be the first in Australia broadly impacted by the direct and indirect effects of climate change, and may thus act as an early warning indicator. Specifically, we expect that the threatened and restricted-range species isolated on high altitude mountain-top "refugia" (particularly in far north Queensland and the Alpine zone) will likely be the first to be impacted by the anticipated direct (i.e., atmospheric warming, drying, altered hydrological regimes, and increased incidence of severe weather events) and indirect (i.e., changes in vegetation communities) effects of climate change. Therefore, these isolated populations should be monitored closely for any fluctuations or reductions in distribution and abundance, as they might provide the first early indications of a deteriorating situation for the genus in general, and for other biota in Australia and perhaps elsewhere in the World.


## CONCLUDING REMARKS

*Euastacus* is Australia's largest freshwater crayfish genus and contains some of the Worlds' most charismatic species. However, although it has been recognised for two decades that these crayfish are directly and appreciably impacted by a range of serious threats, little or nothing has been done to address or rectify this situation. This present assessment is not encouraging, with 80% of *Euastacus* threatened, and 70% of the genus facing serious risk of extinction. If the inaction continues, we can expect a major extinction event in this genus over years-to-decadal timeframes. In order to ensure conservation of these crayfish the documented recommendations for conservation and management for species in this genus need to be adopted and implemented immediately, and the chronic shortage of funding for basic (and inexpensive) research needs to be addressed.


## ACKNOWLEDGEMENTS

Generous assistance with travel costs was extended us by the organising committee and sponsors, which enabled us to present this work at the "Conservation Biology of Freshwater Crayfishes" Symposium at The Crustacean Society Summer Meeting in Ikebukuro, Tokyo (September 2009). Particular thanks are due to Tadashi Kawai, Yakichi Kobayashi and Hirotomo Kumaki for their considerable assistance in facilitating our attendance, and warm hospitality. Many thanks to Alistair Richardson, Todd Walsh and Robert McCormack who kindly provided crayfish images used in our conference presentation. We appreciate the efforts of Kathryn Dawkins (Griffith School of Environment) who kindly volunteered to proof-read all papers in this series. We thank our anonymous reviewers for helpful comments on earlier versions of all three papers in this conservation revision. We extend many, and sincere, thanks to the Editor in Chief Akira Asakura for his patience and assistance during the production of these three papers. These papers are the product of over 20 cumulative years of study by the authors, and we acknowledge the support of many people over this time, in particular our Ph.D. Supervisors, Professors Don F. Gartside (JC) and Clyde H. Wild (JMF), who allowed us to do our thing, and encouraged us to ask the really hard questions. This research project was supported by the Environmental Futures Centre, and the Griffith School of Environment, Griffith University, Gold Coast, Australia.

# REFERENCES

ABS, 2009. Regional population growth, Australia, 2007-08. (Report Number 3218.0, The Australian Bureau of Statistics, Canberra, ACT, Australia). Available online via: http://www.abs.gov.au/AUSSTATS/abs@.nsf/DetailsPage/3218.02007-08?OpenDocument

COUGHRAN, J., 2000. The distribution, habitat and conservation status of *Euastacus gumar* (Decapoda: Parastacidae), in northeastern New South Wales. (BAppSc (Hons) Thesis, School of Resource Science & Management, Southern Cross University, Lismore).

— —, 2006. Biology of the freshwater crayfishes of Northeastern New South Wales, Australia. (Ph.D. Thesis, School of Environmental Science and Management, Southern Cross University, Lismore).

— —, in press. Biology of the Blood crayfish, *Euastacus gumar* Morgan 1997, a small freshwater crayfish from the Richmond Range, northeastern New South Wales. Australian Zoologist.

COUGHRAN, J. & J. M. FURSE, 2010. An assessment of genus *Euastacus* (49 species) *versus* IUCN Red List criteria. Report prepared for the global species conservation assessment of crayfishes for the IUCN Red List of Threatened Species: 1-170. (The International Association of Astacology, Auburn, Alabama, USA. ISBN: 978-0-9805452-1-0).

COUGHRAN, J., R. B. McCORMACK & G. DALY, 2009. Translocation of the Yabby *Cherax destructor* into eastern drainages of New South Wales, Australia. Australian Zoologist, **35**(1): 100-103.

FURSE, J. M., 1999. *Euastacus sulcatus* the Lamington Spiny Crayfish (Crustacea: Decapoda: Parastacidae). (B.Sc. (Honours) Thesis, School of Environmental and Applied Sciences, Griffith University. Gold Coast, Queensland, Australia).

FURSE, J. M. & J. COUGHRAN, 2011a. An assessment of the distribution, biology, threatening processes and conservation status of the freshwater crayfish, genus *Euastacus* (Decapoda: Parastacidae), in continental Australia. I. Biological Background and Current Status. In: A. ASAKURA (ed.), New frontiers in crustacean biology: Proceedings of the TCS Summer Meeting, Tokyo, 20-24 September 2009. Crustaceana Monographs, **15**: 241-252.

— — & — —, 2011b. An assessment of the distribution, biology, threatening processes and conservation status of the freshwater crayfish, genus *Euastacus* (Decapoda: Parastacidae), in continental Australia. II. Threats, conservation assessments and key findings. In: A. ASAKURA (ed.), New frontiers in crustacean biology: Proceedings of the TCS Summer Meeting, Tokyo, 20-24 September 2009. Crustaceana Monographs, **15**: 253-263.

FURSE, J. M. & C. H. WILD, 2002. Prediction of crayfish density from environmental factors for *Euastacus sulcatus* (Crustacea: Decapoda: Parastacidae). Freshwater Crayfish, **13**: 316-329.

— — & — —, 2004. Laboratory moult increment, frequency, and growth in *Euastacus sulcatus*, the Lamington Spiny Crayfish. Freshwater Crayfish, **14**: 205-211.

FURSE, J. M., C. H. WILD & N. N. VILLAMAR, 2004. In-stream and terrestrial movements of *Euastacus sulcatus* in the Gold Coast hinterland: developing and testing a method of accessing freshwater crayfish movements. Freshwater Crayfish, **14**: 213-220.

GILLIGAN, D., R. ROLLS, J. MERRICK, M. LINTERMANS, P. DUNCAN & J. KOHEN, 2007. Scoping the knowledge requirements for Murray crayfish (*Euastacus armatus*). New South Wales Department of Primary Industries, Narrandera, New South Wales, Australia. Fisheries Final Report Series, **89**: 1-107.

HORWITZ, P., 1995. The conservation status of Australian freshwater crayfish: review and update. Freshwater Crayfish, **10**: 70-80.

MERRICK, J. R., 1997. Conservation and field management of the freshwater crayfish, *Euastacus spinifer* (Decapoda: Parastacidae), in the Sydney region, Australia. Proc. Linn. Soc. N.S.W., **118**: 217-255.

MORGAN, G. J., 1986. Freshwater crayfish of the genus *Euastacus* Clark (Decapoda, Parastacidae) from Victoria. Memoirs of the Museum of Victoria, **47**(1): 1-57.

— —, 1988. Freshwater crayfish of the genus *Euastacus* Clark (Decapoda: Parastacidae) from Queensland. Memoirs of the Museum of Victoria, **49**(1): 1-49.

TURVEY, P. & J. R. MERRICK, 1997a. Diet and feeding in the freshwater crayfish, *Euastacus spinifer* (Decapoda: Parastacidae), from the Sydney region, Australia. Proc. Linn. Soc. N.S.W., **118**: 175-185.

— — & — —, 1997b. Growth with age in the freshwater crayfish, *Euastacus spinifer* (Decapoda: Parastacidae), from the Sydney region, Australia. Proc. Linn. Soc. N.S.W., **118**: 205-215.

— — & — —, 1997c. Moult increments and frequency in the freshwater crayfish, *Euastacus spinifer* (Decapoda: Parastacidae), from the Sydney region, Australia. Proc. Linn. Soc. N.S.W., **118**: 188-204.

— — & — —, 1997d. Population structure of the freshwater crayfish *Euastacus spinifer*, (Decapoda: Parastacidae), from the Sydney region, Australia. Proc. Linn. Soc. N.S.W., **118**: 157-174.

— — & — —, 1997e. Reproductive biology of the freshwater crayfish, *Euastacus spinifer* (Decapoda: Parastacidae), from the Sydney region, Australia. Proc. Linn. Soc. N.S.W., **118**: 131-155.

WILD, C. H. & J. M. FURSE, 2004. The relationship between *Euastacus sulcatus* and Temnocephalan spp. (Platyhelminthes) in the Gold Coast hinterland, Queensland. Freshwater Crayfish, **14**: 236-245.

First received 1 December 2009.
Final version accepted 30 January 2010.

# SUPPLEMENTAL INFORMATION ON THE TAXONOMY, SYNONYMY, AND DISTRIBUTION OF *CAMBAROIDES JAPONICUS* (DECAPODA: CAMBARIDAE)

BY


TADASHI KAWAI[1,3]) and VJACHESLAV S. LABAY[2])

[1]) Hokkaido Wakkanai Fisheries Experiment Station, 4-5-15 Suehiro, Wakkanai, Hokkaido 097-0001, Japan

[2]) Sakhalin Research Institute of Fishery and Oceanography, Yuzhno-Sakhalinsk, 693016, Russia


## ABSTRACT


The Japanese endangered freshwater crayfish *Cambaroides japonicus* (De Haan, 1841) is the only indigenous species in Japan. To supplement somewhat insufficient previous works based on dry specimens, *C. japonicus* is taxonomically discussed and illustrated on the basis of newly obtained specimens.


## INTRODUCTION

Taxonomic information on *C. japonicus* has remained insufficient since its original description based on the dry specimens of the type series (Fitzpatrick, 1995) and, subsequently, some fine specific structures of this species were not presented (Kawai & Fitzpatrick, 2004). The present paper provides a brief description and illustrations of this species based on new specimens preserved in alcohol. Further information regarding some problems left unsolved in the taxonomy of *C. japonicus* are discussed, including synonymies, distribution, author name of the taxon, publication dates, and morphological variations.

Fine specific structures were examined in detail and illustrations were prepared from 9 freshly-obtained specimens, collected from Sapporo, Hokkaido, Japan. Measurements of crayfish structures were made to the nearest 0.1 mm with a digital precision caliper and a stereomicroscope, following the methods

---

[3]) Corresponding author; e-mail: kawaita@fishexp.pref.hokkaido.jp

of Hobbs (1981). Abbreviations used in the text are: CBM, Natural History Museum and Institute, Chiba; POCL, post orbital carapace length; TCL, total carapace length; TL, total length.

**Cambaroides japonicus** (De Haan, 1841)
Japanese name: Zarigani or Nihon-zarigani
(figs. 1, 2)

*Astacus japonicus* De Haan, 1841: 164, XIII, fig. 9; Herklots, 1861: 31; Erichson, 1846: 86, 94. Gerstfeldt, 1857: 292-293; Herklots, 1861: 31; Hagen, 1870: 6; Kessler, 1874: 364-365; 1876: 286, 288; Huxley, 1880: 308; Stebbing, 1893: 208; Pierantoni, 1905: 1, 18-19; 1912: 9, 18-19.

*Astacus (Cambaroides) japonicus*: Faxon, 1884: 149; 1885: 126-129, 163; Koelbel, 1892: 651, figs. 1, 2, 4, 5, 7-11; Faxon, 1898: 665; Imaiziumi, 1938: 177; Hobbs, 1974: 17.

*Potamobius (Cambaroides) japonicus*: Ortman, 1902: 286.

*Cambaroides japonicus neglectus* Skorikov, 1907: 116; Entz, 1912: 80.

*Cambaroides japonicus*: Skorikov, 1907: 116; Okada, 1933: 155-15; Urita, 1942: 38-39; Kawai & Fitzpatrick Jr., 2004: 23, 25-26, 28-33.

Description. — Cephalothorax (fig. 1A) subcylindrical: thoracic section of carapace dorsoventrally depressed (maximum width 1.6 times its depth). Rostrum lacking spines or tubercles, tip extending distal margin of antennal scale and ultimate podomere of antennal peduncle, floor (dorsal surface) of rostrum convex, median carina absent, rostrum comprising 18.9% of TCL. Areola 1.9 times as long as wide, comprising 30.6% of TCL (36.7% of POCL). Suborbital ridge and post orbital ridge poorly defined dorsally, caudal end very weak without spines or tubercles.

Antennal scale (fig. 1E) 1.8 times as long as broad, widest at midlength, tip reaching midlength of ultimate podomere of antennular peduncle. Epistome (fig. 1C) with subovate cephalic lobe bearing cephalomedian projection, margin of lobe markedly elevated; fovea of epistome scarcely visible; zygome slightly curved.

Pleura (fig. 1B) at second to fifth abdominal somites rounded. Mesial ramus of uropoda with submedian dorsal ridge, terminating in moderate median spine, barely reaching dorsocaudal edge. Telson with single fixed, corneous spine at each lateral corner; lateral margin slightly tapering to rounded caudal margin, lacking telson notch.

Palm of chela (fig. 1D) with scattered large punctuations on dorsal, lateral, mesial, and ventral surfaces; palm inflated, 1.4 times wider than deep, length of mesial margin 32.3% of lateral margin of chela length; fixed finger with single longitudinal ridge on dorsal surface; opposable surface of fixed finger with single row of 12 tubercles. Length of dactylus 1.8 times length of mesial

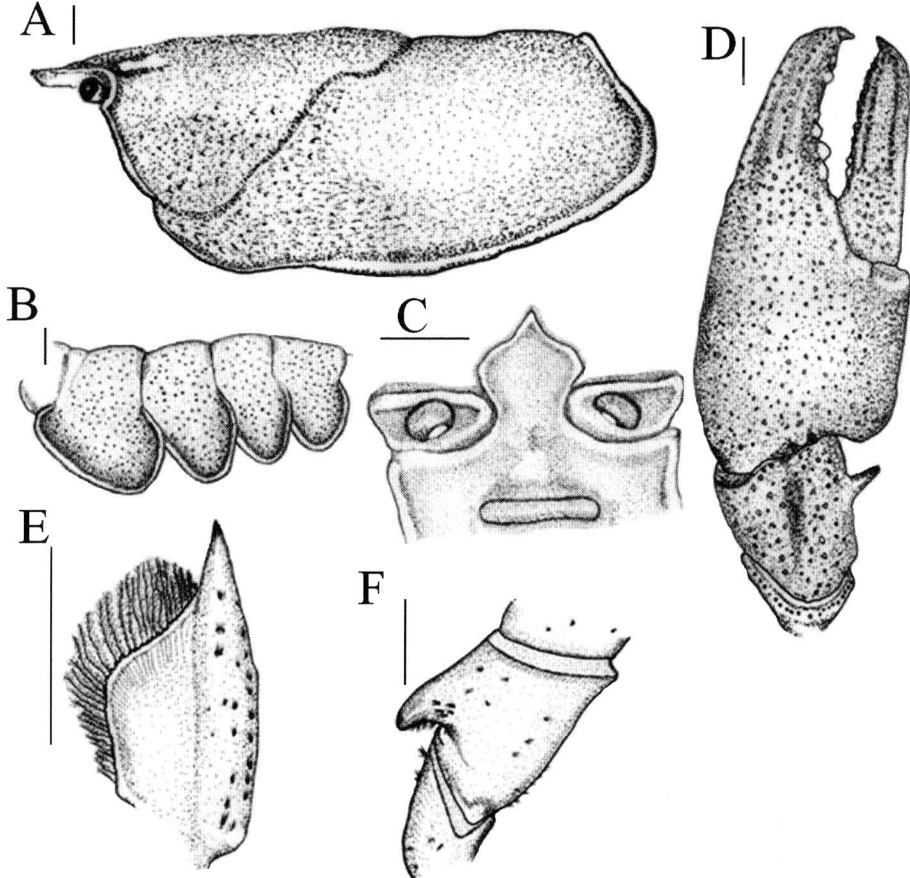

Fig. 1. *Cambaroides japonicus* (De Haan, 1841) male (CBM-2C 10183): A, lateral view of carapace; B, lateral view of second to fifth abdominal segments; C, epistome; D, dorsal view of distal podomeres of cheliped; E, antennal scale; F, ischium of third pereopod. Scales = 2 mm.

margin of palm, dactylus with weak longitudinal ridge dorsally, opposable surface with single row of 10 tubercles.

Carpus of cheliped 1.4 times as long as wide, 1.1 times length of mesial margin of palm; dorsal surface with shallow, wide, slightly oblique sulcus. Merus of cheliped 2.1 times longer than its greatest depth, length 36.9% of TCL (44.2% of TCL), dorsal surface with single subdistal spine. Hooks (fig. 1F) on ischia of second and third pereopods rounded, extending beyond basioischical articulation.

Gonopods (fig. 2C) symmetrical, bases not contiguous, tip extending to edge of basis of second pereopod when abdomen flexed. In ventral aspect, apex directed cephalodistally at about 45° to axis of shaft. Adult male

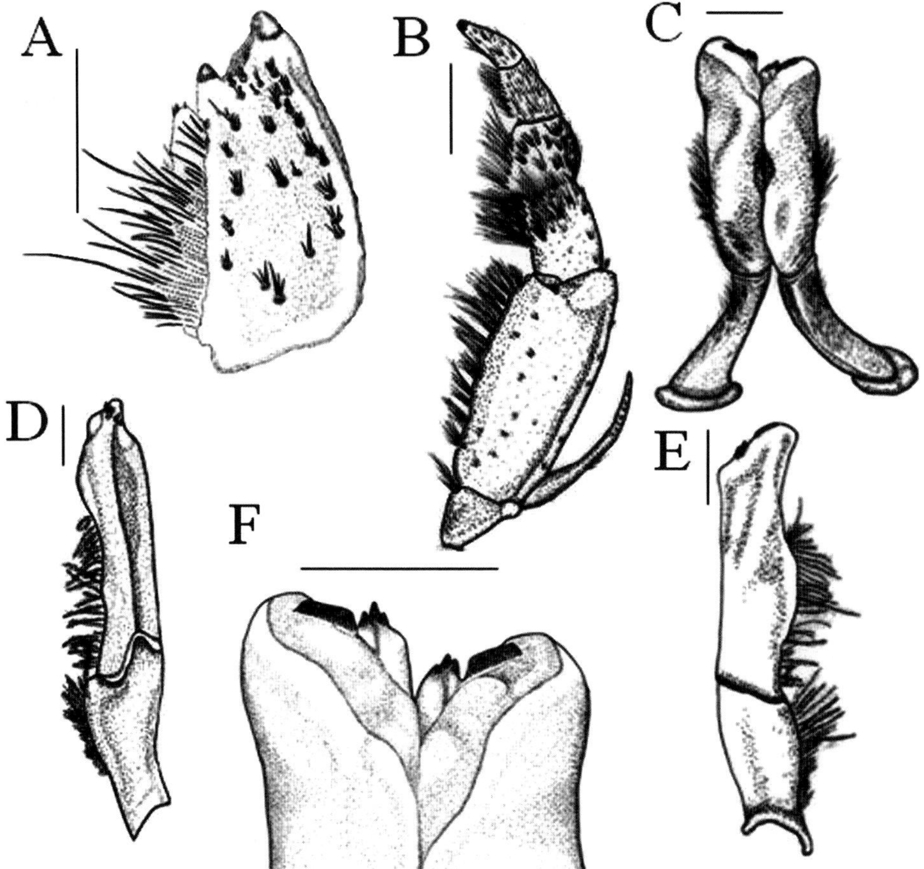

Fig. 2. *Cambaroides japonicus* (De Haan, 1841) male (CBM-2C 10183): A, merus of third maxilliped; B, third maxilliped; C, caudal view of in situ gonopods; D, mesial view of gonopod (first pleopod); E, lateral view of gonopod; F, ventral view of gonopod. Scales = 2 mm.

gonopods (fig. 2D, E) with "juvenile suture", sperm groove along mesial face of endopodite shallow and narrow, open between mesial process and central projection, ending in relatively blunt tips. Apex of gonopod (fig. 2F) sclerotized with more corneous distal portion; lateral portion swollen; cephalic margin with three straight, subacute, stout spines of subequal length directed cephalodistally, and mesial, centrocaudal, and centrocephalic processes, length of three processes about one-tenth of apex; blade like caudal process at midwidth on cephalodistal margin of apex, width about one-three times width of apex. Total length of gonopod 33.1% of TCL (39.6% of POCL).

Annulus ventralis (based on female #1) immovable, symmetrical, rounded in outline, about 1.7 times as long as wide; preannular plates transversely subdivided into subtriangular plates, cephalic margin of anterior portion broadly attached to antecedent sternite and slightly depressed middle section of posterior portion with very shallow fossa without sinus; postannular sclerite subtriangular about two-thirds of annular plate.

Third maxilliped (fig. 2B) with tip of endopodite reaching to distal margin of penultimate podomere of antennal peduncle; basal podomere of exopodite very hirsute, tip extending beyond half of ischium of endopodite; caudolateral corner of merus (fig. 2A) with corneous, single tubercle.

Measurements of specimens are provided in table I.

Specimens examined and geographical range. — Kawai & Fitzpatrick (2004) showed the known range of *Cambaroides japonicus*. The species is

TABLE I

Measurements (mm) of freshly-obtained material of *Cambaroides japonicus* (De Haan)

|  | Male #1 | Female #1 |
| --- | --- | --- |
| Carapace |  |  |
|   Total length | 31.7 | 29.3 |
|   Postorbital length | 26.5 | 24.0 |
|   Length cephalic section | 21.8 | 19.9 |
|   Width | 16.3 | 15.2 |
|   Depth | 10.3 | 10.2 |
|   Length rostrum | 6.0 | 6.3 |
|   Length areola | 9.7 | 8.2 |
|   Width areola | 5.2 | 4.8 |
| Antennal scale |  |  |
|   Length | 4.9 | 3.7 |
|   Width | 2.8 | 1.8 |
| Abdomen |  |  |
|   Length | 32.8 | 30.5 |
|   Width | 14.6 | 15.3 |
| Cheliped |  |  |
|   Length lateral margin chela | 25.4[*] | 20.5 |
|   Length mesial margin chela | 8.2[*] | 7.1 |
|   Width palm | 9.8[*] | 9.9 |
|   Depth palm | 6.8[*] | 6.7 |
|   Length dactylus | 14.7[*] | 11.5 |
|   Length carpus | 8.9[*] | 8.3 |
|   Width carpus | 6.5[*] | 5.6 |
|   Length dorsal margin merus | 11.7[*] | 11.1 |
|   Greatest depth merus | 5.5[*] | 5.7 |
| Gonopod length | 10.5 | N/A |

[*] Left chela (right regenerated).

endemic to the northern part of Japan (Hokkaido and Aomori Prefecture), and more southern records (Akita and Iwate Prefectures) may originate from introductions.

In addition to this, the present study revealed its occurrence at a further 30 localities. Voucher specimens (N = 192), all preserved at the Hirosaki University Museum, were collected at the following localities by the senior author (TK): (1) Sakae, Shimamaki, 16 June 2002, 9M (Male), 6F (Female), 2 ovig. F; (2) Sandomari, Rumoi, 29 July 2002, 3M, 3F; (3) Rarumanai, Eniwa, 18 May 2002, 1M; (4) Shuuparo, Yuubari, 12 Sep. 2002, 1M; (5) Tobetsu, Makubetsu, 23 Sep. 2002, 2M; (6) Nishibetsu, Shibecha, 10 Nov. 2002, 1M; (7) Yagi, Minamikayabe, 20 Oct. 2002, 1M, 1F; (8) Karumai, Atsuma, 4 Nov. 2002, 1M; (9) Rikibiru, Tomomae, 29 July 2002, 10M, 8F; (10) Biya, Setana, 26 Aug. 2004, 13M, 14F; (11) Aidomari, Taisei, 27 Aug. 2004, 17M, 13F; (12) Seseki, Rausu, 19 July 2005, 2M, 1F; (13) Satomi, Shintotsukawa, 5 Nov. 2004, 3M, 2F; (14) Asahi, Rubeshibe, 4 July 2005, 2M; (15) Fukuyama, Tokoro, 5 July 2005, 2F; (16) Oketo, Oketo, 4 July 2005, 2M, 1F; (17) Rasa, Shiranuka, 7 Aug. 2006, 7M, 5F; (18) Bekkai, Bekkai, 12 Aug. 2001, 1M, 4F; (19) Nopporo, Ebetsu, 11 June 2003, 5M, 4F; (20) Reuke, Rumoi, 22 July 2003, 6M, 5F; (21) Oniwaki, Obira, 25 July 2003, 7M, 5F; Apoi, Samani, 12 Oct. 2003, 2M; (22) Toikanbetsu, Horonobe, 19 June 2008, 1M; (23) Kawahigashi, Kitami, 19 May 2008, 1M, 1F; (24) Kamiotoshibe, Esashi, 14 June 2008, 2M, 2F; (25) Chiraibetsu, Sarufutsu, 28 June 2008, 1M, 1F; (26) Utanobori, Esashi, 30 Aug. 2008, 1M, 1F; (27) Hiraganai, Nakatonbetsu, 30 Aug. 2008, 1M, 1F; (28) Higashiyama, Asahikawa, 6 Oct. 2008, 2M, 1F; (29) Ohtani, Kunneppu, 6 Oct. 2008, 1M, 1F; (30) Touei, Shiranuka, 5 May 2009, 1M, 3F, 1 ovig. All crayfish specimens were captured by hand. In addition, 6M and 3F collected by S. Hori, from an unnamed brook at Maruyama, Sapporo, Hokkaido, are also held at the Hirosaki University Museum.

Okada (1933) observed specimens of *C. japonicus* collected from South Sakhalin, Russia, and these specimens are apparently deposited in the University Museum, the University of Tokyo. Urita (1942) thoroughly searched South Sakhalin for *C. japonicus*, however he did not obtain any specimens. The senior author (TK) visited The University Museum, University of Tokyo in 1998, 1999, 2000 to confirm the presence of *C. japonicus* in South Sakhalin, but no specimens of this species from Sakhalin could be found. The junior author (VSL) reviewed numerous Russian papers reporting cambaroid species in Sakhalin State (Sakhalin Island and Kuril Islands, Russia) (Labay, 2005, 2007); nevertheless there was no information on *C. japonicus* from Sakhalin

State. Considering these evidences we propose that, at present, *C. japonicus* could be regarded as an endemic species of Hokkaido and Northern Honshu in the Japanese Archipelago.

Recent surveys conducted in at least 10 brooks in the Noshappu neck of land, on the northern edge of Hokkaido (unpublished data), have shown that *C. japonicus* obligatorily cohabits with the amphipod *Sternomoera yezoensis* Ueno, 1933. The Noshappu Peninsula of Hokkaido extends close to the Krilon Peninsula of Southern Sakhalin, separated by a distance of approximately 40 km. Recently *S. yezoensis* have been found in several streams of southern Sakhalin (Labay, 2003) and although we have not collected *C. japonicus* to date the species may be present in other, as yet unsampled streams of South Sakhalin.

Remarks. — Kessler (1874, 1876) pointed out that *Cambaroides japonicus* collected from Hakodate, Hokkaido, Japan, has a complete excavation on the caudal margin of the telson and it differs from many other specimens of *C. japonicus*, and that the specimens collected from Hakodate would be a new species or variation. Skorikov (1907) mentioned that "I propose temporally only a new name *Cambaroides neglectus* for those having a excavation on the caudal margin of the telson. The telson of *C. japonicus* shown in the original illustration shows to have a excavation, but I could not find such a feature among many Japanese specimens available to me. I am reluctant to establish a new species, instead I propose a new name *C. neglectus* for those with caudal excavation".

As examined to date, the major diagnostic characters (carapace, chela, first pleopod, and annulus ventralis) of *C. japonicus* are basically consistent throughout its distributional range. In contrast, an excavation of the caudal margin of the telson differs substantially between specimens from Hokkaido, a remote island in northern Japan, and those from Aomori Prefecture, Honshu mainland of Japan. Based on observations of the type series and many other samples collected from Japan, Kawai & Fitzpatrick (2004) suggested that *C. japonicus* shows clear geographical variations in morphology of the telson, but that the specimens from Hokkaido and Aomori Prefecture nevertheless belong to the same species. Molecular analysis based on the16S rRNA supported their conclusion (Ahn et al., 2006) and, consequently, we consider the temporal name applied by Skorikov (1907), *C. neglectus*, is no longer valid.

*Cambaroides japonicus* was described in the 'Fauna Japonica' written by the Nederland taxonomist, De Haan of the Leiden Museum (1801-1855). The 'Fauna Japonica' was divided into some fascicles. The De Haan's first fascicle was issued in 1833 and the last supplementary one provided by von Siebold was published in 1850. However, each fascicle had only the indication of the approximate year of issue without an accurate publication date. Faxon (1885: 128) mentioned that the description of *C. japonicus* was published in 1842, whereas Holthuis (1993) showed it to be 1841. Thus, the exact year of description of *C. japonicus* still remained unclear. Holthuis and Sakai (1970) made an effort to learn the exact date of publication for each fascicle for the Crustacean volume (Yamaguchi, 1993). The Department of Internal Affairs in Hague holds a list of the dates on which the installments were sent out. Instead of giving a special grant for the publication, the government of the Netherlands subscribed 10 copies of the Fauna Japonica, which were distributed to several libraries in the country. These lists may provide the most important clue to track down the dates of issue of the publication in question, although the date of distribution recorded in the list might have some time lag and may be later than the actual date of publication. In 1993, Holthuis obtained more accurate information regarding these dates of distribution. The Crustacea volume, in which *C. japonicus* included, was delivered on 28 December 1841. Consequently, we proposed that the year 1841 should be applied as the publication date for the description of *C. japonicus*.

Holthuis and Yamaguchi both used "De Haan" instead of "de Haan", as the latter is the common usage for that name in Nederland (Yamaguchi, 1993). The lectotype of this species was designated in Yamaguchi and Baba (1993: 233, fig. 53b).

## ACKNOWLEDGEMENTS

The authors thank to Dr. Y. Hanamura, Dr. A. Asakura, late Professor J. F. Fitzpatrick, Jr., Mr. Y. Machino, and Professor T. Yamaguchi for their help.

## REFERENCES

AHN, D. H., T. KAWAI, S. J. KIM, H. S. RHO, J. W. JUNG, W. KIM, B. J. LIM, M. S. KIM & G. S. MIN, 2006. Phylogeny of northern hemisphere freshwater crayfishes based on 16S rRNA gene analysis. Korean Journal of Genetics, **28**(2): 185-192.

ENTZ, G., JR., 1914. Über die Flußkrebse Ungarans. Mathemathische und Naturwissenschaftliche Berichte aus Ungarn, Leipzig, **30**: 67-127, 4 plates.

ERICHSON, W. F., 1846. Uebersicht der Arten der Gattung *Astacus*. Archiv für Naturgeschichte, Berlin, **12**: 86-103.

FAXON, W., 1884. Descriptions of new species of *Cambarus*; to which is added a synonymical list of the known species of *Cambarus* and *Astacus*. Proceedings of the American Academy of Arts and Science, **20**: 107-158.

— —, 1885. Revision of the Astacidae. Part 1. The genera *Cambarus* and *Astacus*. Memories of the Museum of Comparative Zoology at Harvard College, **10**: i-vi, 1-186, pls. 1-10.

— —, 1898. Observations on the Astacidae in the United States National Museum and in the Museum of Comparative Zoology, with descriptions of new species. Proceeding of the United States National Museum, **20**: 643-694, pls. 62-70.

FITZPATRICK, J. F., JR., 1995. The Eurasian far-eastern crawfishes: a preliminary overview. Freshwater Crayfish, **8**: 1-11.

GERSTFELDT, G., 1858. Ueber einige zum Thiel neue Arten Platoden, Anneliden, Myriapoden und Crustacea Sibirien's, namentlich series östlichen Theiles und des Amur-Gebietes. Mémores présentés à l'Académie Impérial des Sciences de Saint-Pétersbourg par divers savants et lus dans ses assemblées, (6) **8**: 259-296, 1 plate. (Saint-Pétersbourg).

HAAN, W. DE, 1841. Crustacea. In: PH. F. VON SIEBOLD (1833-1850), Fauna Japonica sive descriptio animalium, quate in itinere per Japoniam, jussu et auspiciis superiorum, qui summum in India Batava Imperium tenent, suscepto, annis 1823-1830 collegit, nois, observationibus et adumbrationibus illustravit (Crustacea): i-xvii, i-xxxi, ix-xvi, 1-243, pls. A-J, L-O, 1-55, circ. tab. 2.

HAGEN, H. A., 1870. Monograph of the North American Astacidae. Illustrated Catalogue of the Museum of Comparative Zoology at Harvard College, **3**: 1-109, pls. 1-11.

HERKLOTS, J. A., 1861. Symbolae carcinologicae. Études sur la classe des Crustacés. I. Catalogue des Crustacés qui ont servi de base au système carcinologique de M. W. De Haan, rédige d'après la collection du Musée des Pays-Bas et les Crustacés de la faune du Japon. Tijdschrift voor Entomologie, **4**: 116-156.

HOBBS, H. H., JR., 1974. Synopsis of the family and genera of crayfishes (Crustacea: Decapoda). Smithsonian Contributions to Zoology, **164**: i-iv, 1-32.

— —, 1981. The crayfishes of Georgia. Smithsonian Contributions to Zoology, **318**: 1-549.

HOLTHUIS, L. B., 1993. The three Dutch authors of von Siebold's Fauna Japanica, with some notes on the artist. In: T. YAMAGUCHI (ed.), Ph. F. von Siebold and natural history of Japan. Crustacea: 689-708, figs. 1-5. (The Carcinological Society of Japan, Tokyo).

HOLTHUIS, L. B. & T. SAKAI, 1970. Ph. F. Von Siebold and Fauna Japonica. A history of early Japanese zoology: 1-323, pls. 1-32, 7 unnumbered pls., 1 map. (Academic Press of Japan, Tokyo).

HUXLEY, T. H., 1880. The crayfish, an introduction to the study of zoology: 1-371. (Kegan Paul & Co., London). (Same pagination at fifth edition in 1889 and sixth edition in 1896.)

IMAIZUMI, R., 1938. Fossil crayfishes from Jehol. Science Reports of the Tohoku Imperial University, Sendai, Japan, Second Series, (Geology) **19**: 173-178, pls. 22, 33.

KAWAI, T. & J. F. FITZPATRICK, JR., 2004. Redescription of *Cambaroides japonicus* (De Haan, 1841) (Crustacea: Decapoda: Cambaridae) with allocation of a type locality and month of collection of types. Proceedings of the Biological Society of Washington, **117**: 23-34.

KESSLER, K., 1874. Die russischen Flusskrebse (vorläufige Mittheilung). Byulleten' Imperatorskogo Moskovskogo Obshchestva Ispytatelei Prirody, **48**: 343-372.

— —, 1876. Russkie rechnye raki. Turdy Russkogo Entomologicheskogo Obshchestva, **8**: 228-320, 5 pls. 1-5.

KOELBEL, K., 1892. Ein neuer ostasiatischer Flusskrebs. Aus den Sitzungsberichten der kaiserl, Akademie der Wissenschaften in Wien, Mathematisch-Naturw Classe, Bd. CI. Abth. 1, **101**: 650-656, pl. 1.

OKADA, Y., 1933. Some observations of Japanese crayfishes. Science Report of Tokyo Bunrika Daigaku, (Biology) **1**(15): 155-158.

ORTMANN, A. E., 1902. The geographical distribution of freshwater Decapods and its bearing upon ancient geography. Proceedings of the American Philosophical Society, **41**: 267-400.

PIERANTONI, U., 1905. *Cirrodrilus cirratus* n. g. n. sp. Parassita dell' *Astacus japonicus*. Annuario del Museo Zoologico Della R. Università di Napoli, (new series) **1**: 1-28.

— —, 1912. Monografia dei Discodrilidae. Annuario del Museo Zoologico Della R. Università di Napoli, (new series) **3**: 1-28.

LABAY, V. S., 2003. *Srenomoera yezoensis* Ueno, 1933 (Crustacea, Amphipoda, Eusiridae) a new species for Russia from fresh waters of the Southern Sakhalin Island. Transactions of Sakhalin Research Institute of Fisheries and Oceanography. Yuzhno-Sakhalinsk, SakhNIRO, **5**: 1-282. [In Russian.]

— —, 2005. Fauna of the Malacostraca (Crustacea) from the fresh and brackish water of Sakhalin Island. In: Flora and fauna of Sakhalin Island (Material of International Sakhalin Island Project). Part 2: 64-87. (Dalnauka, Vladivostok). [In Russia with English summary.]

— —, 2007. Sakhalin river crayfish. In: N. I. ONISCHENKO, V. A. NECHAEV, G. A. VORONOV, S. N. SAFRONOV, V. S. LABAY & Z. V. REVJAKINA (eds.), Red data Sakhalin region: 178-179. (The Sakhalin book publishing house, Yuzhno-Sakhalinsk).

SKORIKOV, A. S., 1907. K sistematike evropeisko-aziatskikh Potamobiidae. Ezhegodnik Zoologischeskogo Muzeya Imperatorskoi Akademii Nauk, **12**: 115-118.

STEBBING, T. R. R., 1893. A history of crustacea, recent malacostraca (The international scientific series): 1-208. (D. Appleton and Company, New York).

URITA, T., 1942. Decapoda crustaceans from Saghalien, Japan. Biographical Society of Japan, **12**(1): 1-78.

YAMAGUCHI, T., 1993. The contributions of von Siebold and H. Bürger to the natural history of Japanese Crustacea. In: T. YAMAGUCHI (ed.), Ph. F. von Siebold and natural history of Japan. Crustacea: 15-44. (The Carcinological Society of Japan, Tokyo).

First received 26 December 2009.
Final version accepted 20 January 2010.

# THE EVOLUTION OF SOCIALITY: PERACARID CRUSTACEANS AS MODEL ORGANISMS

BY

MARTIN THIEL[1])

Facultad Ciencias del Mar, Universidad Católica del Norte, Larrondo 1281, Coquimbo, Chile;
Centro de Estudios Avanzados en Zonas Áridas (CEAZA), Coquimbo, Chile

## ABSTRACT

Sociality in peracarids has evolved via two principal pathways, the parasocial road (aggregations of individuals) and the subsocial road (parents caring for offspring), but the factors governing the evolution of group-living are not well understood. Knowing the benefits and costs of group-living is key to understand the evolution of sociality. Herein, I present examples of experimental studies that examined the benefits and costs of group-living in conspecific aggregations (COGs) and parent–offspring groups (POGs). These studies show that peracarids lend themselves to experimental studies. Furthermore, there is a bias towards studies exploring the benefits and few studies have explicitly tested the costs of group-living. Several studies confirmed that individuals in groups are better protected from environmental hazards (stressful conditions and predators) than solitary individuals in both COGs and POGs. However, in very large groups, a tradeoff emerges and individual benefits decrease. Present evidence suggests that increasing resource competition in larger groups may be responsible for this tradeoff. However, other factors (cannibalism and disease/parasite transmission) may also contribute to the increasing costs in larger groups. At present, it is not clear whether and how tradeoffs differ between COGs and POGs, which appears essential to better understand the evolution of sociality in peracarid crustaceans.

## INTRODUCTION

The evolution of group living is influenced by the tradeoff between the benefits that individuals gain from cohabiting with conspecifics and the costs they pay when living with others. These benefits and costs faced by group-living organisms depend both on intrinsic (life history) and extrinsic factors (environmental conditions). There is also increasing evidence that ecological

---

[1]) e-mail: thiel@ucn.cl

    

factors play a major role in the evolution of group-living (Korb, 2008). For example, in many species individuals gathered in groups are better able than single individuals to cope with high abiotic stress or strong predation pressure. High abundance and quality of food resources might also favor group-living, because the capability of individuals to exploit and monopolize food might increase in large groups (e.g., Whitehouse & Lubin, 2005). However, group-living also generates costs. For example, competition for resources may be higher in groups, and if resources become limited, this can result in increased aggression and/or reduced reproductive potential among group members. Also, parasites and diseases may rapidly spread within large groups (Schmid-Hempel, 1998). The tradeoff between benefits and costs determines the potential for social evolution, which is then mediated by the genetic relatedness among group members (Hamilton, 1964).

Many recent studies have focused on the conditions that led to the evolution of eusocial behavior (e.g., Foster et al., 2006; Boomsma & Franks, 2006). However, important insights on social evolution can also be gained from studying more primitive social organisms (e.g., Costa, 2006). Here it will be instructive to study species that inhabit a wide variety of environments with differing selective pressures (for importance of environmental conditions see, e.g., Schwarz et al., 2007; Korb, 2008). Organisms that occur in groups of different degrees of genetic relatedness will be particularly useful, because this allows incorporating the importance of kinship in the model.

The Peracarida (Crustacea) represent a taxon well suited to examine the conditions that govern the initial stages of social evolution (fig. 1). Members of the peracarids occur in a wide variety of environments from the deep sea to dry deserts, and kinship of peracarid groups varies along two extremes: (i) aggregations of unrelated conspecifics, and (ii) parent–offspring groups of closely related family members. These two types of groups also represent the two divergent pathways (parasocial and subsocial) on the evolutionary road towards higher social behavior. Conspecific aggregations (parasocial) have been observed in a variety of peracarid species from both terrestrial and aquatic environments. Members of these conspecific aggregations (COGs in the following) are not related to each other (non-kin) and there is usually no particular group structure and adhesion. In contrast, in parent–offspring groups (subsocial) closely related individuals (kin) cohabit for extended time periods and membership is exclusive, i.e., unrelated individuals are not admitted to these parent–offspring groups (POGs in the following). Since the species that follow either the subsocial or the parasocial pathways live in the same environments,

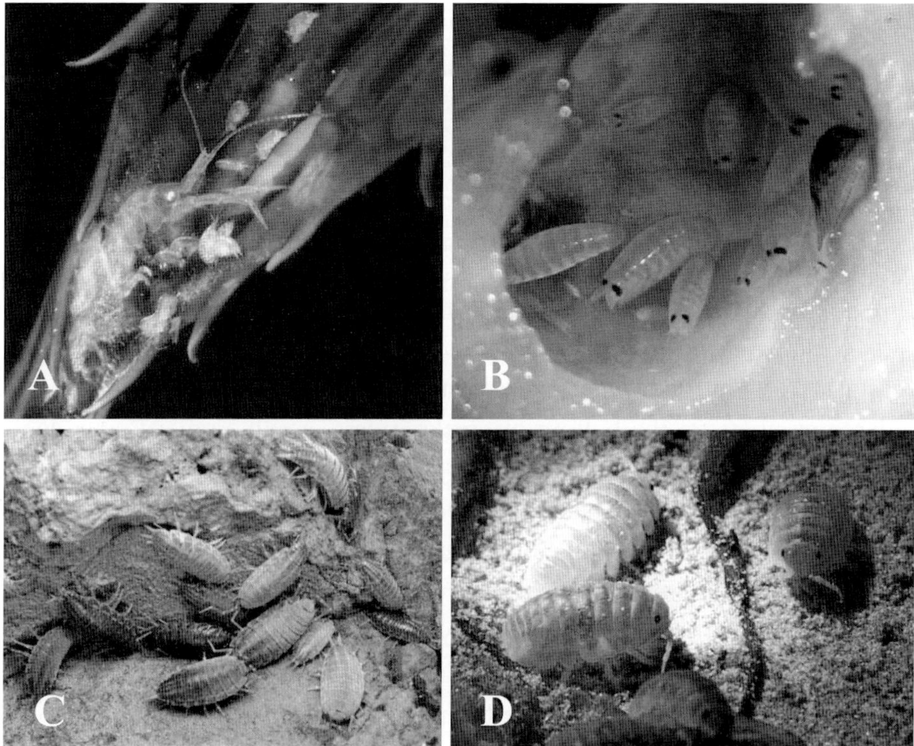

Fig. 1. Amphipods and Isopods in different forms of groups. A, *Peramphithoe femorata* female with offspring in her nest; B, *Parawaldeckia kidderi* in burrow excavated in kelp holdfasts; C, *Porcellio scaber* under rock; D, *Tylos spinulosus* aggregating on food algae in flotsam.

it can be expected that the selective pressures favoring/suppressing cohabitation are similar. However, benefits and costs differ, because group members are either genetically related (subsocial) or not (parasocial). Such differences are expected to have a strong impact on the social evolution of the respective species.

Herein, I will provide examples of peracarid species that live in COGs and POGs where the benefits/costs of group-living have been tested explicitly. Furthermore, I present reports of specific behaviors that might enhance benefits, reduce costs or facilitate group cohesion. For the purpose of this review, the Mysidacea were also included even though their phylogenetic position within the Peracarida is doubtful (Spears et al., 2005). However, mysids share a key life history trait with the true peracarids (direct development) that is known to have a strong influence on social evolution in invertebrates (Costa, 2006). In addition, numerous studies have explicitly tested the benefits and costs of group-living in pelagic mysids (e.g., Ritz, 1994).

## TYPES AND SIZES OF PERACARID GROUPS

Conspecific aggregations. — In the aquatic environment, a wide variety of peracarid aggregations is known. In coastal waters, mysid shoals range from tens to millions of individuals (depending on the species). These schools are usually monospecific, but within aggregations age classes often segregate (O'Brien, 1988; but see Modlin, 1990). Dense aggregations of hyperiid amphipods, swirling in tornado-like swarms (Lobel & Randall, 1986) or apparently cleaning sharks (Whitney & Motta, 2008), have also been reported from coastal waters. In freshwater, isopods and amphipods occasionally aggregate in large numbers on the bottom of creeks, rivers or subterranean cave pools (Allee, 1929; van den Brink et al., 1993; Fenolio & Graening, 2009). In marine soft-bottom habitats, some peracarid species reach extraordinary densities, frequently exceeding 10 000 individuals per $m^2$ (e.g., Drolet & Barbeau, 2009). Similarly, dense aggregations are also found in wood- or algal-dwelling peracarids, where $>10\,000$ individuals may inhabit a volume of $10^3$ cm substratum (Thiel, 2003a). On sandy beaches and on land, (semi-)terrestrial isopods and amphipods gather to the hundreds under algal or plant detritus (e.g., Friend & Richardson, 1986; Jaramillo et al., 2006). Occasionally, peracarid aggregations reach extraordinary densities of hundreds of thousands of individuals (Allee, 1929; Paoletti et al., 2008; Fenolio & Graening, 2009).

Parent–offspring groups. — Females are the nucleus of POGs in most peracarid species. Mothers (in rare cases accompanied by the male) typically continue to cohabit with their offspring after these emerge from the maternal brood pouch. POGs are most common in peracarids that inhabit stable dwellings, such as burrows in sediment and wood, self-constructed tubes, algal nests, or biotic microhabitats (sponges, bivalves, ascidians). In these species, juveniles remain with their parents for extended time periods, sometimes several months (Thiel, 2003b). In some free-living species, females also cohabit with their offspring, but usually for shorter periods (days–weeks). Parents caring for subsequent cohorts of juveniles are known from a few peracarid species, but at present no case of overlapping generations (offspring starting to reproduce in the presence of reproductive parents) has been reported (Thiel, 2007). Brood size in peracarids usually varies from 10-100 eggs, and consequently most POGs have similar sizes.

## BENEFITS AND COSTS OF GROUP-LIVING

Several experimental studies examined the benefits of group-living in COGs and POGs. Benefits can be enhanced survival of group members due to protection against adverse environmental conditions and predators. Furthermore, individuals in aggregations may have higher growth rates due to more efficient exploitation of food resources.

In swarm-living mysids, Ritz (2000) observed that individual food intake is very low in small groups but then increases in parallel with the number of group members (fig. 2A & B). While food intake remains high in larger groups, there is a tendency that benefits decrease with increasing group sizes, most likely due to increase competition (fig. 2A). Very similar observations were reported for terrestrial isopods: single individuals had substantially lower growth rates than individuals in aggregations (fig. 2B). However, in very large aggregations (> 10 individuals) growth rates significantly decreased, albeit not to levels of single individuals (Brockett & Hassall, 2005). These response curves can be resource-dependent, i.e., the optimal group size is positively related to the abundance or quality of resources (e.g., in mysids, Ritz, 2000). It can be expected that group members are sensitive to changes in resource availability and adjust their gregarious behavior accordingly.

Group-living is also an efficient strategy in alleviating stressful conditions. Under dry conditions terrestrial isopods survived much longer in aggregations than as solitary individuals (fig. 3A; Allee, 1926). These isopods are very sensitive to desiccation, and consequently individuals that suffer from water loss are attracted to conspecifics via smell (fig. 3B; Kuenen & Nooteboom,

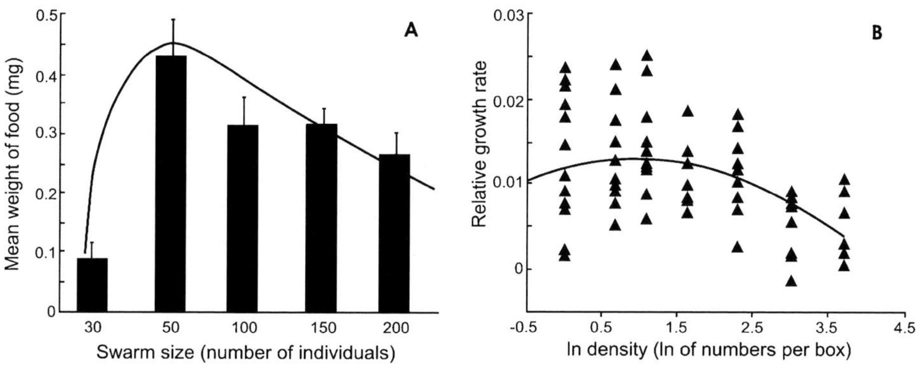


Fig. 2. A, mean dry weight of 10 stomachs of the mysid *Paramesopodopsis rufa* in swarms of different size (after Ritz, 2000); B, growth rates of isopods *Porcellio scaber* maintained at different densities (after Brockett & Hassall, 2005).

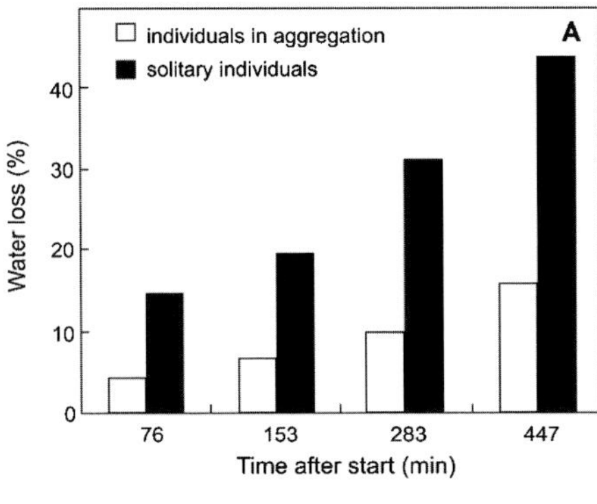
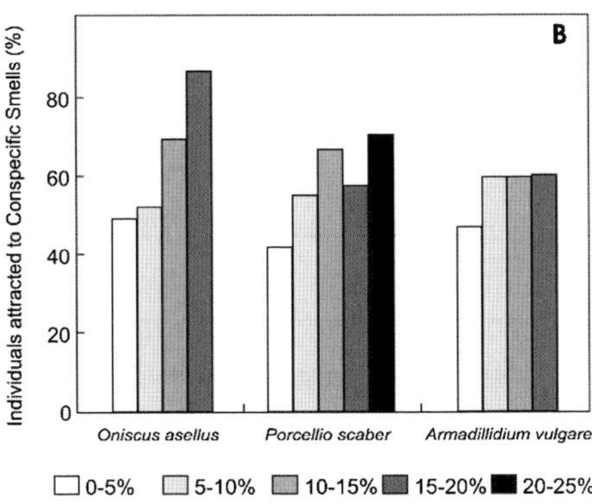

Fig. 3. A, water loss of terrestrial isopods (species not mentioned) that occur in aggregations (10 individuals) or as solitary individuals (data from Allee, 1926). B, proportion of individuals selecting the branch with conspecific odor in a Y-maze (after Kuenen & Nooteboom, 1963).

1963). Many COGs are assumed to form as a strategy to overcome stressful conditions.

In POGs, cohabitation of offspring with mothers was shown to result in enhanced growth and higher survival of the juveniles. Aoki (1997) examined juvenile growth rates in several species of caprellid amphipods. Juveniles were raised as orphans (mothers absent) and with maternal care (mothers present). In species that do not usually exhibit maternal care (*Caprella danilevskii* and *C. subinermis*), growth rates did not differ between the two treatments. However, in two of the three species with maternal care (*C. decipiens*, *C. scaura* and *C. monoceros*), offspring growth rates increased significantly when the mothers were present (fig. 4). The author suggested that in *C. scaura* and *C. monoceros* the females are important as attachment substratum for small offspring, which can not yet efficiently grasp to the substrata (hydroids, algae) on which these species are found in nature (Aoki, 1999). Interestingly, in the intertidal amphipod *Parallorchestes ochotensis*, where juveniles remain for long time periods in the female's brood pouch, juveniles under maternal care grew less than control juveniles without their mother (Kobayashi et al., 2002).

Two experimental studies confirmed that maternal care offers effective protection against predators for small juveniles. In *P. ochotensis*, juvenile survival was high in treatments without predators, regardless of maternal care, but in predator presence juveniles with their mothers had significantly higher survival rates than orphan juveniles (fig. 5A; Kobayashi et al., 2002). Very similar results were obtained for juveniles of the burrow-living amphipod *Leptocheirus pinguis*, where the maternal burrow was instrumental in offering protection against predators (fig. 5B; Thiel, 1999).

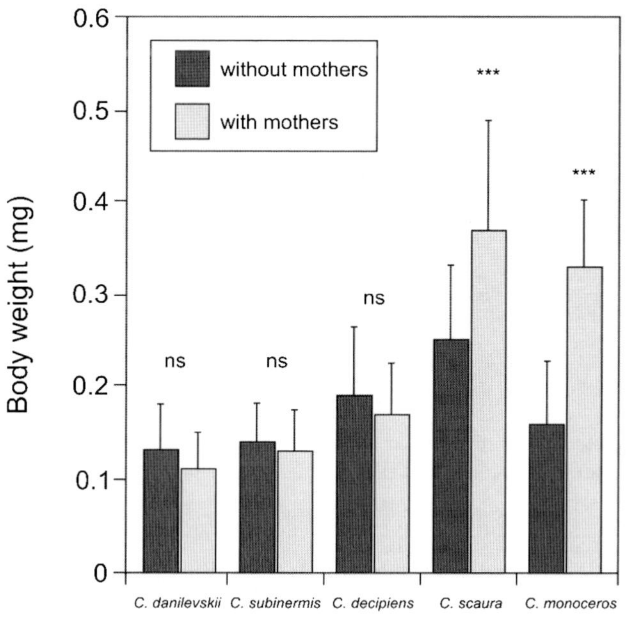



Fig. 4. Growth rates of juvenile caprellids that were reared with and without their mothers (after Aoki, 1997).

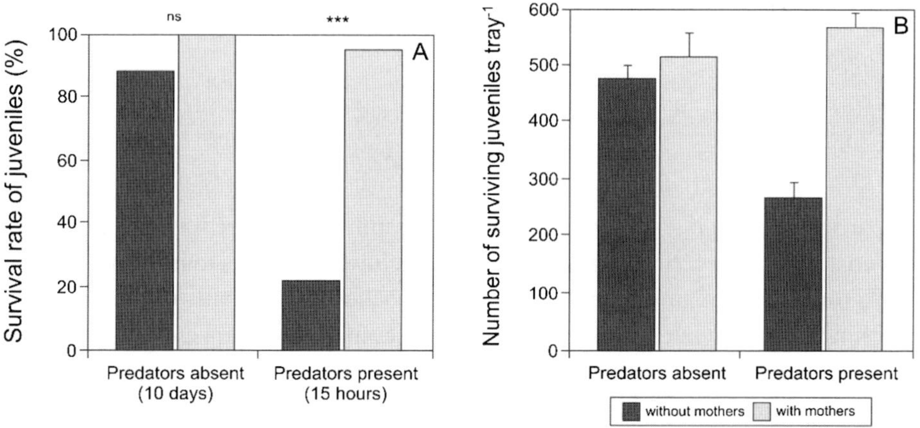



Fig. 5. Survival of juvenile amphipods maintained with and without their mother in presence and absence of predators. A, *Parallorchestes ochotensis*; B, *Leptocheirus pinguis* (A after Kobayashi et al., 2002, and B after Thiel, 1999).

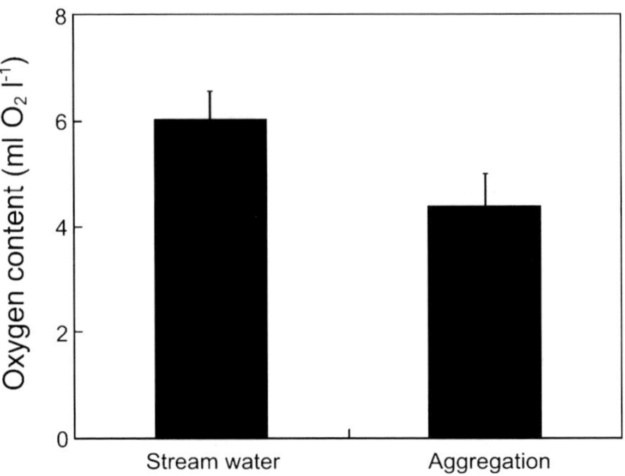

Fig. 6. Oxygen content in stream water above isopod aggregations and of water within dense aggregations of the isopod *Caecidotea communis* (data from Allee, 1929).

Surprisingly, few studies have been conducted to explicitly test the costs of group-living. Costs are generated by resource competition, cannibalism, disease or parasite transmission, contamination, and inbreeding. In very large groups growth or survival may be suppressed, primarily due to competition (see fig. 2A & B). Contamination with excretory products or feces may also cause costs to group members. $CO_2$-concentrations may increase while $O_2$-concentrations are expected to decrease in very large and dense groups. Indication for this comes from data by Allee (1929) who showed that oxygen concentrations decreased in dense COGs of the freshwater isopod *Caecidotea communis* (fig. 6), but it is not known whether this led to the disbandment of the aggregations.

## BEHAVIORAL INTERACTIONS

Particular behaviors may have evolved to enhance the benefits or to reduce the costs of group-living. These behaviors include strategies to initiate or maintain groups. For example, compact separate schools of mysids mingle quickly (within <2 min) after coming into direct contact (Modlin, 1990). Schools are maintained by inward swimming of individuals from the edge (O'Brien, 1988). Individuals that fell behind were sometimes observed to leap forward and swim towards the center of the school (Clutter, 1969). Several authors emphasized the integrated responses of mysids within large schools (Clutter, 1969; O'Brien, 1988; Buskey, 2000). In several pelagic peracarids authors mentioned whorl-like school formations (e.g., Clutter, 1969) or "tornado-like swirls" (Lobel & Randall, 1986), in which individual behaviors need to be highly synchronized to ensure coordinated group movements. Vision is important for school maintenance in mysids (Modlin, 1990; Buskey, 2000). Hydrodynamic and chemical cues have often been inferred, but their role in school cohesion is not well known.

Members of POGs react differently to family members than to unrelated individuals, which enhances family cohesion. Female mysids frequently reincorporate lost embryos and larvae into their marsupium (Wittmann, 1978). While females of some mysid species appear capable of recognizing their own offspring, females in most species do not seem to discriminate between own and unrelated offspring, but they show a preference for larval stages that are similar or more advanced than their own brood (Johnston & Ritz, 2005). In some amphipods the females are also able to recognize their own brood in the marsupium (Patterson et al., 2008). Highly specific family cues have evolved in desert isopods, which ensure that only family members are admitted to the parent–offspring burrow (Linsenmair, 2007).

Aggregation behavior can also be context-dependent. For example, terrestrial isopod aggregate quickly under dry conditions but roam solitarily in moist environments (e.g., Allee, 1926). Aggregations form as soon as two individuals come into direct contact and other individuals then join these aggregations. Kuenen & Nooteboom (1963) showed that individuals of *Oniscus asellus* that had lost a lot of water have a higher tendency to seek out the smell of conspecifics.

While few studies have explicitly examined the costs of group-living in peracarids there are numerous reports on behaviors supposed to reduce these costs. For example, several behaviors appear to have evolved in response to the high risk of cannibalism in peracarid groups. In some species females synchronize brood release, possibly as a mechanism to avoid that their offspring is cannibalized by conspecifics (Johnston & Ritz, 2001). Synchronization appears to be of higher adaptive value in species occurring in small aggregations (hundreds to a few thousand individuals) than in species with large aggregations; whether synchrony is mediated by extrinsic or intrinsic cues is not known at present (Johnston & Ritz, 2001). Habitat segregation between juveniles and adults is considered another strategy to avoid cannibalism (e.g., Jormalainen & Shuster, 1997; McGrath et al., 2007; Taylor, 2008).

In POGs, active cleaning of the brood and removal of feces and dead individuals from the family burrow (Schneider, 1971) probably has evolved as a strategy to keep parasites and diseases at bay. Females of several species have been observed to manipulate their small offspring, which was interpreted as maternal grooming (Thiel, 2007). No grooming or cleaning behavior has been reported from COGs; in fact aggregation sites of terrestrial isopods are known for the accumulation of feces.

## CONCLUSIONS AND OUTLOOK

This review shows that benefits and costs of group-living may influence the evolution of social behaviors in peracarid crustaceans. The few available studies confirm that group-living can improve survival and growth of group-members, both in COGs and in POGs. Similar as for the benefits the costs of group-living appear to affect both the members of COGs and POGs. Enhanced resource competition could be one of the major costs, but anecdotal observations suggest that parasite/disease transmission and cannibalism can additionally affect health and survival of group members. The fact that benefits started to decrease at large group sizes indicates that there are tradeoffs. These tradeoffs can be species-specific and context-dependent and require investigation. The examples presented herein show that peracarids lend themselves to experimental tests of the benefits and costs of group-living. Most species are relatively small and therefore very suitable for small-scale experiments. Since they have direct development, many species can be easily reared in the laboratory.

Studies on the evolution of social behavior often focus on advanced social behaviors (e.g., Korb & Heinze, 2008). Relatively little is known about the environmental conditions that lead to the initial evolution of group-living (for exceptions see, e.g., Tallamy & Wood, 1986; Costa, 2006). To overcome this gap, comparative studies of species from different environments will be particularly instructive. Future studies should thus focus on the mechanisms that drive the social behavior in peracarid groups in terrestrial and aquatic environments. It appears especially desirable to compare the benefits and costs in COGs and POGs to better understand how genetic relatedness (in POGs) or the absence thereof (in COGs) affects the evolution of social behavior.

## ACKNOWLEDGEMENTS

I am very grateful to Akira Asakura for his patience and guidance during the preparation of this manuscript. Furthermore I thank Ivan Hinojosa for the preparation of the figures.

# REFERENCES

ALLEE, W. C., 1926. Studies in animal aggregations: causes and effects of bunching in land isopods. J. Exp. Zool., **45**: 255-277.

— —, 1929. Studies in animal aggregations: natural aggregations of the isopod, *Asellus communis*. Ecology, **10**: 14-36.

AOKI, M., 1997. Comparative study of mother-young association in caprellid amphipods: is maternal care effective? J. Crust. Biol., **17**: 447-458.

AOKI, M. & T. KIKUCHI, 1991. Two types of maternal care for juveniles observed in *Caprella monoceros* Mayer, 1890 and *Caprella decipiens* Mayer, 1890 (Amphipoda: Caprellidae). Hydrobiologia, **223**: 229-237.

BOOMSMA, J. J. & N. R. FRANKS, 2006. Social insects: from selfish genes to self organisation and beyond. Trends Ecol. Evol., **21**: 303-308.

BRINK, F. W. B. VAN DEN, G. VAN DER VELDE & A. BIJ DE VAATE, 1993. Ecological aspects, explosive range extension and impact of a mass invader, *Corophium curvispinum* Sars, 1895 (Crustacea: Amphipoda), in the Lower Rhine (The Netherlands). Oecologia, **93**: 224-232.

BROCKETT, B. F. T. & M. HASSALL, 2005. The existence of an Allee effect in populations of *Porcellio scaber* (Isopoda: Oniscidea). Eur. J. Soil Biol., **41**: 123-127.

BUSKEY, E. J., 2000. Role of vision in the aggregative behavior of the planktonic mysid *Mysidium columbiae*. Mar. Biol., **137**: 257-265.

CLUTTER, R. I., 1969. The microdistribution and social behaviour of some pelagic mysid shrimps. J. exp. mar. Biol. Ecol., **3**: 125-155.

COSTA, J. T., 2006. The other insect societies: 1-767. (The Belknap Press of Harvard University Press, Cambridge, U.S.A.).

DROLET, D. & M. A. BARBEAU, 2009. Differential emigration causes aggregation of the amphipod *Corophium volutator* (Pallas) in tide pools on mudflats of the upper Bay of Fundy, Canada. J. exp. mar. Biol. Ecol., **370**: 41-47.

FENOLIO, D. B. & G. O. GRAENING, 2009. Report of a mass aggregation of isopods in an Ozark cave of Oklahoma with considerations of population sizes of stygobionts. Speleobiology Notes, **1**: 9-11.

FOSTER, K. R., T. WENSELEERS & F. L. W. RATNIEKS, 2001. Kin selection is the key to altruism. Trends Ecol. Evol., **21**: 57-60.

FRIEND, J. A. & A. M. M. RICHARDSON, 1986. Biology of terrestrial amphipods. Annu. Rev. Entomol., **31**: 25-48.

HAMILTON, W. D., 1964. The genetical evolution of social behaviour. I & II. J. Theor. Biol., **7**: 1-52.

JARAMILLO, E., R. DE LA HUZ, C. DUARTE & H. CONTRERAS, 2006. Algal wrack deposits and macroinfaunal arthropods on sandy beaches of the Chilean coast. Rev. Chil. Hist. Nat., **79**: 337-351.

JOHNSTON, N. M. & D. A. RITZ, 2001. Synchronous development and release of broods by the swarming mysids *Anisomysis mixta australis*, *Paramesopodopsis rufa* and *Tenagomysis tasmaniae* (Mysidacea: Crustacea). Mar. Ecol. Prog. Ser., **223**: 225-233.

— — & — —, 2005. Kin recognition and adoption in mysids (Crustacea: Mysidacea). J. Mar. Biol. Ass. UK, **85**: 1441-1447.

JORMALAINEN, V. & S. M. SHUSTER, 1997. Microhabitat segregation and cannibalism in an endangered freshwater isopod, *Thermosphaeroma thermophilum*. Oecologia, **111**: 271-279.

KOBAYASHI, T., S. WADA & H. MUKAI, 2002. Extended maternal care observed in *Parallorchestes ochotensis* (Amphipoda, Gammaridea, Talitroidea, Hyalidae). J. Crust. Biol., **22**: 135-142.

KORB, J., 2008. The ecology of social evolution in termites. In: J. KORB & J. HEINZE (eds.), Ecology of social evolution: 151-174. (Springer Press, Heidelberg).

KORB, J. & J. HEINZE, 2008. The ecology of social life: a synthesis. In: J. KORB & J. HEINZE (eds.), Ecology of social evolution: 245-259. (Springer Press, Heidelberg).

KUENEN, D. J. & H. P. NOOTEBOOM, 1963. Olfactory orientation in some land-isopods (Oniscoidea, Crustacea). Ent. Exp. Appl., **6**: 133-142.

LINSENMAIR, K. E., 2007. Sociobiology of terrestrial isopods. In: J. E. DUFFY & M. THIEL (eds.), Evolutionary ecology of social and sexual systems: crustaceans as model organisms: 339-364. (Oxford University Press, Oxford).

LOBEL, P. S. & J. E. RANDALL, 1986. Swarming behavior of the hyperiid amphipod *Anchylomera blossevilli*. J. Plankton Res., **8**: 253-262.

MCGRATH, K. E., E. T. H. M. PEETERS, J. A. J. BEIJER & M. SCHEFFER, 2007. Habitat-mediated cannibalism and microhabitat restriction in the stream invertebrate *Gammarus pulex*. Hydrobiologia, **589**: 155-164.

MODLIN, R. F., 1990. Observations on the aggregative behavior of *Mysidium columbiae*, the mangrove mysid. Mar. Ecol., **11**: 263-275.

O'BRIEN, D. P., 1988. Direct observations of clustering (schooling and swarming) behaviour in mysids (Crustacea; Mysidacea). Mar. Ecol. Prog. Ser., **42**: 235-246.

PAOLETTI, M. G., A. TSITSILAS, L. J. THOMSON, S. TAITI & P. A. UMINA, 2008. The flood bug, *Australiodillo bifrons* (Isopoda: Armadillidae): a potential pest of cereals in Australia? Appl. Soil Ecol., **39**: 76-83.

PATTERSON, L., J. T. A. DICK & R. W. ELWOOD, 2008. Embryo retrieval and kin recognition in an amphipod (Crustacea). Anim. Behav., **76**: 717-722.

RITZ, D. A., 1994. Social aggregation in pelagic invertebrates. Adv. Mar. Biol., **30**: 155-216.

— —, 2000. Is social aggregation in aquatic crustaceans a strategy to conserve energy? Can. J. Fish. Aquat. Sci., **57**(Suppl. 3): 59-67.

SCHMID-HEMPEL, P., 1998. Parasites in social insects: 1-392. (Princeton University Press, Princeton).

SCHNEIDER, P., 1971. Lebensweise und soziales Verhalten der Wüstenassel *Hemilepistus aphganicus* Borutzky 1958. Zeitschr. Tierpsychol., **29**: 121-133.

SCHWARZ, M. P., M. H. RICHARDS & B. N. DANFORTH, 2007. Changing paradigms in insect social evolution: insicts from halictine and allodapine bees. Annu. Rev. Entomol., **52**: 127-150.

SPEARS, T., R. W. DEBRY, L. G. ABELE & K. CHODYLA, 2005. Peracarid monophyly and interordinal phylogeny inferred from nuclear small-subunit ribosomal DNA sequences (Crustacea: Malacostraca: Peracarida). Proc. Biol. Soc. Wash., **118**: 117-157.

TALLAMY, D. W. & T. K. WOOD, 1986. Convergence patterns in subsocial insects. Annu. Rev. Entomol., **31**: 369-390.

TAYLOR, M. D., 2008. Spatial and temporal patterns of habitat use by three estuarine species of mysid shrimp. Mar. Freshw. Res., **59**: 792-798.

THIEL, M., 1999. Extended parental care in marine amphipods. II. Maternal protection of juveniles from predation. J. exp. mar. Biol. Ecol., **234**: 235-253.

— —, 2003a. Reproductive biology of *Limnoria chilensis*: another boring peracarid species with extended parental care. J. Nat. Hist., **37**: 1713-1726.

— —, 2003b. Extended parental care in crustaceans — an update. Rev. Chil. Hist. Nat., **76**: 205-218.

— —, 2007. Social behavior of parent–offspring groups in crustaceans. In: J. E. DUFFY & M. THIEL (eds.), Evolutionary ecology of social and sexual systems: crustaceans as model organisms: 294-318. (Oxford University Press).

WHITEHOUSE, M. E. A. & Y. LUBIN, 2005. The functions of societies and the evolution of group living: spider societies as a test case. Biol. Rev., **80**: 347-361.

WHITNEY, N. M. & P. J. MOTTA, 2008. Cleaner host posing behavior of whitetip reef sharks (*Triaenodon obesus*) in a swarm of hyperiid amphipods. Coral Reefs, **27**: 363.

WITTMANN, K. J., 1978. Adoption, replacement and identification of young in marine Mysidacea (Crustacea). J. exp. mar. Biol. Ecol., **32**: 259-274.



First received 2 February 2010.
Final version accepted 11 March 2010.

# STABLE ISOTOPE ANALYSIS OF EPIBIOTIC CAPRELLIDS (AMPHIPODA) ON LOGGERHEAD TURTLES PROVIDES EVIDENCE OF TURTLE'S FEEDING HISTORY


BY

TAKASHI HOSONO[1]) and HIROSHI MINAMI

Ecologically Related Species Section, Tropical Tuna Resources Division, National Research Institute of Far Seas Fisheries, Fisheries Research Agency, 5-7-1, Shimizu-orido, Shizuoka 424-8633, Japan



## ABSTRACT

Epibiotic crustacean may act as an external indicator for the feeding history of the host. We analyzed stable isotope ratios, especially carbon isotope ratio ($\delta^{13}C$), of tissues of loggerhead turtles and caprellids (Amphipoda: Caprellidea) collected from the carapace of turtles in pelagic areas and at a nesting beach. There was no significant difference of $\delta^{13}C$ in tissues of turtles sampled in pelagic areas and the nesting beach. These values $\delta^{13}C$ of turtle tissues fell within the range of values of the pelagic food-web components in the western North Pacific. The values of $\delta^{13}C$ of caprellids obtained from turtles at the nesting beach were significantly higher than values found in caprellids sampled from the pelagic area. This indicates that the host turtles spend long enough in the nesting beach area for caprellids to incorporate components of the coastal food-web into their tissues. The results suggest that loggerhead turtles may not feed during the nesting period. Our investigation shows that caprellid $\delta^{13}C$ could function as an external indicator of loggerhead turtle feeding.


## INTRODUCTION

Many epibiotic animals live on the body surface of large marine animals. Most epibiotic animals are non-parasitic species (e.g., barnacles on sea turtles), with a few exceptions of parasitic species such as cyamids on whales (Schell et al., 2000). Parasitic epibionts absorb nutrients directly from their hosts. In contrast, most non-parasitic epibionts utilize detritus in the water column and do not depend on their host for nutrition. But their feeding environment

---

[1]) Corresponding author; e-mail: usujiri3@gmail.com

corresponds to the environment of their hosts because non-parasitic species are attached to their hosts. This unique feature of non-parasitic epibionts may provide useful information on a host's feeding history, when stable isotope in tissues of host and epibiont are analyzed.

Stable isotope analysis has been used in ecological studies to understand foraging habits. Nitrogen isotope ratios ($\delta^{15}N$) reflect the trophic levels of animals within food chains and generally increase about 3 to 4‰ per trophic level (DeNiro & Epstein, 1981; Wada et al., 1987). Therefore $\delta^{15}N$ may be used to investigate the diet of animals and identify the trophic relationships within a food chain. Carbon isotope ratios ($\delta^{13}C$) of animals show the approximate values of the producer of a food chain (DeNiro & Epstein, 1978). In the marine environment, the $\delta^{13}C$ values can indicate inshore versus offshore, or pelagic versus benthic contribution to food intake (France, 1995). Furthermore, when the rate of incorporation of $^{13}C$ or $^{15}N$ into a certain tissue, called turnover rate, are known, the feeding history of particular animals on various time scales can be obtained by comparing somatic tissues with different turnover rates (Hobson et al., 1994). Because epibionts have different turnover rates from their hosts, analysis of both host and epibiont isotopes provides information about the host environment on time scales that cannot be back-calculated from samples of the host's tissues. In the present study, we focus on loggerhead turtles and epibiont caprellids (Amphipoda; Caprellidea) and examine the utility of caprellids as an external indicator for turtle's feeding history.

In the North Pacific, loggerhead turtles range from coastal to pelagic areas through their life cycle. These turtles nest on beaches along the southern coast of the Japanese archipelago (Kamezaki et al., 2003). Juveniles hatch and disperse widely across the North Pacific (Polovina et al., 2006). After maturation, loggerhead turtles return to the Japanese coasts, nest, and re-migrate to foraging areas such as the western North Pacific and the East China sea (Hatase et al., 2002). The nesting period spans several months and female loggerhead turtles nest 2-3 times in the period (Sato et al., 1998).

There is one discussion related to foraging ecology of sea turtles, that is whether sea turtles feed during nesting period (Tanaka et al., 1995; Myers & Hays, 2006). Whether turtles feed or not has important implications for their energy budget in the post-nesting period. A simple and certain method to examine turtle's feeding during the nesting period is to investigate their stomach contents directly in the period. However, this method may stress or kill the turtles. Stable isotope analysis minimizes stress on turtles as only small quantities of tissue samples are required. One of the other merits of using this analysis is to be able to confirm if the turtles feed in coastal area during their nesting period by large changes of stable isotopes ratio, especially $\delta^{13}C$, of turtle tissues, because the food-webs of coastal and pelagic areas are largely different in $\delta^{13}C$ (France, 1995). On the contrary, stable isotopes ratios in turtle tissues are expected to show little change, if turtles do not feed during the nesting period. However to prove non-feeding during the nesting period by using stable isotope, it is also needed to prove that turtles stay at coastal area during the nesting period. In this regard, as epibionts feed independently on ambient detritus, the epibiont isotopic ratios will provide evidence that the host remains within the coastal area.

In the present paper we focus on *Caprella andreae*. The caprellid is a pelagic species mainly observed on the carapace of loggerhead turtles (Aoki, 1995). In general, caprellids strongly depend on their habitat substrates, because caprellids have no swimming larval stage and no swimming appendage in adult stage (Johnson et al., 2001). Like many other caprellids, *C. andreae* is considered to be a detritus feeder (Guerra-Garcia & Figueroa, 2009).

The objective of our study is to examine availability of caprellids on loggerhead turtles as a tool for understanding feeding habits in the turtles. In the present study, we analyzed the isotope ratios of loggerhead turtles sampled from pelagic and nesting areas; isotope ratios of caprellids sampled from these turtles in pelagic and nesting areas; and compare the changes of the isotope ratios in turtles and caprellids between pelagic and nesting areas.

## MATERIAL AND METHODS

Field sampling. — We conducted pelagic longline surveys in the western North Pacific (27°49′N–39°31′N 138°16′E–146°21′E) from June to July in 2007 and 2008, using the research vessel Taikei-maru No. 2. Eleven live loggerhead turtles were caught (Straight carapace length [SCL] = 70.2 ± 7.6 cm, mean ± SD). Samples of muscle tissue and blood were collected by syringes from all these turtles. Blood samples were divided into hemocyte and plasma by centrifuging on board ship (3500 rpm; 10 min). One live turtle that accidentally died on deck, was dissected to collect gut contents. *C. andreae* were collected together with algae from the carapace of the loggerhead turtles by a metallic spatula. Tissue samples, gut contents and caprellid samples were preserved at −10°C on board ship.

Tissue samples were collected from nesting loggerhead turtles at Omaezaki beach (34°36′N 138°11′E) on 17-29 July 2003. On this beach, the nesting period of loggerhead turtles spans from May to September. Our samples were taken at the peak of nesting activity on the beach (Nishimura et al., 1992). We collected muscle tissue, blood samples, and caprellids from nesting turtles (n = 5, SCL: 79.6 ± 4.9 cm mean ± SD) in the same way as in the pelagic survey.

Isotopic analysis. — Caprellid samples and turtle gut contents were sorted by species under a binocular microscope. Caprellid specimens were pooled by individual turtles, 10-20 caprellids were randomly selected, and combined together for the isotope analysis. Blood samples, caprellid samples and prey species obtained from the gut contents were dried at 60°C and ground to a fine powder. Muscle samples were delipidized with chloroform: methanol (2:1) solution. Stable isotope ratios of carbon and nitrogen were measured with a MAT 252 mass spectrometer coupled with an element analyzer (EA1110). Isotope ratios are expressed as per mil deviations from the standard as defined by the following equation: $\delta X = \{(R\ sample/R\ standard) - 1\} \times 1000$, where X means $^{13}C$ or $^{15}N$, and R means $^{13}C/^{12}C$ or $^{15}N/^{14}N$. The Vienna Pee Dee Belemnite and atmospheric nitrogen were used as standards of carbon and nitrogen isotopes respectively. The analytical precision for the isotopic analyses was $\leqslant 0.1‰$ for both $\delta^{13}C$ and $\delta^{15}N$.

Turnover rate and discrimination factor in loggerhead turtle were reported in plasma and hemocyte from the rearing experiments using juveniles (SCL: 9.0-13.1 cm) (table I; Reich et al., 2008). Minami (2004) also reared loggerhead turtles (SCL: 19.9-42.6 cm) and identified turnover rate and discrimination factor (table I). We used the results of these two studies to back-calculate

TABLE I

Reported turnover rates and discrimination factors in a loggerhead sea turtle. Turnover rates are shown in half-life period, which are calculated by ln(2)/kst, where kst is the fractional rate of isotope incorporation of a tissue

| Reference | Carbon | | | Nitrogen | | |
|---|---|---|---|---|---|---|
| | Plasma | Hemocyte | Muscle | Plasma | Hemocyte | Muscle |
| Reich et al. (2008) | | | | | | |
| Discrimination factor | −0.38 | 1.53 | – | 1.5 | 0.16 | – |
| Turnover rate (Days) | 27.4 | 27.8 | – | 15.6 | 25.2 | – |
| Minami (2004) | | | | | | |
| Discrimination factor | −1.7 | 0.4 | 0.7 | 1.1 | −0.7 | 1.5 |
| Turnover rate (Days) | 29.6 | 74.9 | 65.4 | 53.6 | 121.1 | 115.5 |

feeding history of the turtles with the formula: $\delta X_{prey} = \delta X_{tissue} - \Delta X_{tissue}$, where $\Delta X_{tissue}$ indicates a discrimination factor specific to a tissue.

## RESULTS

Mean values of $\delta^{13}C$ of turtle tissues collected in pelagic area were $-19.6‰ \pm 0.5$, $-18.7‰ \pm 0.3$ and $-18.1‰ \pm 0.8$ (mean $\pm$ SD, plasma, hemocyte and muscle, respectively; fig. 1), and $\delta^{15}N$ were $9.9‰ \pm 0.3$, $9.0‰ \pm 0.5$ and $11.3‰ \pm 0.7$. Goose barnacles, fish eggs, and digested materials were

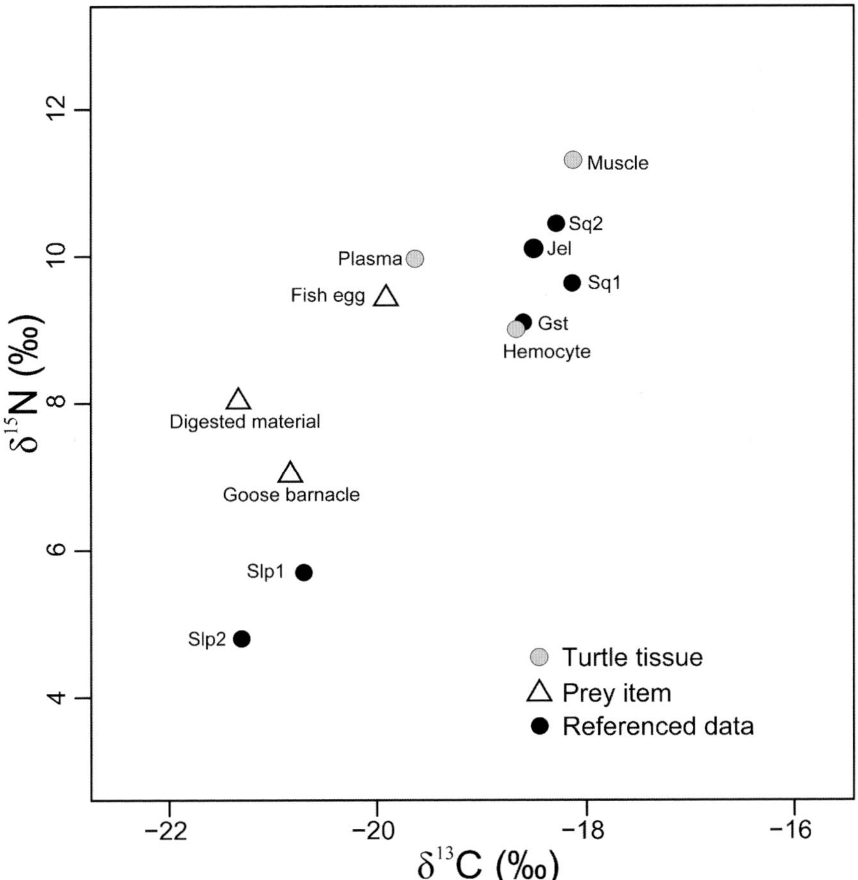


Fig. 1. Stable isotopes ratios of turtle issues and prey items collected from the pelagic area; Sq1: *Onychoteuthis banksii*, Sq2: *Gonatopsis borealis* cited from Minami (1994); Gst: *Pterotrachea* sp., Jel: *Beroe cucumis*, Slp1: *Salpa fusiformis*, Spl2: unidentified salp cited from Hatase et al. (2002).

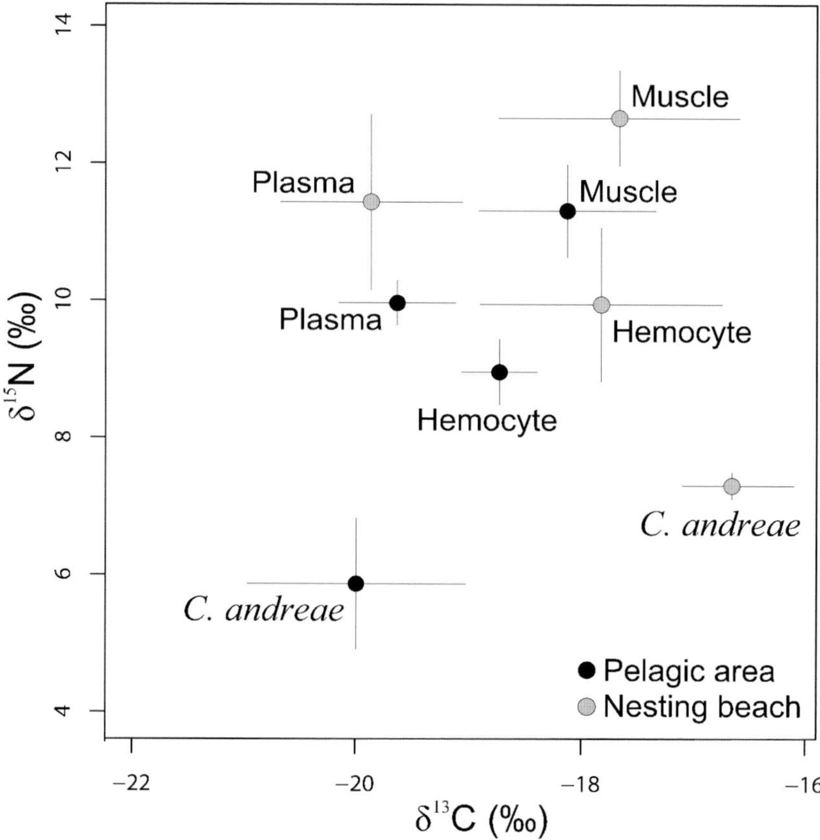


Fig. 2. Stable isotope ratios of turtle tissues and *Caprella andreae* sampled from the pelagic area and from the nesting beach.

found in the turtle gut contents. The mean isotope ratios of these prey items ranged from $-21.3$ to $-19.9‰$ in carbon, and 7.1 to 9.4‰ in nitrogen (fig. 1).

In samples taken from the nesting beach, mean values of $\delta^{13}C$ of turtle tissues were $-19.9‰ \pm 0.8$, $-17.8‰ \pm 1.1$ and $-17.7‰ \pm 1.1$ (mean $\pm$ SD, plasma, hemocyte and muscle, respectively; fig. 2), and $\delta^{15}N$ were 11.4‰ $\pm$ 1.3, 9.9‰ $\pm$ 1.1 and 12.7‰ $\pm$ 0.7. The values of $\delta^{13}C$ in nesting-beach turtle samples were not significantly different from turtle samples taken in the pelagic area (Wilcoxon Rank Sum test, $p > 0.05$). The values of $\delta^{15}N$ in turtle tissue sampled at the nesting beach were significantly higher than pelagic samples (Wilcoxon Rank Sum test, $p = 0.003, 0.032$ and $0.006$ in hemocyte, plasma and muscle, respectively), but the difference was small (0.99-1.4‰). Nesting beach caprellid $\delta^{13}C$ was significantly higher than samples from the pelagic area ($-20.0‰ \pm 0.9$ and $-16.7‰ \pm 0.4$ in pelagic area and nesting

TABLE II

Stable isotope ratio of preys back-calculated using turtle tissues isotope ratios
and tissue-specified discrimination factors

|  | Reich et al. (2008) | | Minami (2004) | |
| --- | --- | --- | --- | --- |
|  | $\delta^{13}$C | $\delta^{15}$N | $\delta^{13}$C | $\delta^{15}$N |
| Pelagic area |  |  |  |  |
| Plasma | −19.3 | 8.5 | −17.9 | 8.9 |
| Hemocyte | −20.3 | 8.8 | −19.1 | 9.7 |
| Muscle | − | − | −18.8 | 9.8 |
| Nesting beach |  |  |  |  |
| Plasma | −19.5 | 9.9 | −18.2 | 10.3 |
| Hemocyte | −19.4 | 9.8 | −18.2 | 10.7 |
| Muscle | − | − | −18.4 | 11.2 |

beach; Wilcoxon Rank Sum test, $p < 0.001$; fig. 2), while $\delta^{15}$N were not significantly different between areas (5.9‰ ± 1.0 and 7.3‰ ± 0.2 in pelagic area and nesting beach, respectively; Wilcoxon Rank Sum test, $p = 0.053$).

Prey isotope ratios were back-calculated to acquire turtle's feeding history at different time scales using the isotope ratios of turtle tissues and the tissue-specified discrimination factors. Based on the values of the discrimination factor reported by Reich et al. (2008), the estimated carbon isotope ratios of prey were similar between tissues and areas, Carbon isotope values ranged from −20.3 to −19.3‰. While, nitrogen isotope ratios were estimated to range from 8.5 to 8.8‰ in the pelagic area, and 9.8 to 9.9‰ in the nesting beach area (table II). Similar pattern of estimates were obtained when the discrimination factors reported in Minami (2004) were applied ($\delta^{13}$C: 17.9-19.1‰ in pelagic and 18.2-18.4‰ in nesting beach, $\delta^{15}$N: 8.9-9.8‰ in pelagic and 10.3-11.2‰ in nesting beach; table II).

DISCUSSION

Loggerhead turtles feed on gelatinous planktons and cephalopods in pelagic foraging areas (Hatase et al., 2002; Parker et al., 2005). Information on stable isotope ratios of animals in pelagic ecosystems in the western North Pacific is limited. Minami (1994) analyzed isotope ratios of Cephalopods *Onychoteuthis banksii* and *Gonatopsis borealis* in the western North Pacific (39°N–42°N 155°E) in 1991. The author reported $\delta^{13}$C of these species were around −18‰ (fig. 1). Hatase et al. (2002) reported $\delta^{13}$C of potential loggerhead turtle prey, in western North Pacific (36°30′N–42°20′ N 143°30′E–144°E), ranged −21.3 to −18.5‰ (fig. 1). These values for $\delta^{13}C$ are similar to ratios back-calculated from the tissues and gut contents of turtles collected in the pelagic area in the present study. This means the turtles feed organism in the pelagic food-web in the western North Pacific. The values of $\delta^{13}C$ of turtle tissues from the nesting beach were not significantly different from those in the pelagic area. This indicates that turtle tissues at the nesting beach continue to reflect the isotope value of pelagic food items ingested before migration to the nesting beach.

Back-calculated isotope values of preys did not change among the tissues of the nesting turtles, and were similar to reported values of pelagic food items. Turnover rate for carbon isotope in loggerhead turtle were reported to be 27.4-29.6 days in plasma and 27.8-74.9 days in hemocyte (table I). Nesting turtles have feeding record reflecting pelagic food items even in plasma reflecting feeding previous one month, suggests two conceivable explanations on feeing of loggerhead turtles in nesting period. First is that loggerhead turtles do not feed during nesting season. A second possibility is that the turtles sampled had recently arrived in coastal waters. If the duration of stay were shorter than the turnover rate, then isotopic components derived from coastal food sources would not be sufficiently incorporated into turtle tissue. The isotope ratios of caprellids provides information in relation to duration of stay of the host turtle in the coastal area. The values of $\delta^{13}C$ of caprellids collected from the nesting turtles were higher than values for caprellids collected from pelagic turtles (−16.7‰ and −20.0‰ in nesting beach and pelagic area). In coastal food-web, and $\delta^{13}C$ of caprellids were reported to be −15.2‰ at the inland sea of Japan (Takai et al., 2002), and −17.7‰ to −16.2‰ in Gokasyo bay, the Pacific coast of Japan (Yokoyama & Ishihi, 2007). The value of caprellid $\delta^{13}C$ collected from the turtles in nesting beach was similar to coastal food-web values. This indicates that the host turtles spend enough time in the coastal area for caprellids to incorporate isotopic components of coastal food-web into tissue. This result supports the explanation that loggerhead turtle do not feed during nesting season.

Possibility of non-feeding during nesting period are also suggested from increasing of turtle $\delta^{15}N$ at nesting beach. It is generally accepted that tissues of fasting animals become enriched in $\delta^{15}N$ following nutritional restriction (Hobson et al., 1993; Oelbermann & Scheu, 2002; Cherel et al., 2005). Hobson et al. (1993) reported that fasting geese showed significantly higher $\delta^{15}N$ values compared to non-fasting geese but $\delta^{13}C$ values were unchanged. This can well explain the different behavior of $\delta^{15}N$ and $\delta^{13}C$ in turtle tissues at the two area. But it is also noted that the effect of nutritional restriction could be dependent of species-specific differences in physiological response (Williams et al., 2007) or largely a function of the level of nutritional stress (Kempster et al., 2007). Further studies on the isotopic values of sea turtles during fasting would be needed.

Although stable isotope analysis is a powerful tool for studying animal's feeding habits, it is difficult to verify non-feeding using only the animal's own tissues. We revealed that caprellid $\delta^{13}C$ on the carapace of loggerhead turtles showed different changes from that in turtle tissues between pelagic foraging area and coastal nesting beach. This suggests that loggerhead turtles do not feed during nesting period. To verify the non-feeding of the loggerhead turtles during nesting period based on the results of the stable isotope analysis, it is also required to present a specific length of stay of the host turtles in the coastal area. This could be achieved by revealing turnover rate in caprellid. Although the present study focused on caprellids as an epibiont, stable isotope analysis of other turtle epiobionts such as barnacles, gammarids and flotsam crabs would also provide additional knowledge on the host behavior and feeding patterns.


## ACKNOWLEDGMENTS

We thank the captain and the crews of R/V Taikei-maru for their cooperation and L. Yashige for assistance on board ship. We are grateful to K. Yokota and T. Nobetsu for assistance in field sampling at the nesting beach. K. Yokawa provided helpful comments on the early version of our manuscript. This research was conducted under the project study entitled "Understanding of dynamic in marine ecosystem and feeding habits in marine animals using stable isotope analysis", which is funded by the National Research Institute of Far Seas Fisheries, Fisheries Research Agency, Japan.


## REFERENCES

AOKI, M. & T. KIKUCHI, 1995. Notes on *Caprella andreae* Mayer, 1890 (Crustacea, Amphipoda) from the carapace of loggerhead sea turtles in the East China Sea and in Kyushu, Japan. Proc. Jap. Soc. System. Zool., **53**: 54-61.

CHEREL, Y., K. A. HOBSON & S. HASSANI, 2005. Isotopic discrimination between food and blood and feathers of captive penguins: implications for dietary studies in the wild. Physiol. Biochem. Zool., **78**: 106-115.

DENIRO, M. J. & S. EPSTEIN, 1978. Influence of diet on the distribution of carbon isotopes in animals. Geochim. Cosmochim. Acta., **42**: 495-506.

— — & — —, 1981. Influence of diet on the distribution of nitrogen isotopes in animals. Geochim. Cosmochim. Acta., **45**: 341-351.

FRANCE, R. L., 1995. Carbon-13 enrichment in benthic compared to planktonic algae: foodweb implications. Mar. Ecol. Prog. Ser., **124**: 307-312.

GUERRA-GARCIA, J. M. & J. M. D. FIGUEROA, 2009. What do caprellids (Crustacea: Amphipoda) feed on? Mar. Biol., **156**: 1881-1890.

HATASE, H., N. TAKAI, Y. MATZUZAWA, W. SAKAMOTO, K. OMUTA, K. GOTO, N. ARAI & T. FUJIWARA, 2002. Size-related differences in feeding habitat use of adult female loggerhead turtles *Caretta caretta* around Japan determined by stable isotope analyses and satellite telemetry. Mar. Ecol. Prog. Ser., **233**: 273-281.

HOBSON, K. A., R. T. ALISAUSKAS & R. G. CLARK, 1993. Stable-nitrogen isotope enrichment in avian tissues due to fasting and nutritional stress: implications for isotopic analysis of diet. Condor, **95**: 388-394.

HOBSON, K. A., J. F. PIATT & J. PITOCHELLI, 1994. Using stable isotope to determine seabird trophic relationships. J. Anim. Ecol., **63**: 786-798.

JOHNSON, W. S., M. STEVENS & L. WATLING, 2001. Reproduction and development of marine peracaridans. Adv. Mar. Biol., **39**: 105-260.

KAMZAKI, N., K. MATZUZAWA, O. ABE, H. ASAKAWA, T. FUJII & K. GOTO, 2003. Loggerhead turtles nesting in Japan. In: A. BOLTEN & B. WITHERINGTON (eds.), Loggerhead sea turtles: 210-217. (Smithsonian Institution Press, Washington).

KEMPSTER, B., L. ZANETTE, F. J. LONGSTAFFE, S. A. MacDOUGALL-SHACKLETON, J. C. WINGFIELD & M. CLINCHY, 2007. Do stable isotopes reflect nutritional stress? Results from a laboratory experiment on song sparrows. Oecolog., **151**: 365-371.

MINAMI, H., 1994. The biogeochemical and ecological study of four species of shearwaters in the Pacific using stable carbon and nitrogen isotope analysis. (Fisheries Science, Ph.D. Thesis, Hokkaido University, Hakodate). [In Japanese.]

— —, 2004. Stable isotope analysis. Report on the program on the international fishery resources survey: 1-18. (Cetacean group and ecosystem sub group, Fisheries Agency, Japan). [In Japanese.]

MYERS, A. E. & G. C. HAYS, 2006. Do leatherback turtles *Dermochelys coriacea* forage during the breeding season? A combination of data-logging devices provide new insights. Mar. Ecol. Prog. Ser., **322**: 259-267.

NISHIMURA, W., I. MATSUURA & T. TAKATSUKA, 1992. Time lag of the nesting of *Caretta caretta* between Fukiage-hama and Omaezaki. Umigame News Lett., **12**: 8-14. [In Japanese.]

OELBERMANN, K. & S. SCHEU, 2002. Stable isotope enrichment ($\delta^{15}$N and $\delta^{13}$C) in a generalist predator (*Pardosa lugubris*, Araneae: Lycosidae): effects of prey quality. Oecolog., **130**: 337-344.

PARKER, D. M., W. J. COOKE & G. H. BALAZS, 2005. Diet of oceanic loggerhead sea turtles (*Caretta caretta*) in the central North Pacific. Fish. Bull., **103**(1): 142-152.

POLOVINA, J. J., I. UCHIDA, G. BALAZS, E. A. HOWELL, D. PARKER & P. DUTTON, 2006. The Kuroshio extension bifurcation region: a pelagic hotspot for juvenile loggerhead sea turtles. Deep Sea Res. II, **53**: 326-339.

REICH, K. J., K. A. BJORNDAL & C. M. DEL RIO, 2008. Effects of growth and tissue type on the kinetics of $^{13}$C and $^{15}$N incorporation in a rapidly growing ectotherm. Oecolog., **155**: 651-663.

SATO, K., Y. MATSUZAWA, H. TANAKA, T. BANDO, S. MINAMAKWA, W. SAKAMOTO & Y. NAITO, 1998. Internesting intervals for loggerhead turtles, *Caretta caretta*, and green turtles, *Chelonia mydas*, are affected by temperature. Can. J. Zool., **76**: 1651-1662.

SCHELL, D. M., V. J. ROWNTREE & C. J. PFEIFFER, 2000. Stable-isotope and electron-microscopic evidence that cyamids (Crustacea: Amphipoda) feed on whale skin. Can. J. Zool., **78**: 721-727.

TAKAI, N., Y. MISHIMA, A. YOROZU & A. HOSHIKA, 2002. Carbon sources for demersal fish in the western Seto Inland Sea, Japan, examined by $\delta^{13}$C and $\delta^{15}$N analyses. Limnol. Oceanogr., **47**(3): 730-741.

TANAKA, H., K. SATO, Y. MATSUZAWA, W. SAKAMOTO, Y. NAITO & K. KUROYANAGI, 1995. Analysis of possibility of feeding of loggerhead turtles during internesting periods based on stomach temperature measurements. Bull. Jap. Soc. Sci. Fish., **61**(3): 339-345. [In Japanese.]

WADA, E., M. TERAZAKI, Y. KABAYA & T. NEMOTO, 1987. $^{15}$N and $^{13}$C abundances in the Antarctic Ocean with emphasis on the biogeochemical structure of the food web. Deep Sea Res., **34**(5-6): 829-841.

WILLIAMS, C. T., C. L. BUCK, J. SEARS & A. S. KITAYSKY, 2007. Effects of nutritional restriction on nitrogen and carbon stable isotopes in growing seabirds. Oecolog., **153**: 11-18.

YOKOYAMA, H. & Y. ISHIHI, 2007. Variation in food sources of the macrobenthos along a land–sea transect: a stable isotope study. Mar. Ecol. Prog. Ser., **346**: 127-141.

First received 15 November 2009.
Final version accepted 30 January 2010.

# PERACARIDS FROM THREE LOW-ENERGY FINE-SAND BEACHES OF MEXICO: WESTERN COAST OF GULF OF CALIFORNIA

BY


GUADALUPE TORRES[1,3]) and JIM LOWRY[2])

[1]) CICIMAR — IPN, Av. IPN s/n Col. Playa el Conchalito; CIBNOR, Mar Bermejo No. 195, Col. Playa Palo de Santa Rita, La Paz BCS 23096, Mexico

[2]) Division of Invertebrate Zoology, Australian Museum, 6 College Street, Sydney, NSW 2010, Australia



## ABSTRACT

Taxonomic knowledge of species from low-energy sandy beaches is scarce in the literature; in spite of the fact, many phyla inhabit sandy shores. Here, is present peracarid species found from concurrent field sampling on intertidal and subtidal zones on low-energy beaches. We found 59 species of orders Amphipoda, Isopoda, Tanaidacea and Cumacea. Dominant orders were amphipods (39 species) and isopods (9 species). For first time 18 species were cited for Gulf of California beaches and 16 for mexican pacific coast. There were 16 common species shared between different beaches. Cirolanid isopods dominated intertidal and amphipods, cumaceans and tanaids dominated subtidal. Although, tanaids where the more dense group ($1191$ ind·m$^{-2}$). Both, number of species and abundance increase from long beaches with fine mineral sands and high organic and silt clay content to small beach with coarse calcareous sand and high slope. Slope varies seasonally according with physical regime of the bay. Patterns of seasonal variation and across shore distribution could not be definite because depend of many variables not studied here, but according with our results some species are distributed both intertidal and subtidal zones.



## RESUMEN

El conocimiento de especies provenientes de playas de baja energía es escaso en la literatura a pesar de que muchos grupos están presentes en las playas arenosas. En este trabajo se presentan las especies de peracaridos encontrados durante un muestreo concurrente en el intermareal y submareal de playas de baja energía. Encontramos 59 especies de los órdenes Amphipoda, Isopoda, Tanaidacea y Cumacea. Los grupos dominantes fueron anfípodos (39 especies) e isópodos (9 especies). Se citan por primera vez 18 especies para playas del Golfo de California y 16 para la costa del pacífico mexicano. Dieciséis especies estuvieron presentes



[3]) Corresponding author; e-mail: gmtorres@ipn.mx

en todos los tipos de playa estudiadas. Los isópodos cirolánidos dominaron en el intermareal y en el submareal los anfípodos, cumáceos y tanaidáceos. Aunque los tanaidáceos fueron el grupo con mayor densidad (1191 ind·m$^{-2}$). Tanto el número de especies como la abundancia aumentan desde playas largas con arena fina y alto contenido de materia orgánica y límo-arcilla hacia la playa pequeña con arena de calcio, gruesa y con pendiente mayor. La pendiente varía estacionalmente de acucrdo al régimen ambiental de la bahía. Las tendencias de las especies en su variación estacional y distribución a través de la playa no fueron claramente definidas ya que dependen de muchas variables no estudiadas aquí, aunque se observaron algunas especies distribuidas tanto en el intermareal como en el submareal.

## INTRODUCTION

Globally low-energy sandy beaches are a frequent feature of coastal environment. However, detail ecological studies on embayed sandy beaches are uncommon in the literature (Eliot et al., 2006). This highlights the need to undertake more detailed research to analyze communities on this kind of sandy beaches.

The Gulf of California, a semi-enclosed sea on the Pacific coast of Mexico, is one of the most biologically diverse regions in the world with 1032 species (80% of the fauna are invertebrates), macrocrustacea, and in particular Peracarida and Decapoda (Hendrickx et al., 2002).

Peracarids are among the most diverse and numerically dominant organisms of sandy beaches (Schlacher et al., 2008). However, despite the research conducted on taxonomy and systematics (Barnard, 1969, 1979; Brusca, 1980; Donath-Hernández, 1985; Brusca, Coelho & Taiti, 2001), peracarids of the Gulf of California are not well known yet, with the exception of the Isopoda for which more taxonomic and ecological information is available than the other orders (Hendrickx et al., 2002). Peracarids associated with sand (67 species reported to entire Gulf of California) are almost certainly underestimate and reflects the absence of detailed study of the small macrofauna associated with sandy-shores (Hendrickx et al., 2002).

In this article, we present the species found during concurrent field sampling performed on three low-energy sandy beaches on Bahía de La Paz (Mexico) and some notes about seasonal variation and across-shore distribution.

## MATERIAL AND METHODS

Three low-energy sandy beaches located on the west coast of Gulf of California, on Bahía La Paz (24°06′N–24°48′N 110°19′W–110°44′W), were

Fig. 1. Map showing the location of the three sandy beaches studied on Bahía de la Paz (Mexico): Balandra, Conchalito and Mogote.

sampled during summer 2003 (10, 17 October 2003), winter 2004 (18, 20 February 2004) and summer (17, 24 February 2006) and winter 2006 (26, 28 September 2006) (fig. 1). Seven stations were located along a transect extending from intertidal to subtidal, in water depths of 0.5 m, 1.0 m, 1.5 m and 3.0 m (fig. 2). Peracarids samples were collected by corers 15 cm diameter (0.018 m$^2$) penetrating 20 cm deep into the substrate, and dredge (last two stations) fig. 2. The sediment was sieved through 1 mm mesh and the residue was preserved in 5% neutralized formalin. The individuals were later sorted from the sediments under a microscope into 70% ethanol. Amphipod families were sorted, employing the interactive keys in the DELTA system (Lowry & Springthorpe, 2001). Isopods families followed taxonomic keys

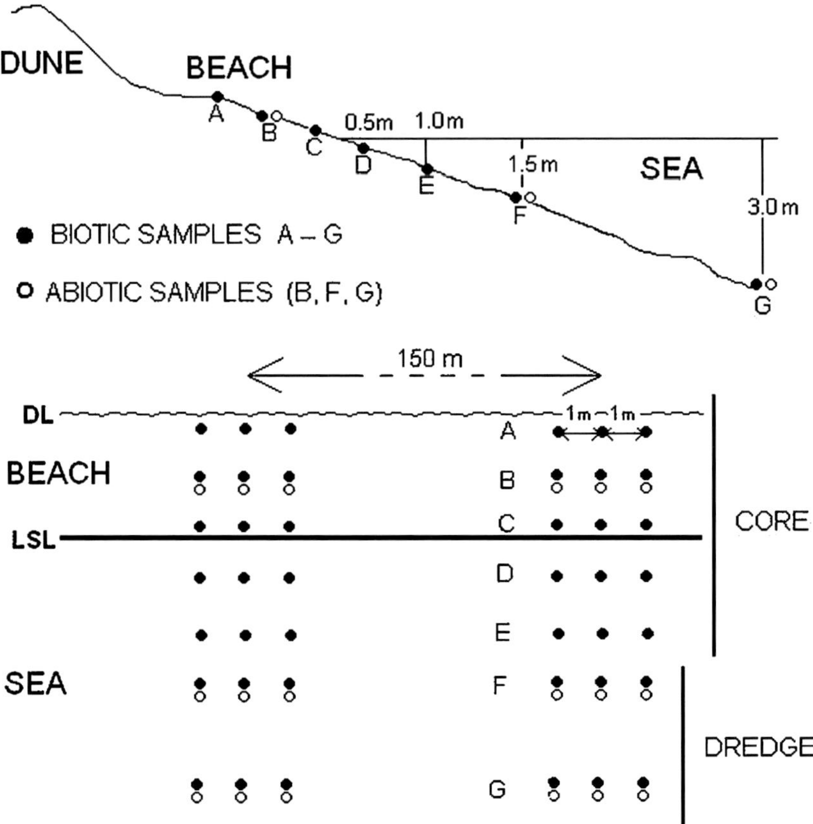



Fig. 2. Sampling design showing the position of transects and stations on the three sandy beaches studied in Bahía de la Paz (Mexico): Balandra, Conchalito and Mogote. LSL: lower swash limit. DL: drift line.

of Brusca, Coelho & Taiti (2001); Cumaceans followed Donath-Hernández (1985) and Tanaids with interactive keys in DELTA system Lowry (1999 onwards). Amphipods species, using many keys mainly from East Pacific coast (Barnard, 1969, 1973, 1979; Barnard & Barnard, 1982, 1985; Barnard & Karaman, 1991; Blake et al., 1995); Isopods, keys of Brusca et al. (1995) and Brusca, Coelho & Taiti (2001); Cumaceans, keys in Donath-Hernández (1985); Tanaids, interactive keys in DELTA system Lowry (1999).

Additionally, at each sampling campaign, were collected three replicate samples for sediment analysis (permeability, organic matter content and mean particle size) in stations B, F and G (fig. 2). Permeability was estimated by Holmes & McIntyre's (1971) method; organic matter content by Schollenberger's (1945) chromic oxidation method and mean particle size following

Folk (1974). Slope of the beaches, were calculated by Emery's method (1961) and tidal height was consulted from the chart datum (González et al., 2006).

## RESULTS

The abiotic data for three low-energy sandy beaches studied are given in table I. Both seasons, Balandra was the smallest (210 m length) and steepest beach with highest permeability and lowest silt clay content. Mogote was the largest (13 000 m length) and gentlest beach with lowest permeability and grain size. Conchalito (1000 m length), longer than Balandra, had coarsest grain size and the highest organic matter and silt clay content. The beach slope was steeper in summer than in winter and mean tidal level was always highest in summer.

Throughout the study period were collected 5163 benthic peracarids belonged to 34 families, 41 genera and 59 species. Amphipods were the dominant taxa (39 species), followed by isopods (9 species), cumaceans (7 species) and tanaids (4 species). Among them, 18 species were cited for the first time for Gulf of California beaches: *Aoroides* cf. *californica*, *Tritella* cf. *tenuissima*, *Caprella mendaz*, *Resupinus visendus*, *Resupinus coloni*, *Pardalisca tenuipes*, *Photis lacia*, *Foxiphalus golfensis*, *Grandiphoxus* cf. *grandis*, *Rhepoxynius* species "C", *Rhepoxynius homocuspidatus*, *Rhepoxynius* cf. *lucubrans*, *Eudevenopus honduranus*, *Tritella laevis*, *Campylaspis biplicata*, *Excirolana chamensis*, *Synidotea laevidorsalis*, *Gnorimosphaeroma oregonense*. Species mentioned above were cited for the first time for the Pacific coast of Mexico too, except *Photis lacia* (Barnard, 1970) and *Eudevenopus honduranus* (Escobar et al., 2002). The main taxonomic groups were present in the three beaches, but in different proportion (fig. 3). Dominant order on Balandra was Amphipoda (46%) and on both Conchalito and Mogote was Isopoda (74% and 52% respectively) (fig. 3). There were 16 common species between three beaches and according with abundance, tanaids presented the highest density in Balandra (1191 ind·m$^{-2}$) (table II). The abundant common specie changed in Conchalito and Mogote: in summer 2003 were an isopod and a tanaid; next season, winter 2004, were amphipod and isopod; in winter 2006, were an isopod and an amphipod respectively; and finally, summer 2006, were isopods for both beaches (table II). Balandra showed the highest abundance of the almost half of common species (7), in Conchalito were 5 and in Mogote 3 (fig. 4). The caprellid, *Tritella* cf. *tenuissima*, was found in the same proportion between Balandra and Mogote (fig. 4).

## TABLE I

Seasonal physical characteristics (mean ± standard error) of low-energy sandy beaches studied at Bahía de La Paz (Mexico) during 2003-2006: Balandra, Conchalito and Mogote

| | Balandra Summer | Conchalito Summer | Mogote Summer | Balandra Winter | Conchalito Winter | Mogote Winter |
|---|---|---|---|---|---|---|
| Beach slope (°) | 5.25 ± 0.35 | 3.50 ± 0.22 | 2.31 ± 0.06 | 3.89 ± 0.00 | 2.68 ± 0.09 | 1.82 ± 0.01 |
| Tidal level (cm) | 50.1 ± 15.60 | 65.4 ± 16.99 | 65.4 ± 16.99 | 47.7 ± 19.10 | 45.5 ± 15.85 | 45.5 ± 15.86 |
| Permeability (m/seg) | 3.4 ± 0.30 | 2.4 ± 0.36 | 2.1 ± 0.17 | 4.8 ± 0.92 | 2.1 ± 0.28 | 1.8 ± 0.12 |
| Organic matter (%) | 0.146 ± 0.017 | 0.24 ± 0.041 | 0.247 ± 0.030 | 0.208 ± 0.100 | 0.26± 0.044 | 0.08 ± 0.014 |
| Silt-clay (%) | 0.047 ± 0.011 | 2.1 ± 1.110 | 0.57 ± 0.257 | 0.060 ± 0.012 | 0.80 ± 0.200 | 0.28 ± 0.066 |
| Grain (mm) | 0.25 ± 0.022 | 0.34 ± 0.018 | 0.23 ± 0.009 | 0.34 ± 0.028 | 0.34 ± 0.022 | 0.22 ± 0.014 |
| Textural group | medium-fine | medium | fine | medium | medium | fine |

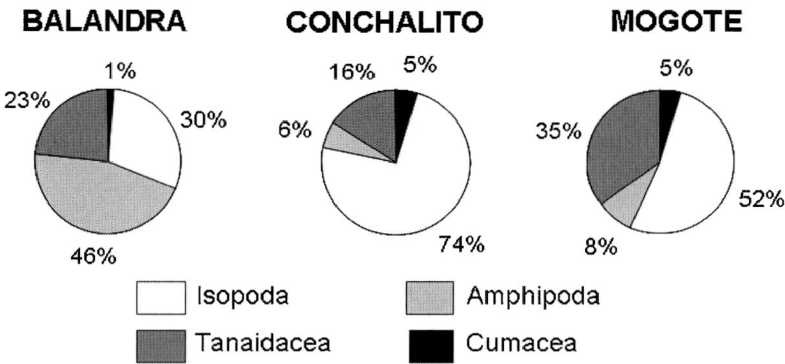


Fig. 3. Percentages of peracarid orders found on the three sandy beaches studied in Bahía de la Paz (Mexico): Balandra, Conchalito and Mogote.

Across-shore distribution of common species showed cumaceans, tanaids and amphipods as subtidal species mainly (fig. 5). However, there were exceptions, as the distribution of tanaids in Mogote (intertidal in summer) and the distribution of *Gnorimosphaeroma oregonense* in Conchalito (intertidal and subtidal) (fig. 5). There were more species inhabiting subtidal that intertidal and in Balandra the distribution of the species is wider than in the other two beaches (fig. 5).

Each beach had other species. Balandra, 14 particular species (*Cumella* n. sp., *Synidotea laevidorsalis*, *Tritella laevis*, *Aoroides* cf. *californica*, *Batea transversa*, *Bemlos tehuecos*, *Foxiphalus golfensis*, *Foxiphalus obtusidens*, *Harpiniopsis* sp., *Melita dentata*, *Melita* sp., *Uruthoe* sp., *Batea* sp., *Orchestia* sp.); Conchalito, five particular species (*Campylaspis biplicata*, *Munna* sp., *Phyllodurus abdominalis*, *Tritella* sp. "A", *Uristes* sp.) and Mogote, three particular species (*Pardalisca tenuipes*, Dexaminidae, *Resupinus coloni*).

## DISCUSSION

This is the first report of peracarid species of low-energy sandy beaches on Bahía de La Paz, in the southwestern part of the Gulf of California. Beaches have different peracarid assemblage and share 16 common species. In therms of both species and abundance of individuals, Balandra beach shelters highest number of species in an environment with steep slope and high permeability. By contrast, Mogote beach shelters a poorest assemblage in an environment with gentle slope and small grain size. Subtidal zone shelters a higher number of species than intertidal one. The slope showed seasonal variations, although most of the species distribution is similar during both seasons.

TABLE II

Mean seasonal density of common species (ind·m$^{-2}$) shared between three sandy beaches studied at Bahía de La Paz (Mexico) during 2003-2006. Balandra (B); Conchalito (C); Mogote (M); summer (s); winter (w); 2003 (03); 2004 (04); 2006 (06). *Abbreviatons, see fig. 4

| Species | * | Bs03 | Bw04 | Bw06 | Bs06 | Cs03 | Cw04 | Cw06 | Cs06 | Ms03 | Mw04 | Mw06 | Ms06 |
|---|---|---|---|---|---|---|---|---|---|---|---|---|---|
| *Cyclaspis bituberculata* Donath-Hernández, 1988 | Cb | | 3 | 3 | 3 | | 16 | | 1 | | 1 | | |
| *Cyclaspis* n. sp. | Cn | | 9 | 8 | 1 | | | | 3 | | 1 | | |
| *Excirolana braziliensis* Richardson, 1912 | Eb | 442 | 112 | 94 | 304 | 3 | | 1 | | | | | 7 |
| *Gnorimosphaeroma oregonense* (Dana, 1853)[1,2] | Go | 49 | 20 | 18 | 7 | 3 | 11 | 66 | 358 | | | 3 | |
| *Excirolana mayana* Ives, 1891 | Em | 30 | 7 | 1 | 7 | 26 | 16 | 46 | 90 | 1 | 3 | | |
| *Excirolana chamensis* Brusca & Weinberg, 1987[1,2] | Ec | 7 | 19 | 11 | 1 | 12 | 3 | | | 7 | 36 | 11 | 97 |
| *Pseudosphyrapus* sp.[1,2] | Ps | 493 | 326 | 163 | 1191 | | 1 | 50 | | 27 | | 3 | 3 |
| Kalliapseudidae[1,2] | Ka | | 1 | | 1 | | | | 3 | 1 | | | |
| *Eudevenopus honduranus* Thomas & Barnard, 1983[2] | Eh | 1 | 73 | 140 | 75 | | 1 | | | 3 | | | 12 |
| *Ampelisca pugetica* Stimpson, 1864 | Ap | 1 | | 9 | 1 | | 8 | | | | 1 | | |
| *Photis lacia* Barnard, 1962[2] | Pl | | 1 | 16 | 1 | | 47 | | 1 | | 1 | | |
| *Photis californica* Stout, 1913 | Pc | 3 | 7 | 1 | 1 | | 3 | | | | | | 15 |
| *Rephoxynius heterocuspidatus* (Barnard, 1960) | Rh | 3 | 30 | 233 | 47 | | | 4 | | 3 | 34 | | 11 |
| *Rhepoxynius* species C Barnard & Barnard, 1982[1,2] | Rc | 12 | 51 | 53 | 35 | | | | | 1 | 1 | | |
| *Tritella* cf. *tenuissima* Dougherty & Steinberg, 1953[1,2] | Tt | | 3 | | | | 1 | | | | | 3 | |
| *Rhachotropis* sp. | Ra | 1 | | | 1 | 15 | | | | | | | 20 |

[1] Cited for the first time in beaches from the Gulf of California.
[2] Cited for the first time in Pacific Mexican Coast.

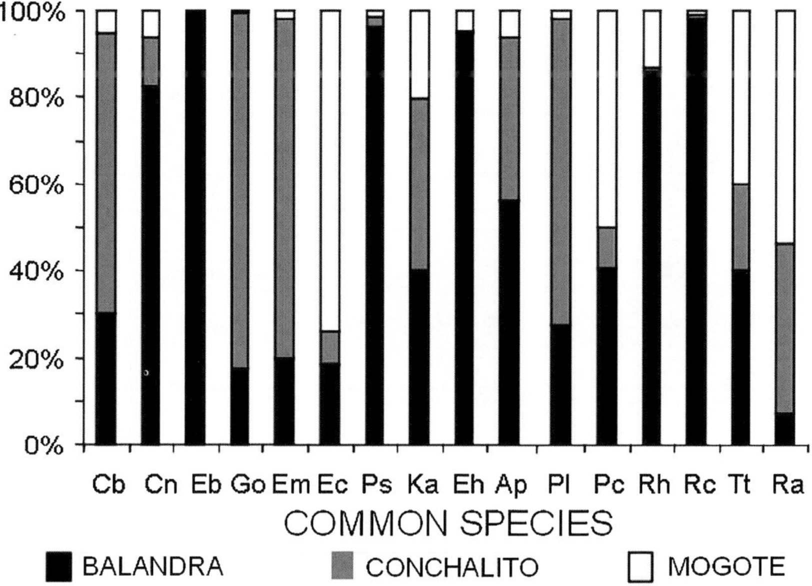


Fig. 4. Percentages of common species shared between the three sandy beaches studied on Bahía de la Paz (Mexico): Balandra, Conchalito and Mogote.

All beaches studied here backed by a dune system, are stable and supply by sediments from the very shallow subtidal (Álvarez-Arellano et al., 1997). However, the origin of their sediments is different. Mogote belonged to a sandy barrier supplied by northwest mineral rocks (Álvarez-Arellano et al., 1997). Calcium carbonate and mineral sediments from a neighbor lagoon (Cruz-Orozco et al., 1989) supply Conchalito and Balandra sands are totally of calcium carbonate since come from fossil coral banks (Álvarez-Arellano et al., 1997). Such sedimentary differences result in their physical qualities as permeability, organic matter and silt clay content. McLachlan & Brown (2006) explain characteristics related to sediments on interstitial environment. First, there are two main types of sands: calcium carbonate (of marine origin) and mineral (of cliff erosion). Calcium carbonate sands have irregular shape and slow sink in water while mineral sands tended to be regular and sink quickly. On the other hand, granulometry of sand defines the interstitial system (i.e., nature and size of sand), and the process of water filtration controls its dynamics. That is, beaches with coarser sands filter large volumes of water at fast rates and consequently have elevated flush and oxygenation; by contrast, fine sands filter smaller volumes of water at low rates. Furthermore, surplus organic input will lead anaerobic conditions. Last, fine sands tended to reduce its resistance to sharing depending of its content of water and in saturated

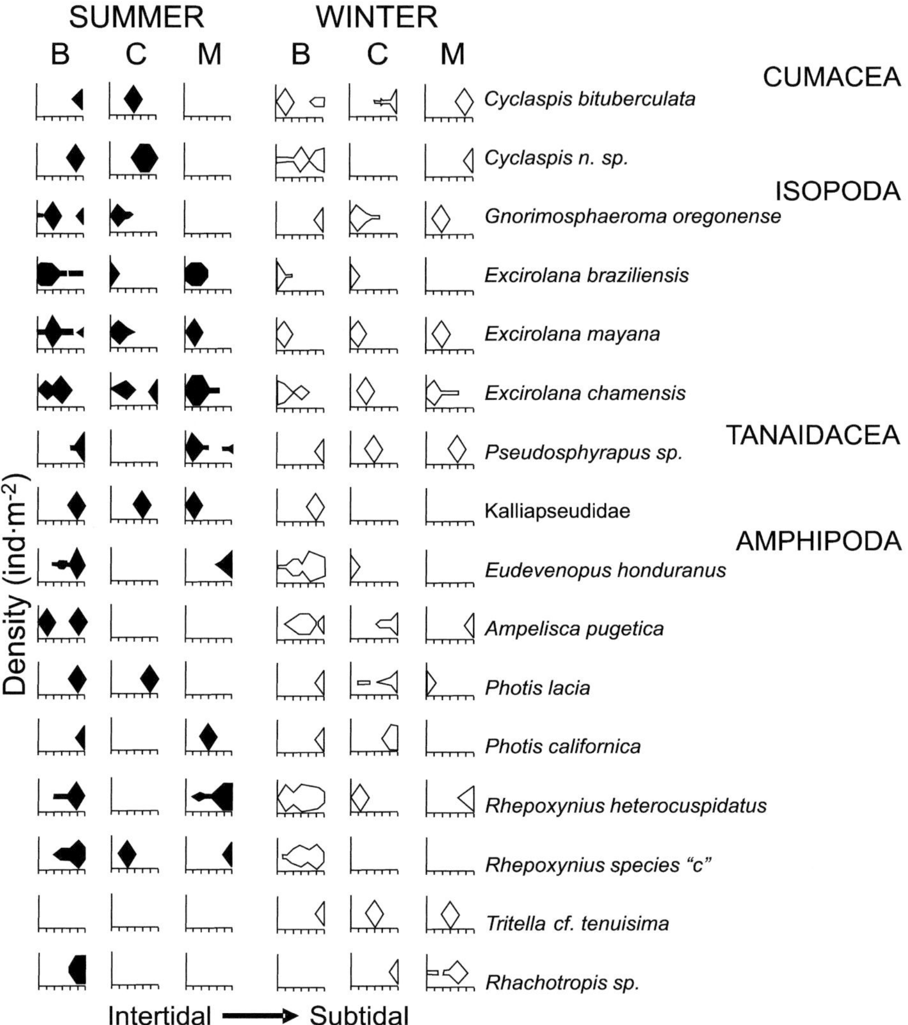


Fig. 5. Across shore distribution of peracarid species found on the three sandy beaches studied in Bahía de la Paz (Mexico): Balandra, Conchalito and Mogote.

condition the reduction is highest and the burrowing of animals is impossible. We sampled peracarids on all sites because can establish populations on all beach types due to their high motion, rapid burrowing, and their necessity of vigorous swash action (McLachlan & Brown, 2006). Peracarids, as tanaids, find on sandy shores adequate conditions to living since they can select particles for both feeding and tube construction (Krasnow & Taghon, 1997). In this sense, all beaches are environments to peracarids. However, Balandra has better environmental conditions because has coarse, sharply, calcareous sand

and elevated flush and oxygenation. Conchalito and Mogote are not better than Balandra because have highest organic matter and silt clay contents and this implicate anaerobic conditions.

In a preliminary description of the sandy-beach fauna along Mexican coastline was established that Mexican sandy-beach fauna was impoverished in terms of number of species (Dexter, 1976). We disagree with this work because the species richness found was higher (59) than previously reported (three). Our differences could be due to Dexter's study is a snapshot. Jaramillo et al. (1995) and Brazeiro & Defeo (1996) explain that sampling strategies affect observed trends on sandy beaches. Our sampling took place during two seasons and the same number or years while Dexter sampled only once. Peracarid fauna found in this assemblage (59 species) represent almost 88% of 67 species reported to sandy shores on entire Gulf of California (Hendrickx et al., 2002).

Subtidal peracarid species are more numerous than intertidal due to on intertidal of sandy beaches exist a swash climate (i.e., water movement over the beach face) that limits the settlement of organisms (Defeo & McLachlan, 2005).

Our results show seasonal differences in slope, which agree with our expectations because Bahía de La Paz has marked spatial and seasonal variability of oceanographic conditions by variations in local wind regime and solar radiation (Jiménez-Illescas et al., 1997). In spite of the fact, most of the time tidal currents control coastal processes (Brambila-Miranda, 1989), strong northerly, northwesterly winds erode beach shores in winter, and strong tidal currents in summer carry sediments to the beaches and increase the budget of sediments on beaches shores (Brambila-Miranda, 1989; Obeso-Nieblas et al., 2002). Variability between seasons in distribution of peracarid assemblage was not clear because it was not a continuous response to the intertidal gradient of types of beaches. In across shore distribution of macrofauna physical factors, taxonomic and physiological differences in the environment and fauna play a role (McLachlan & Brown, 2006).

We concluded that there are few other studies of Gulf of California with which the results of the present investigation can be compared. Even if peracarid assemblage can vary seasonally and across shore, in this work we have not elements to give one explanation, so this factor requires further study whit population studies. It is necessary too a detailed systematic study to define new species and biogeography of all species found.

Low-energy beaches on Bahía de La Paz have different peracarid assemblages that increase in both number of species and abundances from long beaches with fine sands and high organic and silt-clay content to short beach with coarse calcareous sand and high slope. Even though peracarids increased from higher to lower beach levels zonation patterns were not distinguished.

Some specimens of Amphipoda and Isopoda orders are in the Invertebrate Collection of Australian Museum with the numbers of reference: P.78860-P.78864 and P.78991-P.79016.

## ACKNOWLEDGEMENTS

We express our appreciation to Dr. Stephen J. Keable and Dr. Lauren Hughes (Marine Invertebrate Section, Australian Museum) for making the hospitality and resources of the laboratory available to us during the identification of specimens. We specially thank Dr. Daniel Roccatagliata for his valuable suggestions for improvement the previous identification of cumaceans. We also thank anonymous referees for critical readings and useful comments. The first author is grateful for the financial support granted by EDI, COFAA, IPN and CIBNOR-CONACYT, through fellowship (Grant 200917 at AUMUS, Australia).

## REFERENCES

ÁLVAREZ-ARELLANO, A. D., H. ROJAS-SORIANO & J. J. PRIETO-MENDOZA, 1997. Geología de la Bahía de La Paz y áreas adyacentes. In: R. J. URBÁN-RAMÍREZ & M. RAMÍREZ-RODRÍGUEZ (eds.), La Bahía de La Paz, investigación y desarrollo: 13-29. (Universidad Autónoma de Baja California Sur, Centro Interdisciplinario de Ciencias Marinas, México y SCRIPPS Institute, La Paz, Baja California Sur).

BARNARD, J. L., 1969. A biological survey of Bahía de los Ángeles of Gulf of California, Mexico, IV. Benthic Amphipoda (Crustacea). Trans. San Diego Soc. Nat. Hist., **15**(13): 175-228.

— —, 1973. Revision of Corophiidae and related families (Amphipoda). Smith. Contrib. Zool., **151**: 1-27.

— —, 1979. Littoral gammaridean Amphipoda from the Gulf of California and the Galapagos Islands. Smith. Contrib. Zool., **271**: 1-149.

BARNARD, J. L. & CH. M. BARNARD, 1982. Revision of *Foxiphalus* and *Eobrolgus* (Crustacea: Amphipoda: Phoxocephalidae) from American oceans, Smith. Contrib. Zool., **372**: 1-35.

— — & — —, 1985. The genus *Rhepoxynius* (Crustacea: Amphipoda: Phoxocephalidae) in American seas, Smith. Contrib. Zool., **357**: 1-49.

BARNARD, J. L. & G. S. KARAMAN, 1991. The families and genera of marine gammaridean Amphipoda (except marine gamaroids). Rec. Austr. Mus., (supplement) **13**: 1-866.

BLAKE, J. A., L. WATLING & P. H. SCOTT, 1995. Taxonomic atlas of the benthic fauna of the Santa Maria Basin and the Western Santa Barbara Channel, California USA, **12**: 1-247.

BRAMBILA-MIRANDA, E. S., 1989. Variación morfológica, textural y mineralógica en el período de marzo de 1986 a febrero de 1987 de las playas del sureste de la Bahía de La Paz, B.C.S.: 1-59. (Tesis Licenciatura, Universidad Autónoma de Baja California Sur, Mexico).

BRAZEIRO, A. & O. DEFEO, 1996. Macrofauna zonation in microtidal sandy beaches: is it possible to identify patterns in such variable environments? Estuar. Coast. Shelf. Sci., **42**: 523-536.

BRUSCA, R. C., 1980. Common Intertidal Invertebrates of the Gulf of California: 1-513. (University of Arizona Press).

BRUSCA, R. C., V. COELHO & S. TAITI, 2001. A guide to the coastal isopods of California. Electronic publication: http://tolweb.org/notes/?note_id=3004

BRUSCA, R. C., R. WETZER & S. C. FRANCE, 1995. Cirolanidae (Crustacea: Isopoda: Flabellifera) of the tropical eastern Pacific. Proc. San Diego Soc. Nat. Hist., **30**: 1-96.

CRUZ-OROZCO, R. A., P. ROJO-GARCÍA, L. GODÍNEZ-ORTA & E. NAVA-SÁNCHEZ, 1989. Topografia, hidrología y sedimentos de los márgenes de la laguna de La Paz, B.C.S. Rev. Inv. Cient. UABC, **1**(3): 3-16.

DEFEO, O. & A. MCLACHLAN, 2005. Patterns, processes and regulatory mechanisms in sandy beach macrofauna: a multi-scale analysis. Mar. Ecol. Prog. Ser., **295**: 1-20.

DEXTER, D., 1976. The effect sandy-beach fauna of Mexico. South. Nat., **20**(4): 479-485.

DONATH-HERNÁNDEZ, F. E., 1985. Cumáceos (Crustacea, Peracarida de Baja California y del Golfo de California: Sistemática, aspectos ecológicos y biogeraía: 1-124. (Tesis Maestría, Centro de Investigación Científica y de Educación Superior de Ensenada, Mexico).

ELIOT, M. J., A. TRAVERS & I. ELIOT, 2006. Morphology of a low-energy beach, Como beach, western Australia. Journ. Coast. Res., **22**(1): 63-77.

EMERY, K. O., 1961. A simple method of measuring beach profiles. Limnol. Ocean., **6**: 90-93.

ESCOBAR-BRIONES, E., I. WINFIELD, M. ORTIZ, R. GASCA & E. SUÁREZ, 2002. Biodiversidad, Taxonomía y Biogeografía de Artrópodos de México: Hacia una síntesis del conocimiento. Facultad de Ciencias, UNAM Mexico, **III**: 341-371.

FOLK, R. L., 1974. Petrology of sedimentary rocks: 1-182. (Hemphill Publishing Company, Austin).

GONZÁLEZ, J. I., R. SOTO & J. OCHOA, 2006. Predicción de Mareas en México. Oceanografía Física. CICESE. http://oceanografia.cicese.mx/predmar/

HENDRIX, M. E., R. C. BRUSCA & G. RAMÍREZ-RESÉNDIZ, 2002. Biodiversity of macrocrustaceans in the Gulf of California, Mexico. Contributions to the Study of East Pacific Crustaceans, **1**: 349-368.

HOLMES, N. A. & A. D. MCINTYRE (eds.), 1971. Methods for the study of marine benthos. IBP Handbook, **16**: 1-387.

JARAMILLO, E., A. MCLACHLAN & J. DUGAN, 1995. Total sample area and estimates of species richness in exposed sandy beaches. Mar. Ecol. Prog. Ser., **119**: 311-314.

JIMÉNEZ-ILLESCAS, A. R., M. OBESO-NIEBLAS & D. A. SALAS-DE LEÓN, 1997. Oceanografía física de la Bahía de La Paz, B. C. S. In: R. J. URBÁN-RAMÍREZ & M. RAMÍREZ-RODRÍGUEZ (eds.), La Bahía de La Paz, investigación y desarrollo: 31-42. (Universidad Autónoma de Baja California Sur, Centro Interdisciplinario de Ciencias Marinas, México y SCRIPPS Institute, La Paz, Baja California Sur).

KRASNOW, L. D. & G. L. TAGHON, 1997. Rate of tube building and sediment particle size selection during tube construction by the tanaid crustacean, *Leptochelia dubia*, Estuaries, **20**(3) (1997): 534-546.

LOWRY, J. K., 1999-onwards. 'Crustacea, the higher taxa: description, identification, and information retrieval.' Version: 2 October 1999. http://crustacea.net/
LOWRY, J. K. & R. T. SPRINGTHORPE, 2001. Amphipoda: families. Version 1: 2 September 2001. http://www.crustacea.net/
McLACHLAN, A. & A. BROWN, 2006. The ecology of sandy shores: 1-373. (Academic Press, Burlington).
OBESO-NIEBLAS, M., J. H. GAVIÑO RODRIGUEZ, A. JIMÉNEZ-ILLESCAS & B. SHIRASAGO-GERMÁN, 2002. Simulación numérica de la circulación por marea y viento del noroeste y sur en la Bahía de La Paz, Baja California Sur. Oceánides, **17**(1): 1-12.
SCHLACHER, T. A., D. S. SCHOEMAN, J. DUGAN, M. LASTRA, A. JONES, F. SCAPINI & A. MACLACHLAN, 2008. Sandy beach ecosystems: key features, sampling issues, management challenges and climate change impacts. Mar. Ecol., **29** (1): 70-90.
SCHOLLENBERGER, C. J., 1945. Determination of soil organic matter. Soil Sci., **59**: 53-56.

First received 16 November 2009.
Final version accepted 4 March 2010.

# TEMPORAL AND SPATIAL VARIATIONS IN FUNCTIONAL-TRAIT COMPOSITION (FUNCTIONAL DIVERSITY) OF MACROCRUSTACEAN COMMUNITIES IN SEAGRASS MEADOWS

BY


KATSUMASA YAMADA[1,3]), MASAKAZU HORI[1]), MASAHIRO NAKAOKA[2])
and MASAMI HAMAGUCHI[1])

[1]) National Research Institutes of Fisheries and Environment of Inland Sea, 2-17-5 Maruishi, Hatsukaichi, Hiroshima 739-0452, Japan

[2]) Field Science Center for Northern Biosphere, Akkeshi Marine Station, Hokkaido University, 5 Aikappu, Akkeshi-cho, Akkeshi-gun, Hokkaido 088-1113, Japan


## ABSTRACT


Functional diversity (FD) is the value and range of functional traits and ecosystem function of organisms present in a community. In marine macrocrustacean communities, FD has traditionally been expressed by taxonomic composition, and patterns of taxonomic richness (e.g., species richness [SR]) are considered relevant to FD. However, SR is not always used to express FD because there may be overlaps in species' functional roles (i.e., functional redundancy) due to multiple-traits of many macrocrustacean species. This study described the spatiotemporal variations of FD of a macrocrustacean communities using the data collected by 2 spatiotemporal sampling designs in seagrass meadows at the Akkeshi-ko estuary (northeastern Japan) and Tokyo Bay (central part of Japan) along Pacific coast. FDs were measured using ecological multiple-traits such as occurrence, life type, feeding type, and size, a total of 32 categories in 4 functional traits. FD variation was significantly different among areas (tens of km$^2$). Areas separated into 2 types based on FD variations, i.e., high-FD areas and low-FD areas. Spatiotemporal variations in FD and SR were compared to determine whether SR patterns are related to variations in FD. The pattern of spatiotemporal variations was different between FD and SR. High SR but low FD (i.e., high functional redundancy) and low SR but high FD (i.e., low functional redundancy) were occasionally found. These results suggest that SR could not be represented as FD in the macrocrustacean community. Evaluation of the functional redundancy in the macrocrustacean community is important to detect patterns of variation in ecosystem function and to identify the ecological determinants of FD.


---


[3]) Corresponding author; e-mail: yamadaka@affrc.go.jp

## INTRODUCTION

Functional diversity (FD) is the value and range of functional traits and ecosystem function of organisms present in a community (e.g., Díaz & Cabido, 2001; Petchey et al., 2004). Understanding FD spatial and temporal variations is important in explaining patterns of ecological-trait components, which can further leads to explaining assembly and/or ecosystem processes (Wright et al., 2006; Petchey et al., 2007; Sasaki et al., 2009). FD variation in a community depends on degrees of overlap in species' functional roles (i.e., functional redundancy) (e.g., Rosenfeld, 2002; Loreau, 2004). For example, loss of a few species with unique functions may lead to FD changes in the community, whereas loss of a few functionally-redundant species will not have this effect. Understanding how species richness (SR) is related to FD is, therefore, needed to detect ecosystem consequences by SR changing.

There have been few attempts to measure FD in marine macrocrustacean assemblages (e.g., Decapoda, Amphipoda, Mysidacea, Isopoda, and Tanaidacea) (Boström et al., 2006; Yamada et al., 2007b). FD was measured in previous studies by grouping species based on few or several dominant ecological traits. On the other hand, it has been suggested that the use of multiple categorical traits (which are associated directly and/or indirectly with ecological function) may be more appropriate to explain patterns of ecological-trait components in communities (Petchey et al., 2007; Cianciaruso et al., 2009; Sasaki et al., 2009). However, definitions of multiple categorical traits are difficult for many macrocrustacean species. For example, categories of ecological feeding type are difficult to discern in many amphipods because they are often feeding generalists (i.e., opportunistic feeders) that contribute to both grazing and detritus food chains. Consequently, FD of a marine macrocrustacean community has traditionally been addressed by describing taxonomic composition of the assemblage, and emergent patterns are considered relevant to putative FDs. However, taxonomic indices (e.g., SR) do not always represent FD (Rosenfeld, 2002; Loreau, 2004; Wright et al., 2006) because of overlaps in species' functional roles (i.e., functional redundancy) due to multiple-traits of many macrocrustacean species.

In this study, we firstly described FD spatiotemporal variations in macrocrustacean communities in seagrass meadows, which are hotspots of biodiversity in coastal areas. FD measurement was based on multiple traits (Botta-Dukát, 2005; De Bello et al., 2005; Lavorel et al., 2008) of macrocrustacean species, such as occurrence, life type, feeding type, and size, a total of 32 categories in 4 functional traits (table I). To determine spatiotemporal variations

TABLE I

Macro-crustacean functional traits and their categories used in the analysis. Almost species belong to more than one trait category (multiple membership)

| Macro-crustacean functional trait | Trait categories |
| --- | --- |
| 1. Occurrence | Abyssal; Marine; Brackish; Fresh water; Terrestrial and littoral |
| 2. Life type | Bbrore; Commensal; Epi-infauna; Epifauna; Infauna; Interstitia; Pelagic; Phreatic; Periphytic; Substrata (e.g., Rock); Streams; Terrestrial; Swarm; Nest builder; Live in the shell of gastropoda; Tube |
| 3. Feeding type | Detrivore; Predator; Planktivore; Scavenger; Suspension feeder; Grass grazer; Algae (seaweed) grazer |
| 4. Size | Large (adult >30 mm); Middle (adult 10-30 mm); Small (adult <10 mm) |

in FD, we collected data using 2 types of spatiotemporal sampling designs in 2 case studies (Nakaoka et al., 2007; Yamada et al., 2007b): (1) in Case I, we made measurements of the variations on local scales (several kilometres: in the Akkeshi-ko estuary); (2) in Case II, we made measurements on a regional scale (tens of kilometres: in Tokyo Bay).

Second, we compared spatiotemporal variations in FD and SR in order to assess the overlap in species' functional roles (functional redundancy) in the communities. If the spatiotemporal variations and patterns of fluctuations between FD and SR show the same trend, variation in SR can be represented as that in FD (i.e., non functional redundancy). On the other hand, if spatiotemporal variations were different between them, functional redundancy may cause in the community. Further, degree of functional redundancy was also evaluated from the relationship between SR and FD (Díaz & Cabido, 2001; Petchey et al., 2007; Cianciaruso et al., 2009). For example, if high SR but low FD, high functional redundancy may occur in the community. Conversely, low SR but high FD indicates low functional redundancy.

## MATERIALS AND METHODS

### Study areas and sampling procedure

We measured spatiotemporal variations in FD and SR in 2 case studies (field sampling designs). In Case I, measurements were made in the Akkeshi-ko estuary (North-eastern Japan; fig. 1), which is semi-enclosed (32 km$^2$) and connected to Akkeshi Bay by a narrow channel (ca. 500 m wide). Seagrass beds were continuous in this estuary (ca. 12.6 km$^2$ in areal extent) (Mukai,

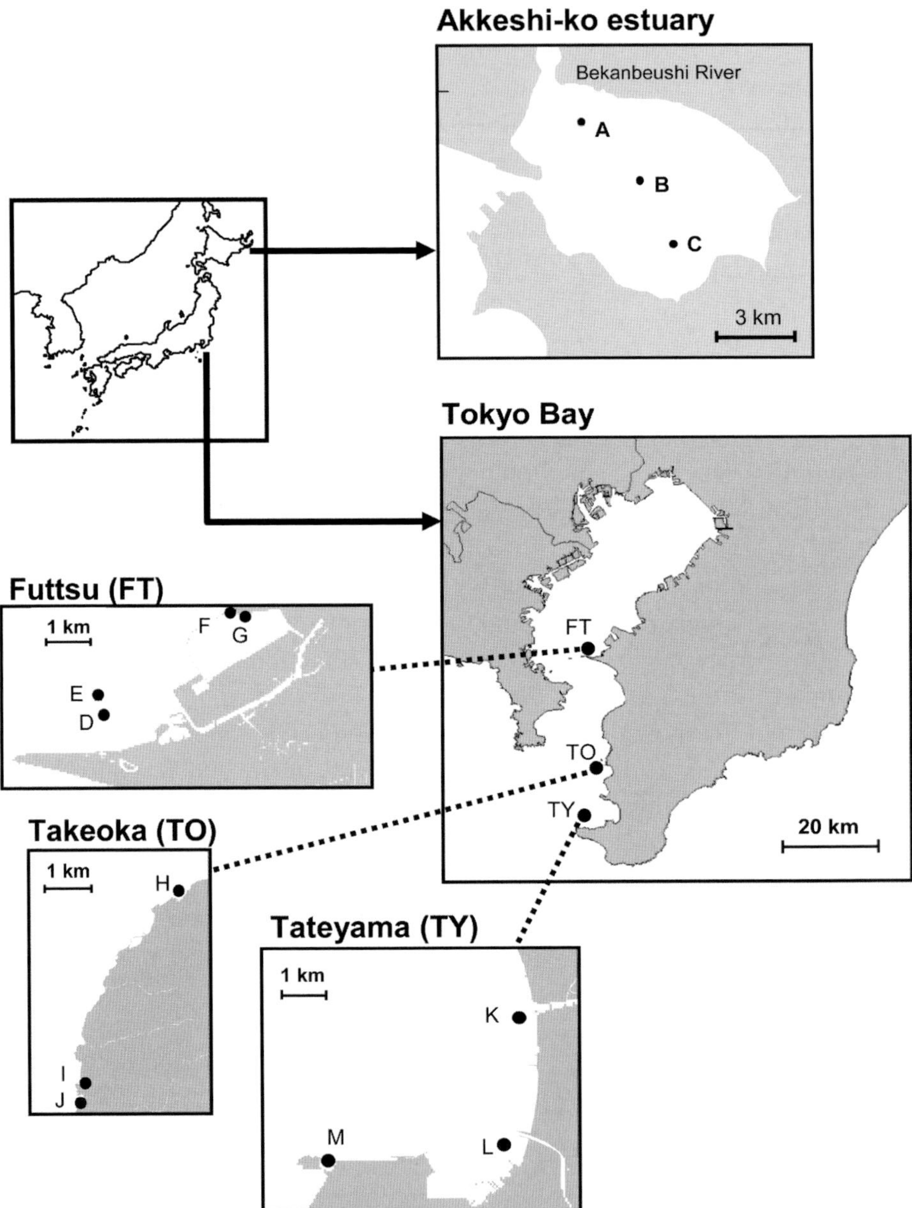


Fig. 1. Three study sites in the Akkeshi-ko estuary (St. A–C), and 10 study sites (St. D–M) in 3 areas (FT, TO, and TY) of Tokyo Bay. Each site had dense vegetation of *Zostera* species (Nakaoka et al., 2007; Yamada et al., 2007b).

2006). Three sites (St. A–C; depth, 0.9-1.5 m) with dense vegetation of seagrass (*Zostera marina*) were selected for the collection of macrocrustacean community samples (fig. 1). Physiological and biological environmental conditions (e.g., river input, tides and currents, and seagrass vegetation) at each site are summarized in Mukai (2006), Yamada et al. (2007a, b, 2010), Tanaka et al. (2008) and Yamada (2008). Quantitative sampling was conducted in June, September, and November in 2004. Three replicate samples were collected with an epibenthic sledge (height, 40 cm; width, 60 cm; mesh size, 500 $\mu$m) on each sampling occasion at each site. The sledge was towed horizontally for a distance of 40 m. The actual tow lengths were measured with a global positioning system (GPS) plotter (JLU-128; JRC).

In Case II, hierarchical spatial variations in FD and SR were evaluated in Tokyo Bay (Central Japan; fig. 1) (Nakaoka et al., 2007; T. Era, K. Yamada and M. Nakaoka, unpubl.), which is also semi-enclosed, but broader (922 km$^2$) than Akkeshi-ko estuary (Case I). Seagrass beds were patchy or continuously developed (ca. <100 m$^2$-1.8 km$^2$) at these sites (Shouji & Hasegawa, 2004; Nakaoka et al., 2007; Yamakita & Nakaoka, 2009). Three areas (Futtsu [FT], Takaoka [TO], and Tateyama [TY]) with seagrass vegetation (*Z. marina* and *Z. caulescens*) were selected, and within each, 3-4 sites (St. D–M; depth, 0.7-3.8 m) were selected for the collection of macrocrustacean community samples (fig. 1). Quantitative sampling was conducted in summer (June–August), 2006. Physiological and biological environmental conditions at each site are summarized in Unoki (1985), Shouji & Hasegawa (2004) and Furukawa & Okada (2006). Three replicate samples were collected from each site using an epibenthic sledge (equipment used in Case I). The sledge was towed horizontally for a distance of 20-140 m. The actual length of each tow was measured with a GPS plotter.

A total of 57 samples (Akkeshi-ko estuary, 3 sites × 3 months × 3 replicates; Tokyo Bay, 10 sites [in 3 areas] × 3 replicates) was collected and preserved in a 10% formalin seawater solution. In the laboratory, the species (or genus) of macrocrustacean community members were identified, and the numbers of individuals were counted.

## Macrocrustacean functional traits and numerical analyses

In analyses of FD, traits must be linked to the function of interest (Petchey & Gaston, 2002; Petchey et al., 2004, 2007). Therefore, we first selected the functional traits of macrocrustacean species that are considered key determinants of their role in seagrass ecosystems. A total of 32 categories in four ecological traits: habitat, life type, feeding type, and body size, a total of 32 categories were evaluated (table I), according to mainly Bousfield (1973), Biernbaum (1979), Nelson (1980), Nishimura (1995), Yamada et al. (2007b, 2010), Yamada (2008) and Guerra-García & Tierno de Figueroa (2009) and literature therein. When we were unable to identify specimens down to the species level, we assigned functional categories by generic traits.

Each value for the traits was recorded in a separate column of the trait category (i.e., a value of 0 or 1 was recorded in a separate column of the trait category). A "fuzzy coding approach" (Bremner et al., 2003; Bady et al., 2005; Tillin et al., 2006; Sasaki et al., 2009) was applied because species showed multiple memberships in different categories in which a trait was divided, and the sum of the scores for a species across that trait was set to 1. For example, with regard to the trait of feeding type, when a species is both a detritivore and a suspension feeder, the trait category have a score of 0.5 as a detritivore and 0.5 as a suspension feeder for this species.

For measurement of FD, pairwise dissimilarity between species $i$ and $j$ was calculated as Euclidean distance ($d_{ij}$):

$$d_{ij} = \sqrt{\sum_{k=1}^{h} (q_{ik} - q_{jk})^2},\tag{1}$$

where $q_{ik}$ and $q_{jk}$ are the scores for trait category $k$ ($1 \leqslant k \leqslant h$) for species $i$ and $j$, respectively. FD was then calculated for each sampling using an adaptation of the index of species dissimilarity (Rao, 1982; Bady et al., 2005; Botta-Dukát, 2005; Lavorel et al., 2008; Sasaki et al., 2009):

$$FD = \sum_{i=1}^{N} \sum_{j=1}^{N} d_{ij} P_i P_j,\tag{2}$$

where $P_i$ and $P_j$ are the proportional abundances of species $i$ and $j$, respectively. Thus, FD is the sum of dissimilarities in trait space among all possible pairs of species weighted by the product of the species' relative abundances.

Spatial and temporal variations in FD and SR in Case I (Akkeshi-ko estuary) were evaluated by two-way factorial analysis of variance (ANOVA), and spatial variations in Case II (Tokyo Bay) were evaluated by nested ANOVA.

## RESULTS

### Case I: Akkeshi-ko estuary

A total of 40 species were captured at the Akkeshi-ko estuary (table II). Changes in FD were in a low level (0.05-0.36). Spatiotemporal variations differed between FD and SR (table III, fig. 2). Main effects of month and site were significant for SR, but the interaction term was not. On the other hand, the temporal variation in FD and interaction between month and site were significant, but spatial variation was not (table III). Functional redundancy (high SR, low FD) and reduced functional redundancy (low SR, high FD) were observed occasionally. For example, high functional redundancy was observed at St. A and B in September and at all sites in December; low functional redundancy was observed at St. A in June and St. C in September.

### Case II: Tokyo Bay

A total of 83 species were captured at the Tokyo Bay (table IV). Spatial variations were different between FD and SR (table V, fig. 3). Spatial variation in SR was significant at the site level, but not at the area level. In contrast, spatial variation in FD was significant at the area level, but not at the site level (table V). Areas were of 2 types based on FD variations. FT and TO were high-FD areas (0.28-0.52), whereas TY was a low-FD area (0.17-0.23). High functional redundancy and low functional redundancy occurred at regional and local scales. For example, high functional redundancy occurred at St. L and M in the TO area, and low functional redundancy occurred in the FT area (St. D–G).

## DISCUSSION

We showed spatiotemporal variations in the FD and SR of a macrocrustacean community in seagrass beds. Spatiotemporal patterns of variation differed between FD and SR in both case studies (tables III & V), indicating that SR was not representative of FD in these communities. In other words, our key finding highlights that even if the SR of a macrocrustacean community is high, FD is not always high, as suggested by theoretical and empirical studies for other terrestrial and aquatic faunal communities (e.g., Micheli & Halpern, 2005; Wright et al., 2006; Petchey et al., 2007).

## TABLE II

Mean densities (inds · m$^{-2}$) and temporal and spatial occurrence patterns for 40 species collected in Akkeshi-ko estuary. ++: >100 inds · m$^{-2}$, +: <100 inds · m$^{-2}$

| Species | Mean density (inds · m$^{-2}$) | A June | A Sep. | A Dec. | B June | B Sep. | B Dec. | C June | C Sep. | C Dec. |
|---|---|---|---|---|---|---|---|---|---|---|
| **Decapoda** | | | | | | | | | | |
| *Crangon* sp. | 27.36 | ++ | + | + | + | + | + | + | + | + |
| *Spirontocaris ochotensis* | 1.20 | | | | + | | + | + | + | + |
| *Heptacarpus grebnitzkii* | 0.91 | | + | + | + | + | + | | | + |
| *Pandalopsis pacifica* | 0.84 | | | | + | + | + | | + | + |
| *Pandalus kessleri* | 0.19 | | | | | + | + | | | |
| *Heptacarpus rectirostris* | 0.19 | + | | | + | | | + | | |
| *Eualus leptognathus* | 0.01 | | | | + | + | | | | |
| *Lebbeus speciosus* | 0.01 | | | | | | + | | | |
| **Amphipoda** | | | | | | | | | | |
| *Pontogeneia rostrata* | 617.50 | + | ++ | ++ | ++ | + | ++ | + | ++ | + |
| *Caprella bispinosa* | 242.24 | | | | ++ | + | | ++ | | + |
| *Ampithoe* spp. | 73.59 | + | ++ | | + | + | + | | | + |
| *Grandidierlla* spp. | 59.88 | ++ | ++ | + | + | + | | + | + | + |
| *Corophium* spp. | 53.30 | ++ | ++ | + | + | | + | + | | |
| *Caprella penantis* | 28.89 | | ++ | + | | | | | | + |
| *Pleustes panopla* | 24.69 | | + | | | | ++ | + | | + |
| *Aoroides* spp. | 17.90 | + | + | | + | | + | + | | + |
| *Photis reinhardi* | 16.82 | | ++ | | | | | + | | |
| *Caprella gigantochir* | 15.56 | | ++ | | | + | + | | | |
| *Hyale* spp. | 4.38 | + | + | + | + | | | | | |
| *Caprella kroyeri* | 3.95 | | | | | | | + | | |
| *Metaphoxus* sp. | 2.84 | | | | | | | | + | + |
| *Gammaroposis japonica* | 2.78 | + | | | + | | | | | |
| *Synchelidium lenorstalum* | 2.13 | | | | + | | + | + | | + |
| *Caprella scaura* | 1.23 | | | | | | + | | | |
| *Orchomene* sp. | 1.05 | | | | + | | | + | | + |
| *Caprella polyacantha* | 0.56 | | | | | + | | | | |
| *Metopa* sp. | 0.56 | | | | | + | | + | | + |
| *Pleusirus secowns* | 0.52 | | | | | | | + | | |
| *Caprella laeviuscula* | 0.37 | | | | | | | | | + |
| *Ischyrocerus anguipes* | 0.19 | | | | | + | | | | |
| *Allorchestes* sp. | 0.12 | | | | | | | | | + |
| *Urothoe grimaldii japonica* | 0.12 | | | | | | | | | + |
| **Mysida** | | | | | | | | | | |
| *Neomysis awatschensis* | 1751.33 | ++ | ++ | ++ | + | ++ | ++ | ++ | ++ | ++ |
| *Neomysis mirabilis* | 1405.15 | + | ++ | ++ | ++ | ++ | ++ | + | ++ | ++ |
| *Acanthomysis schrenčki* | 31.20 | | | | + | + | ++ | + | | + |
| *Neomysis czerniawskii* | 28.67 | | + | + | + | ++ | + | + | + | + |
| *Exacanthomysis japonica* | 26.51 | | ++ | | + | + | + | | | |

TABLE II

(Continued)

| Species | Mean density (inds · m$^{-2}$) | A | | | B | | | C | | |
|---|---|---|---|---|---|---|---|---|---|---|
| | | June | Sep. | Dec. | June | Sep. | Dec. | June | Sep. | Dec. |
| Isopoda and Tanaidacea | | | | | | | | | | |
| *Cymodoce japonica* | 34.86 | + | + | + | | ++ | + | + | ++ | + |
| *Paranthura japonica* | 2.72 | | | | + | | + | | + | + |
| *Idotea ochotensis* | 0.92 | | | + | + | + | + | + | + | + |

TABLE III

Results of two-way ANOVA testing temporal and spatial variation in species richness and functional diversity ($FD_{\mathrm{Rao}}$) of macro-crustacean communities in the Akkeshi-ko estuary

| Factor | df | MS | F | P |
|---|---|---|---|---|
| Species richness (*SR*) | | | | |
| Month | 2 | 70.037 | 11.67 | 0.001 |
| Site | 2 | 84.704 | 14.12 | <0.001 |
| Month × Site | 4 | 9.037 | 1.51 | 0.242 |
| Error | 18 | 6.000 | | |
| Functional diversity ($FD_{\mathrm{Rao}}$) | | | | |
| Month | 2 | 0.016 | 4.99 | 0.019 |
| Site | 2 | 0.004 | 1.35 | 0.285 |
| Month × Site | 4 | 0.330 | 10.23 | <0.001 |
| Error | 18 | 0.003 | | |

FDs of macrocrustacean communities in seagrass beds were lower (maximum $FD_{\mathrm{Rao}}$, 0.52) than those of other faunal communities such as coastal fish, avian, and rangeland plant communities (Micheli & Halpern, 2005; Petchey et al., 2007; Sasaki et al., 2009). Such lower FDs of macrocrustacean communities may be due to multiple traits of the macrocrustacean species. Further, spatiotemporal variations in FD were different from those of SR in both our case studies (tables III & V), indicating the occurrence of functionally redundant species, but also changes in the degree of functional redundancy. Degree of functional redundancy is, therefore, a critical variable in the interpretation of FD variation in macrocrustacean communities (Rosenfeld, 2002; Loreau, 2004; Petchey et al., 2007; Flynn et al., 2009).

In Tokyo Bay, variation in FD was significantly different among areas (table V). For example, areas FT and TO were high FDs (0.28-0.52), whereas values in area TY were low (0.17-0.23). In the Akkeshi-ko estuary, the range of values was low (0.05-0.36) and comparable to that in area TY in Tokyo Bay,

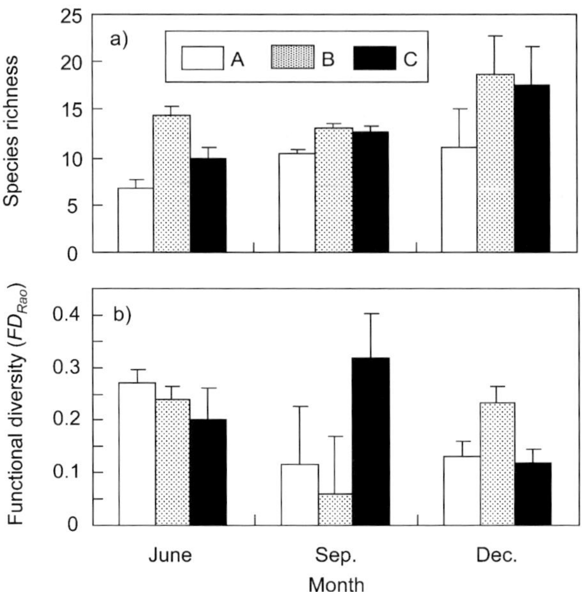

Fig. 2. Spatiotemporal variations in (a) species richness and (b) functional diversity of macrocrustacean communities of seagrass meadow at 3 sites (St. A–C) in 3 seasons (months) in Akkeshi-ko estuary. Values are means ± SD.

which had approximately the same areal extent. This similarity of FD variations between area TY (Tokyo Bay) and the Akkeshi-ko estuary separated by ca. 1000 km suggests that environmental differences between large geographic regions (e.g., latitudinal differences) is not major effects on FD variations. In contrast, there were significant FD variations at area scales within Tokyo Bay, suggesting that different environmental-factors among areas (tens of km$^2$) are more important for determine variation of FD, i.e., 2 levels (phases) of FD variations (i.e., high-FD at FT and FO or low-FD at TY and the Akkeshi-ko). However, this study could not identify the regional environment-factors and evaluate whether the environmental factors change temporally among areas.

Although a variety of techniques based on a comprehensive range of ecological traits has recently been developed to measure FD (Petchey & Gaston, 2002; Wright et al., 2006), there may be many cases in which the knowledge of ecological traits (which are directly associated with ecological functions) to estimate FD is critically limiting. This limitation is particularly adapted for aquatic macrocrustacean faunas because unsatisfactory examinations conducted for determining the ecological traits of each macrocrustacean species. FD measurement of macrocrustacean communities using multiple ecological traits contributes to understanding assembly rules and/or ecosystem processes

## TABLE IV

Mean densities (inds · m$^{-2}$) and spatial occurrence patterns for 83 species collected in 3 areas in Tokyo Bay. ++: >100 inds · m$^{-2}$, +: <100 inds · m$^{-2}$

| Species | Mean density (inds · m$^{-2}$) | Futti (FT) | | | | Takeoka (TO) | | | Tateyama (TY) | | |
|---|---|---|---|---|---|---|---|---|---|---|---|
| | | D | E | F | G | H | I | J | K | L | M |
| **Decapoda** | | | | | | | | | | | |
| *Heptacarpus pandaloides* | 21.38 | + | + | + | + | + | + | + | ++ | + | + |
| *Crangon* spp. | 0.87 | + | + | + | + | + | | + | + | | |
| *Heptacarpus futilirostris* | 0.54 | | | | | + | + | + | | | + |
| *Heptacarpus geniculatus* | 0.20 | | | + | | + | + | | | | + |
| *Palaemon pacificus* | 0.13 | | | | | | | + | | | + |
| *Latreutes planirostris* | 0.12 | | | | | + | | | | + | |
| *Latreutes acicularis* | 0.04 | | | | | + | + | | | + | + |
| *Heptacarpus rectirostris* | 0.04 | | | + | | | | + | | | |
| *Hippolyte* sp. | 0.01 | | | | | | | | | + | + |
| *Eualus leptognathus* | 0.01 | | | | + | + | | | | | |
| *Heptacarpus* sp. | 0.01 | | | | | | | | + | + | |
| **Amphipoda** | | | | | | | | | | | |
| *Cerapus* spp. | 207.28 | | | | | + | + | + | + | + | ++ |
| *Jassa* spp. | 134.46 | ++ | ++ | ++ | ++ | ++ | + | + | + | + | + |
| *Caprella kroyeri* | 28.74 | + | ++ | + | + | + | + | + | + | | |
| *Caprella scaura* | 22.35 | | | | | | | | | + | + |
| *Pontogeneia rostrata* | 19.81 | + | + | + | + | + | + | + | + | + | + |
| *Podocerus* sp. | 17.82 | | | + | + | + | + | + | | | ++ |
| *Caprella penantis* | 15.05 | + | | + | + | + | + | + | + | + | + |
| *Maxillipius* sp. | 14.94 | | | | | + | + | + | | + | ++ |
| *Gammaropsis* spp. | 14.35 | + | + | + | + | + | + | + | + | + | + |
| *Ampithoe* spp. | 9.74 | + | + | + | + | + | + | + | | | + |
| *Parapleustes* spp. | 9.36 | + | + | + | + | + | + | + | | + | + |
| *Aoroides* sp. | 9.30 | + | + | + | + | + | + | + | | + | + |
| *Monoculodes* spp. | 7.99 | + | + | + | + | + | + | + | + | | |
| *Corophium* sp. | 6.56 | | | + | + | + | | | | | |
| *Paradexamine* spp. | 6.13 | + | + | + | + | + | + | + | | + | + |
| *Pleustes* sp. | 4.01 | + | + | + | + | + | + | + | | + | + |
| *Stegocephalus* sp. | 2.31 | + | + | + | | + | + | | | | + |
| *Caprella tsugarensis* | 1.83 | | | + | | + | + | + | + | | |
| *Melita* sp. | 1.34 | + | + | + | + | + | + | + | | | |
| *Parhyale* sp. | 1.15 | + | | | | | | + | | | + |
| *Phoxocephalus* spp. | 0.99 | | | | + | | | | + | | |
| *Gitanopsis* sp. | 0.57 | + | + | | + | | | + | | | + |
| *Pereionotus* sp. | 0.37 | | | | | | | | | | + |
| *Dexamine* sp. | 0.30 | | | | | | | + | | | |
| *Caprella verrucosa* | 0.22 | | | | | + | + | | | | + |
| *Hyale* sp. | 0.21 | | + | | | | | + | | | + |
| *Ampelisca* spp. | 0.12 | | | + | + | + | + | | | + | |
| *Allorchestes* sp. | 0.12 | | | | | | | + | | | + |
| *Najna* sp. | 0.09 | | | | | | | + | | | + |
| *Caprella brevirostris* | 0.06 | | | | | + | | | | | |
| *Caprella danilevskii* | 0.05 | | | | | | | + | | | |
| *Parahyalella* sp. | 0.03 | | | | | | | + | | | |
| *Protomima* spp. | 0.02 | | | + | | | | | | + | |
| *Urothoe* sp. | 0.01 | + | | | | | | | | | |

## TABLE IV
### (Continued)

| Species | Mean density (inds · m$^{-2}$) | Futti (FT) | | | | Takeoka (TO) | | | Tateyama (TY) | | |
|---|---|---|---|---|---|---|---|---|---|---|---|
| | | D | E | F | G | H | I | J | K | L | M |
| **Mysida** | | | | | | | | | | | |
| *Acanthomysis milsiikurii* | 1606.34 | | | | | | | | ++ | + | |
| *Acanthomysis* sp. | 863.92 | | | | | + | + | | ++ | | |
| *Acanthomysis tamurai* | 518.57 | | | | | | | | ++ | | |
| *Siriella watasei* | 92.59 | | | + | + | + | + | + | ++ | + | + |
| *Nipponnomysis calcarata* | 16.57 | + | + | + | + | + | + | + | | | |
| *Siriella longipes* | 13.92 | | + | + | | + | + | + | | + | |
| *Acanthomysis nakazatoi* | 13.33 | | | | | | | | ++ | | |
| *Nipponnomysis takitai* | 5.25 | | | | | + | | | | | |
| *Anisomysis* sp. | 4.72 | | | | | + | | + | | | + |
| *Nipponnomysis fusca* | 3.57 | | | | | + | + | | | + | + |
| *Iiela ohshimai* | 1.21 | + | | | | | | | | | |
| *Nipponnomysis ornata* | 0.98 | | + | | | | | | | | + |
| *Mysidopsis japonica* | 0.68 | | + | + | | + | | | | + | |
| *Nipponomysis eriopedes* | 0.67 | | | | | + | + | + | | | + |
| *Neomysis awatscheinsis* | 0.43 | | | + | | | | | | | + |
| *Paracanthomysis hispida* | 0.37 | | | | | | | | | | + |
| *Nipponomysis perminuta* | 0.33 | | | | | | | | | | + |
| *Mysidopsis surugae* | 0.31 | | + | | + | | | | | | |
| *Siriella nodosa* | 0.12 | | | | | | | | | | + |
| *Nipponomysis surugensis* | 0.10 | | | | | | + | | | | |
| *Siriella japonica izuensis* | 0.10 | | | | | | + | | | | |
| *Nipponomysis longvura* | 0.07 | | | | | | | | | | + |
| *Hypererythrops zimmei* | 0.05 | | | | | | + | | | | |
| *Paracanthomysis kurilensi* | 0.05 | | | | | | + | | | | |
| *Siriella lingvura* | 0.03 | | | | | | + | | | | |
| **Isopoda and Tanaidacea** | | | | | | | | | | | |
| *Zeuox normani* | 45.65 | + | ++ | + | + | + | + | + | + | + | ++ |
| *Janiropsis* spp. | 3.65 | | | + | | | | + | | | + |
| *Synidotea laevidorsalis* | 3.11 | | | | | + | + | | + | + | + |
| *Holotelson tuberculatus* | 1.53 | | | | | + | + | + | | | + |
| *Cymodoce japonica* | 0.47 | + | | | | | | | | | + |
| *Paranthura japonica* | 0.21 | + | | + | + | | | | | | + |
| *Cleantioides planicauda* | 0.11 | | | | | + | | + | | | |
| *Leptosphaeroma gottschei* | 0.10 | | | | | | + | + | | | |
| *Cleantiella strasseni* | 0.04 | | | + | | + | | | | | + |
| *Symmius planus* | 0.03 | | | | | + | | | | | |
| *Aega* sp. | 0.03 | | | | | | | + | | + | |
| *Gnathia* sp. | 0.01 | | | | | + | | | | | |
| *Excorallana yamamuroae* | 0.01 | | | | | | + | | | | |

in diverse communities and to generalization of the concept of FD in aquatic
ecology (Wright et al., 2006; Petchey et al., 2007; Sasaki et al., 2009; Yamada
et al., 2010).

TABLE V

Results of nested ANOVA testing spatial variation in species richness and functional diversity ($FD_{Rao}$) of the macro-crustacean communities in Tokyo Bay

| Factor | df | MS | F | P |
|---|---|---|---|---|
| **Species richness ($SR$)** | | | | |
| Area | 2 | 264.261 | 2.28 | 0.173 |
| Site (Area) | 7 | 115.778 | 7.82 | <0.001 |
| Error | 20 | | | |
| **Functional diversity ($FD_{Rao}$)** | | | | |
| Area | 2 | 0.148 | 31.35 | <0.001 |
| Site (Area) | 7 | 0.005 | 1.50 | 0.223 |
| Error | 20 | | | |

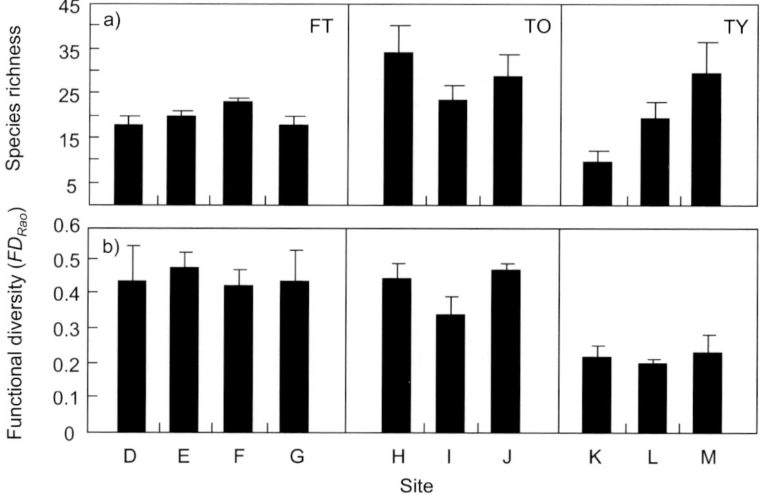


Fig. 3. Spatial variations in (a) species richness and (b) functional diversity of macrocrustacean communities of seagrass meadow at 10 sites (St. D–M) in 3 areas (FT, TO, and TY) of Tokyo Bay. Values are means ± SD.

## ACKNOWLEDGEMENTS

We gratefully thank T. Era, S. Katoh, M. Ishii, Y. Ohfuji, N. H. Kumagai, W. Kentaro, T. Yamakita and N. Whanpetch for assistance with the field collection. T. Sasaki, H. Shudo, G. Yoshida, T. Takano and members of FEIS, Fishery Research Agency Japan, helped with analysis. This research was partly supported by the grants-in-aids from Ministry of Environment, Chiba Prefecture and the Ministry of Education, Science, Culture and Sports, Japan to MN (No. 18201043), and a Grant-in-Aid by Akkeshi Town for Scientific

Research of Lake Akkeshi and Bekanbeushi Wetland and the Sasagawa Scientific Research Grant from the Japan Science Society to KY.

# REFERENCES


BADY, P., S. DOLEDEC, C. FESL, S. GAYRAUD, M. BACCHI & F. SCHOLL, 2005. Use of invertebrate traits for the biomonitoring of European large rivers: the effects of sampling effort on genus richness and functional diversity. Freshw. Biol., **50**: 159-173.

BIENBAUM, C. K., 1979. Influence of sedimentary factors on the distribution of benthic amphipods of Fisher Island Sound, Connecticut. J. Exp. Mar. Biol. Ecol., **38**: 201-224.

BOSTRÖM, C., K. O'BRIEN, C. ROOS & J. EKEBOM, 2006. Environmental variables explaining structural and functional diversity of seagrass meadows in an archipelago landscape. J. Exp. Mar. Biol. Ecol., **335**: 52-73.

BOTTA-DUKAT, Z., 2005. Rao's quadratic entropy as a measure of functional diversity based on multiple traits. J. Veg. Sci., **16**: 533-540.

BOUSFIELD, E. L., 1973. Shallow-water gammaridean Amphipoda of New England: 1-312. (Cornell University, New York).

BREMNER, J., S. I. ROGERS & C. L. J. FRID, 2003. Assessing functional diversity in marine benthic ecosystems: a comparison of approaches. Mar. Ecol. Prog. Ser., **254**: 11-25.

CIANCIARUSO, M. V., M. A. BATALHA, K. J. GASTON & O. L. PETCHEY, 2009. Including intraspecific variability in functional diversity. Ecology, **90**: 81-89.

DE BELLO, F., J. LEPS & M. T. SEBASTIA, 2005. Predictive value of plant traits to grazing along a climatic gradient in the Mediterranean. J. Appl. Ecol., **42**: 824-833.

DIAZ, S. & M. CABIDO, 2001. Vive la difference: plant functional diversity matters to ecosystem processes. Trends Ecol. Evol., **16**: 646-655.

FLYNN, D., M. GOGOL-PROKURAT, T. NOGEIRE, N. MOLINARI, B. RICHERS, B. LIN, N. SIMPSON, M. MAYFIELD & F. DECLERCK, 2009. Loss of functional diversity under land use intensification across multiple taxa. Ecol. Let., **12**: 22-33.

FURUKAWA, K. & T. OKADA, 2006. Tokyo Bay: its environmental status — past, present, and future. In: E. WOLANSKI (ed.), The environment in Asia Pacific harbors: 15-34. (Springer, Dordrecht).

GUERRA-GARCÍA J. M. & J. M. TIERNO DE FIGUEROA, 2009. What do caprellids (Crustacea: Amphipoda) feed on? Mar. Biol., **156**: 1881-1890.

LAVOREL, S., K. GRIGULIS, S. MCINTYRE, N. S. G. WILLIAMS, D. GARDEN, J. DORROUGH, S. BERMAN, F. QUETIER, A. THEBAULT & A. BONIS, 2008. Assessing functional diversity in the field: methodology matters! Func. Ecol., **22**: 134-147.

LOREAU, M., 2004. Does functional redundancy exist? Oikos, **104**: 606-611.

MICHELI, F. & S. HALPERN, 2005. Low functional redundancy in coastal marine assemblages. Ecol. Let., **8**: 391-400.

MUKAI, H., 2006. Seagrass bed ecosystem in a total system of land- and coastal marine systems. Monthly Kaiyo, **37**: 148-155. [In Japanese.]

NAKAOKA, M., K. WATANABE & T. ERA, 2007. Evaluation of relationships between biodiversity and ecosystem functions in coastal seas: a case study of seagrass beds in Tokyo Bay. Jpn. J. Benth., **62**: 82-87. [In Japanese with English abstract.]

NELSON, W. G., 1980. Reproductive patterns in gammaridean amphipods. Sarsia, **65**: 61-71.

NISHIMURA, S., 1995. Guide to seashore animals of Japan with color pictures and keys, **II**: 1-663. (Hoikusha, Osaka). [In Japanese.]

PETCHEY, O. L., K. L. EVANS, I. S. FISHBURN & K. J. GASTON, 2007. Low functional diversity and no redundancy in British avian assemblages. J. Anim. Ecol., **76**: 977-985.

PETCHEY, O. L. & K. J. GASTON, 2002. Functional diversity (FD), species richness and community composition. Ecol. Lett., **5**: 402-411.

PETCHEY, O. L., A. HECTOR & K. J. GASTON, 2004. How do different measures of functional diversity perform? Ecology, **85**: 847-857.

RAO, C. R., 1982. Diversity and dissimilarity coefficients: a unified approach. Theor. Pop. Biol., **21**: 24-43.

ROSENFELD, J. S., 2002. Functional redundancy in ecology and conservation. Oikos, **98**: 156-162.

SASAKI, T., S. OKUBO, T. OKAYASU, U. JAMSRAN, T. OHKURO & K. TAKEUCHI, 2009. Two-phase functional redundancy in plant communities along a grazing gradient in Mongolian rangelands. Ecology, **90**: 2598-2608.

SYOUJI, Y. & K. HASEGAWA, 2004. Distribution of eelgrass *Zostera marina* of coastal sea waters in Chiba Prefecture. Bull. Chiba Pref. Fish. Res. Cen., **3**: 77-86.

TANAKA, Y., T. MIYAJIMA, K. YAMADA, M. HORI, N. HASEGAWA, Y. UMEZAWA & I. KOIKE, 2008. Specific growth rate as a determinant of carbon isotopic composition of a temperate seagrass *Zostera marina*. Aquat. Bot., **89**: 331-336.

TILLIN, H. M., J. G. HIDDINK, S. JENNINGS & M. J. KAISER, 2006. Chronic bottom trawling alters the functional composition of benthic invertebrate communities on a sea-basin scale. Mar. Ecol. Prog. Ser., **18**: 31-45.

UNOKI, S., 1985. Tokyo Bay: 335-387. In: S. UNOKI (ed.), Coastal oceanography of Japanese islands. (Tokai University Press, Tokyo). [In Japanese.]

WRIGHT, J. P., C. G. JONES, B. BOEKEN & M. SHACHAK, 2006. Predictability of ecosystem engineering effects on species richness across environmental variability and spatial scales. J. Ecol., **94**: 815-824.

YAMADA, K., 2008. Functional responses and effects of macrofaunal community at a seagrass meadow in northeastern Japan: 146 pp. (Ph.D. Thesis, Chiba Univ. Chiba, Japan).

YAMADA, K., M. HORI, Y. TANAKA, N. HASEGAWA & M. NAKAOKA, 2007b. Temporal and spatial macrofaunal community changes along a salinity gradient in seagrass meadows of Akkeshi-ko estuary and Akkeshi Bay, northern Japan. Hydrobiologia, **592**: 345-358.

— —, — —, — —, — — & — —, 2010. Contribution of different functional groups to the diet of major predatory fishes at a seagrass meadow in northeastern Japan. Est. Coast, Shelf Sci., **86**: 71-82.

YAMADA, K., K. TAKAHASHI, C. VALLET, S. TAGUCHI & T. TODA, 2007a. Distribution, life history, and production of three species of *Neomysis* in Akkeshi-ko estuary, northern Japan. Mar. Biol., **150**: 905-917.

YAMAKITA, T. & M. NAKAOKA, 2009. Scale dependency in seagrass dynamics: how does the neighboring effect vary with grain of observation? Popul. Ecol., **51**: 33-40.

First received 19 December 2009.
Final version accepted 13 March 2010.

# SPATIAL DIFFERENCES IN STABLE ISOTOPE SIGNATURES OF CRUSTACEANS IN BRACKISH LAKE SYSTEMS, WESTERN JAPAN

BY


KENGO KURATA[1,3]), MASAHIRO HORINOUCHI[1]) and DAVID L. DETTMAN[2])

[1]) Research Center for Coastal Lagoon Environments, Shimane University, 1060 Nishikawatsucho, Matsue 690-8504, Japan

[2]) Department of Geosciences, University of Arizona, Tucson, Arizona 85721, U.S.A.


## ABSTRACT


Samples of several species of crustaceans and potential food sources from the Hii River basin, including Lake Shinji, Lake Nakaumi, the Honjo area and the Sakai Strait, were collected and measured for stable isotope ratios. Carbon and nitrogen stable isotope ratios of the potential food materials, suspended solids, surface sediment, attached matter, marsh plants, seagrass and seaweed, varied significantly among the water bodies. The $\delta^{13}$C values of the potential food materials in L. Shinji were generally low. While samples collected from L. Nakaumi were generally higher in $\delta^{13}$C, some showed unexpected high $\delta^{13}$C values. Samples at Iya, on the southern coast of L. Nakaumi, had high carbon isotope ratios, with extreme values in the green algae *Ulva* sp. ($-6.7$ to $-5.8‰$). Carbon stable isotope ratios of crustaceans also differed among the water bodies. The $\delta^{13}$C values of the samples from L. Shinji were lower ($-24.2$ to $-17.7‰$) than those from L. Nakaumi ($-14.4$ to $-10.8‰$) with the exception of Sphaeromatidae. Carbon isotope ratios of suspended organic matter in the two lakes are different, indicating that carbon stable isotope ratios of crustaceans strongly reflect the ratios of primary producers in nearby water masses in Lakes Shinji and Nakaumi.


## INTRODUCTION

Carbon and nitrogen stable isotope ratios have been used to study the relationship between consumers and food materials in many aquatic food webs (e.g., Yamada & Yoshioka, 1999; Fry, 2006; Yokoyama, 2008). Organic matter in estuarine environments can come from a variety of sources making isotope studies complex (e.g., Cifuentes et al., 1988; Yokoyama & Ishihi, 2007). In coastal brackish water systems carbon isotope ratios of benthic organisms

---


[3]) Corresponding author; e-mail: kengo@soc.shimane-u.ac.jp

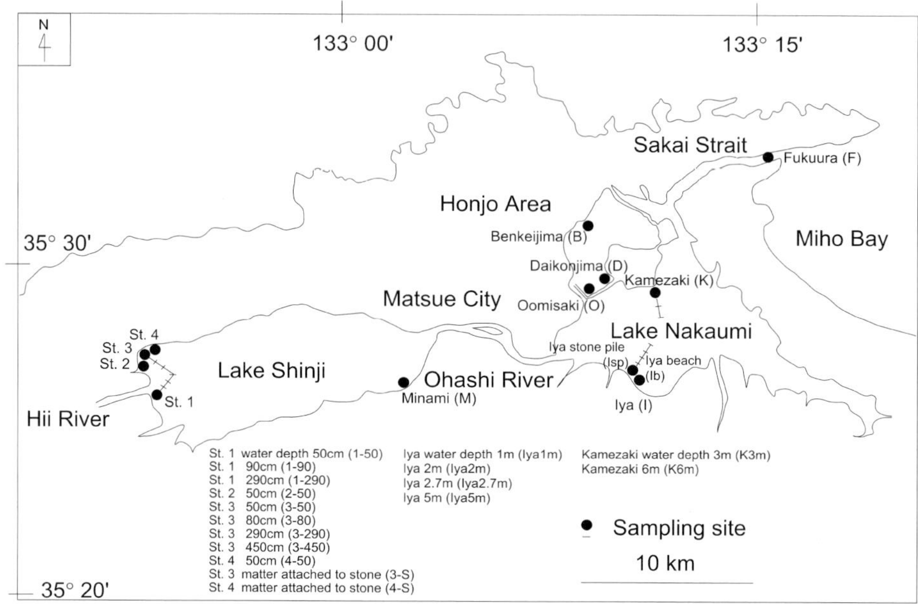


Fig. 1. Study sites and sampling stations.

should reflect the food sources, such as autochthonous primary producers, phytoplankton and benthic microalgae, as well as allochthonous organic matter derived from coastal or riverine vegetation. Benthic invertebrates in the Gulf of Lions, north-western Mediterranean Sea, showed a wide range in $\delta^{13}$C, suggesting different diets and organic sources (Darnaude et al., 2004). Nyunja et al. (2009) showed that $\delta^{13}$C values of potential carbon sources varied from mangrove sites to seagrass sites in Gazi Bay, Kenya. These results indicated that both autochthonous and allochthonous organic matter could play a part of role in nutrition for benthic invertebrates.

Lake Shinji, Lake Nakaumi, and the Ohashi River are the largest brackish water habitat in Japan (fig. 1). Because a large salinity gradient is observed from Lake Shinji to Lake Nakaumi, we set out to investigate the position of crustaceans in the food chain and changes in local food sources across the salinity gradient. Could the salinity variation in this system affect the use of food sources by a large group of macroinvertebrates? We used the relationship between environmental gradients and stable isotope signatures of benthic invertebrates to look at this question. Gradients in carbon stable isotope ratios in estuarine systems have been reported for particulate organic matter (Thornton & McManus, 1994; Canuel et al., 1995), sedimentary organic carbon (Chanton & Lewis, 2002), and benthic organism (Canuel et al., 1995;

Chanton & Lewis, 2002). There are, however, no studies concerning stable isotope ratios of benthic invertebrates inhabiting the Shinji-Nakaumi brackish water system so far.

In this study we collected samples from Hii River basin, including Lake Shinji, Lake Nakaumi, the Honjo area and the Sakai Strait, and measured stable isotope ratios of potential food sources and several species of crustaceans.

## MATERIAL AND METHODS

### Study sites and sampling

The Hii River basin, located in Shimane and Tottori Prefectures, has two shallow brackish lakes with markedly different salinities (fig. 1). The Honjo area, formerly a northwestern part of Lake Nakaumi, is surrounded by dikes and forms a semi-enclosed water body. The Hii River is the main input river to Lake Shinji and the Sakai Strait is the only outlet of Lake Nakaumi. The salinity of Lake Shinji ranges from oligohaline to mesohaline, and that of Lake Nakaumi ranges from mesohaline to polyhaline.

Samples were collected from 23 stations in this brackish water system (10 stations in Lake Shinji, 12 stations in Lake Nakaumi including the Honjo area and 1 station in the Sakai Strait) from May to August in 2004 and 2005 (fig. 1). Sediment samples were taken using either an acrylic resin core in shallow areas or an Ekman-birge grab sampler from a boat. Sediments were sieved with a 1 mm mesh sieve to collect infauna. Epifauna were collected at cobble beds and concrete structures using a scraper and tweezers. Phytal organisms in macrophytes were collected with a 1 mm mesh sieve. Vascular plants, macroalgae, suspended solids, surface sediments and attached matter on hard substrates were also collected.

### Measurements

Crustaceans, vascular plants and macroalgae were sorted, washed to remove any visible attached matter, and put into glass vials in the laboratory. Suspended solids were collected on glass fiber filter (Whatman GF/F) by filtration of surface water taken at the sampling stations. Surface sediments and attached matter were treated with 1N HCl to remove carbonates. All samples were freeze-dried and the entire sample was then ground and transferred to a tin container. Isotope ratios ($\delta^{13}$C and $\delta^{15}$N) were measured on a continuous-flow gas-ratio mass spectrometer (Finnigan Delta PlusXL). Samples were combusted using an elemental analyzer (Costech) coupled to the mass spectrometer. Standardization is based on NBS-22 and USGS-24 for $\delta^{13}$C, and IAEA-N-1 and IAEA-N-2 for $\delta^{15}$N. Precision is better than $\pm 0.06$ for $\delta^{13}$C and $\pm 0.2$ for $\delta^{15}$N ($1\sigma$), based on repeated internal standards. Stable isotope ratios are expressed as relative values (‰) as defined by the following equation:

$$\delta^{13}\text{C}, \delta^{15}\text{N} = (\text{R}_{\text{sample}} - \text{R}_{\text{standard}})/\text{R}_{\text{standard}} * 1000,$$

where $\text{R} = {}^{13}\text{C}/{}^{12}\text{C}$ or ${}^{15}\text{N}/{}^{14}\text{N}$. $\text{R}_{\text{standard}}$ is VPDB for carbon and AIR for nitrogen.

## RESULTS

### 1. Potential food materials

Carbon and nitrogen stable isotope ratios of the potential food materials, including suspended solids, surface sediment, attached matter, marsh plant, seagrass and seaweed, varied among the water bodies (fig. 2). With the exception of green algae and vascular plants L. Shinji samples were lower in $\delta^{13}$C than samples of the Nakaumi/Honjo/Sakai group. Vascular plant differences were masked by the large $\delta^{13}$C difference between seagrasses and other near-shore plant taxa. *Ulva* sp. from the southern coast of L. Nakaumi showed extremely high $\delta^{13}$C. Nitrogen isotope ratios also tended to be higher in L. Nakaumi than in L. Shinji, although the L. Shinji samples included the most positive and most negative measured values.

### 2. Crustaceans

Carbon and nitrogen stable isotope ratios of crustaceans are shown plotted by location and species (fig. 3). When viewed by order the samples have very similar compositions. For $\delta^{15}$N: isopods range from 5.3 to 12.4‰, mean 9.8‰ (n = 31); decapods range from 8.3 to 10.9‰, mean 10.2‰ (n = 9); amphipods range from 5.8 to 11.9‰, mean 9.3‰ (n = 20); and mysids range from 8.2 to 11.3‰, mean 9.8‰ (n = 4). Means and ranges for carbon isotopes are: $-22.5$ to $-8.4$‰ for isopods (mean $-14.6$‰, n = 31); $-23.1$ to $-10.8$‰ for decapods (mean $-14.9$‰, n = 9); $-24.2$ to $-10.9$‰ for amphipods (mean $-16.4$‰, n = 25); and $-22.7$ to $-19.5$‰ for mysids (mean $-21.1$‰, n = 4). Variable sample size makes the data difficult to compare for the entire brackish lake system, particularly as regional variability within the system has a very

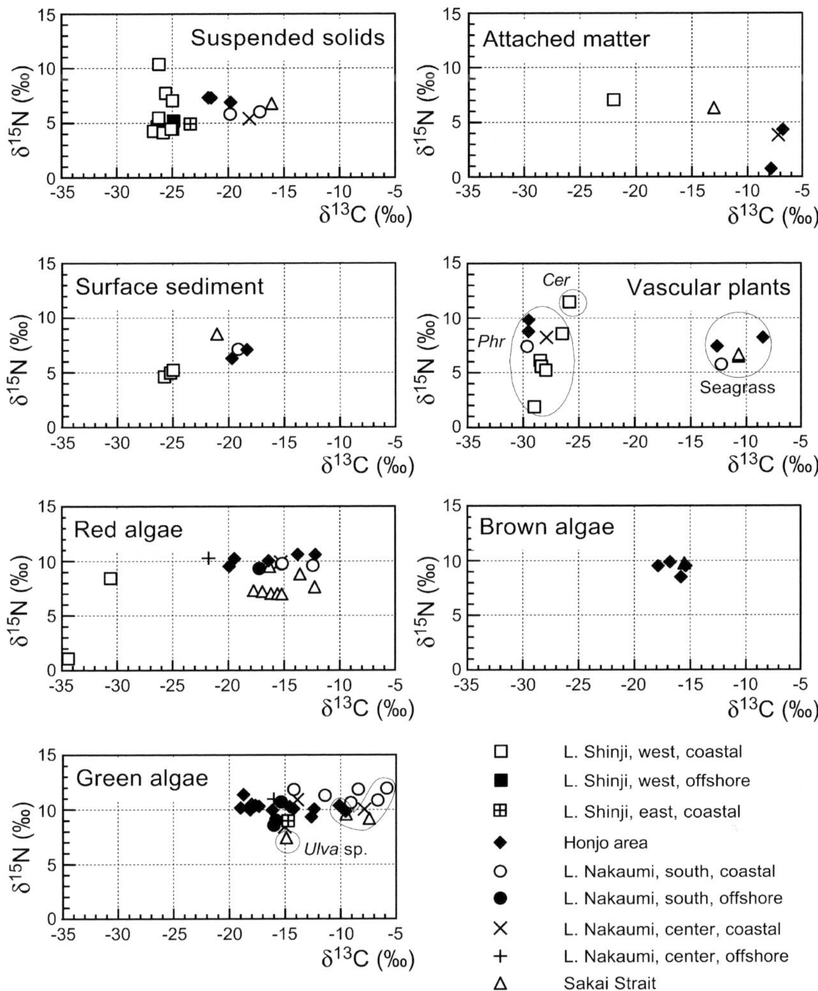



Fig. 2. Relationships between carbon and nitrogen stable isotope ratios of potential food materials (suspended solids, attached matter, surface sediment, vascular plants and macroalgae). Seagrass includes *Ruppia maritima*, *Zostera japonica* and *Zostera marina*. Abbreviations: *Phr*, *Phragmites australis*; *Cer*, *Ceratophyllum demersum*. *Ulva* sp. samples are surrounded in the green algae figure.

strong effect on the carbon isotope ratios. In general samples from L. Shinji are notably more negative in carbon isotope ratio. Nitrogen isotope ratios do not show a clear regional pattern. Amphipods are the only group that shows a combined C and N isotope trend within the data, with L. Shinji trending low in both $\delta^{13}C$ and $\delta^{15}N$ and the L. Nakaumi and marine samples having higher C and N isotope ratios.

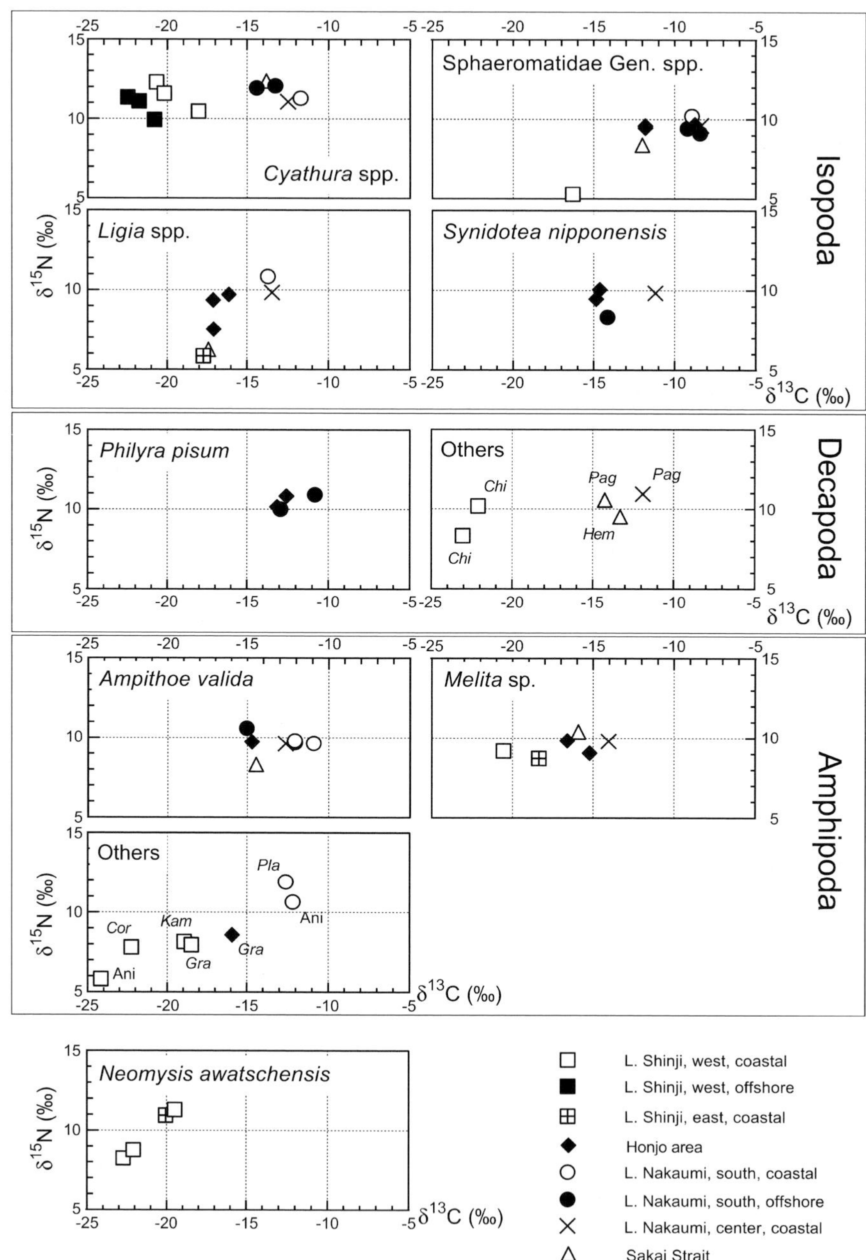



Fig. 3. Relationships between carbon and nitrogen stable isotope ratios of crustaceans (Isopoda, Decapoda, Amphipoda and Mysidacea). Other species of Decapoda: *Chi*, *Chiromantes dehaani*; *Hem*, *Hemigrapsus penicillatus*; *Pag*, *Pagurus minutus*. Other species of Amphipoda: Ani, Anisogammaridae gen. sp.; *Cor*, *Corophium* sp.; *Kam*, *Kamaka* sp.; *Gra*, *Grandidierella japonica*; *Pla*, *Platorchestia platensis*.

## DISCUSSION

### 1. Food web

Although isotopic fractionations between consumer and food sources can vary due to a number of variables, including feeding habit, habitat, and tissue measured (McCutchan et al., 2003; Vanderklift et al., 2003), in general there is an increase in both $\delta^{13}$C and $\delta^{15}$N with each step up the food chain. It has been suggested that a difference of one trophic level is reflected in an increase of 2 to 3‰ in N isotope ratios and a slight increase in carbon (Peterson et al., 1985; Vanderklift et al., 2003). Although the small crustacean species of this study are thought to be omnivorous, the relatively elevated $\delta^{15}$N values of *Cyathura* spp. from both L. Shinji and L. Nakaumi suggest that it may be feeding at a higher trophic level than most other crustaceans and has more of a role as a secondary consumer in the lakes.

For other taxa of crustaceans the case is less clear because of the large variability in the nitrogen isotope ratio of potential food materials. It is difficult to generalize about the nitrogen isotope ratio at the base of the food chain because of the variability apparent even when limiting the question to one body of water. Each water body has a good deal of heterogeneity in both regional distribution of $\delta^{15}$N values and between types of food material. For example, L. Nakaumi foods range from 3.8 to 11.9‰, and L. Shinji food sources are extremely variable, from 1.1 to 11.5‰. In addition different classes of food sources have different and variable isotopic signatures; sediments and suspended particulates are highly variable while red and green algae are much less variable but have different mean $\delta^{15}$N values. This makes it very difficult to place different taxa in trophic order because of the variation in the system.

A number of studies have used carbon and nitrogen isotope ratios to quantify food sources in some of small crustaceans studied in this paper. In the laboratory *Amphithoe valida* fed with fresh *Ulva* sp. ($\delta^{13}$C $= -14.6$‰) showed a similar $\delta^{13}$C value ($-15.5$‰) (Macko et al., 1982). It was suggested that macroalgae such as *Ulva* sp. could be one of the food sources for amphipods, and that the similarity in carbon stable isotope signatures would reflect the close relationship between potential food sources and crustaceans. Sakurai et al. (2007) studied the food sources of the amphipod *Anisogammarus pugettensis*, which preferred leaf litter to macroalgae such as *Ulva pertusa* in a feeding experiment. Carbon isotope ratios were used to quantify the amount of leaf litter in its diet at 31% of its food sources.

Gut contents analysis of the three mysid species *Tenagomysis tasmaniae* Fenton, *Anisomysis mixta australis* (Zimmer) and *Paramesopodopsis rufa*

Fenton revealed that all have an omnivorous diet (Fenton, 1996). Feeding experiments suggest that the mysid shrimp *Gastrosaccus brevifissura*, which is a key crustacean species in most South African estuaries, is able to feed efficiently on both settled and resuspended benthic microalgae. This observation was supported by stable isotopc analysis ($\delta^{13}$C and $\delta^{15}$N), which showed that benthic microalgae contribute 68 and 24% to the total diet of *G. brevifissura* in winter and summer, respectively (Kibirige et al., 2003).

When comparing food source $\delta^{13}$C values to the various taxa of crustaceans in this study the trend from more negative values in L. Shinji to more elevated values in L. Nakaumi that is clearest in the sediment and particulate data is well replicated in the animal tissues (fig. 4). Although this result is consistent with an omnivorous diet for these taxa, the variability in the food dataset prevents any clear conclusions about diet specificity. Any change in the use of food sources or trophic level for crustaceans across the environmental gradient of the lake systems is not visible above the noise in the system. Some general observations can be made, however, when looking at the regional patterns within the study area.

2. Spatial variations of carbon stable isotope ratios of potential food materials

Many estuarine systems are characterized by a large range in the $\delta^{13}$C values in organic matter. Cloern et al. (2002) showed the isotopic composition of primary producers collected from San Francisco Bay estuarine system was highly variable. 52 species and 109 specimens of macroalgae collected in Hiroshima Bay showed a broad range of $\delta^{13}$C values, and isotopic fractionation may have decreased, leading to high $\delta^{13}$C values, in the innermost bay due to the fast growth of seaweed and high productivity (Takai et al., 2001). *Ulva* sp. and *Gracilaria* sp. collected at an artificial tidal flat in Osaka Bay showed carbon stable isotope ratios with $-10.5 \pm 1.1‰$ and $-15.8 \pm 1.6‰$ respectively (Matsuo et al., 2009).

A portion of the samples collected from L. Nakaumi, especially at Iya, showed unexpected high $\delta^{13}$C values (fig. 4). Most extreme was *Ulva* sp., one of the potential food materials for crustaceans, collected from Iya (the southern coast of L. Nakaumi) which had $\delta^{13}$C values of $-6.7$ to $-5.8‰$. The fractionation of carbon isotope ratios in phytoplankton decreases when dissolved inorganic carbon becomes limiting for photosynthesis. This can occur during algal blooms in the summer, resulting in elevated carbon stable isotope ratios of primary producers (see literature cited in Yamada & Yoshioka, 1999).

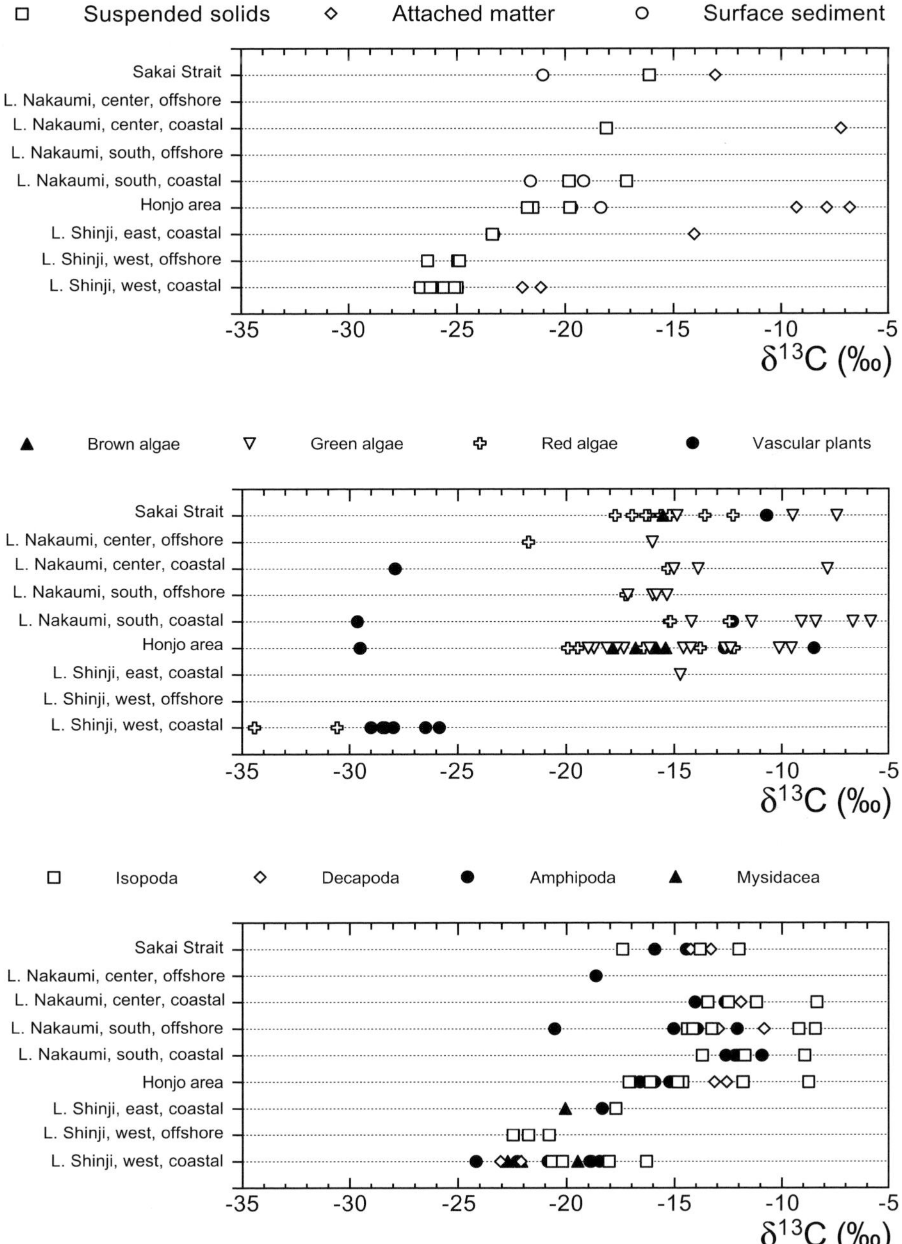



Fig. 4. Spatial variation of carbon stable isotope ratios of potential food materials (suspended solids, attached matter, surface sediment, macroalgae and vascular plants) and crustaceans (Isopoda, Decapoda, Amphipoda and Mysidacea). Note that five data of Amphipoda that are not presented in fig. 3 due to lack of nitrogen isotope ratios are plotted here.

In addition, the presence of seagrass, *Zostera japonica*, in our study area and seagrass-derived detritus may influence the carbon stable isotope ratios of sediments and suspended particulates. Seagrasses measured in this study have $\delta^{13}$C values from $-12.7$ to $-8.5$‰. The seagrass microenvironment can also affcct attachcd algac. Epiphytic algae in an eelgrass *Zostera marina* bed had higher $\delta^{13}$C values ($-11.3$‰), very similar to those of the eelgrass (Jaschinski et al., 2008). Koike et al. (1989) and Takai (2005) also showed that attached microalgae in seagrass beds have elevated $\delta^{13}$C values. Therefore seagrass and other organics derived from the seagrass environment may be another source for suspended solids with high $\delta^{13}$C values at Iya in L. Nakaumi in this study.

### 3. Spatial variations of carbon stable isotope ratios of crustaceans

With very rare exceptions there is no overlap between the carbon isotope ratios of crustaceans collected in L. Shinji and those from L. Nakaumi. There is some overlap in the intermediate salinity environment of the Honjo area (fig. 4). Other studies have shown that crustaceans can have a very wide range of $\delta^{13}$C values, tied to the local carbon sources. In the inner Kiel Fjord where salinity ranges between 10 and 20 psu, carbon isotope ratios ranged from $-17.1$‰ for the isopod *Idotea baltica* to $-23.9$‰ for the amphipod *Amphitoe rubricata* (Jaschinski et al., 2008). In Osaka Bay, *Ampithoe* sp. had $\delta^{13}$C of $-10.6$‰ and *Pagurus minutus* had $\delta^{13}$C with $-13.0 \pm 0.2$‰ (Matsuo et al., 2009). Carbon stable isotope ratios of the samples from L. Shinji were lower ($-24.2$ to $-16.3$‰) than those from L. Nakaumi ($-15.0$ to $-8.4$‰) with the two exceptions of amphipods (fig. 4).

Carbon stable isotope ratios of crustaceans varied strongly among stations in L. Nakaumi. Samples taken from Iya showed higher $\delta^{13}$C values, suggesting that the crustaceans feed on food sources with higher $\delta^{13}$C values, perhaps derived from seagrass, seagrass-microalgae, or high productivity events. Other studies have shown that small crustaceans inhabiting seagrass beds had higher $\delta^{13}$C values, $-14$ to $-10$‰, indicating that they may have used *Zostera* and epiphytic organic matter as food sources (Takai, 2005). The large differences in L. Nakaumi crustaceans show that they are strongly dependant on food sources available within a limited habitat range.

The spatial limits of foraging ranges can be very restricted. The $\delta^{13}$C values of crustaceans collected from an artificial tidal flat in Osaka Bay were very similar to seaweed or benthic microalgae rather than suspended solids, indicating food sources were not suspended particulates, but rather locally derived seaweed or benthic microalgae (Matsuo et al., 2009). Guest et al.

(2004) demonstrated that carbon movement and its assimilation by two crab species in adjacent saltmarsh and mangrove habitats occurred at scales of less than 30 m. They also suggested that the results may differ among invertebrates which have different mobilities and/or feeding habits. Kanaya et al. (2007) suggested that the spatial limit of foraging for macrozoobenthos occurred at a scale of 200 m within a brackish lagoon. Carbon isotope ratios of suspended organic matter in the two lakes of this study are quite different, indicating that carbon stable isotope ratios of crustaceans strongly reflect the local primary producers.

In addition, the carbon stable isotope ratios of Sphaeromatidae were invariably higher than those of other crustaceans in the same body of water. Although this could suggest that these animals feed higher on the food chain, this is not supported by the relatively low $\delta^{15}$N values. Similarly many samples of green algae are notably higher in $\delta^{13}$C values than other types of organic matter from the same area. This offset in a possible food source may explain the higher carbon stable isotope ratios of Sphaeromatidae, suggesting that green alga is a portion of its diet.

## 4. Brackish lake systems

Brackish systems are often associated with a significant gradient in carbon isotope ratios of organic matter and associated consumers. Peterson et al. (1985) traced the carbon and sulfur isotope ratios of the ribbed mussels along a land-to-sea gradient in Great Sippewissett Marsh, Falmouth, Massachusetts, showing that both marsh grass detritus and phytoplankton were consumed by the mussels with a changing proportion according to location. In San Francisco Bay, the suspension-feeding bivalve *Potamocorbula amurensis* showed a spatial gradient of carbon stable isotope ratio from the area influenced by riverine inputs to the seaward portion of the bay (Canuel et al., 1995). Analysis of particulate organic matter indicated that phytoplankton is important throughout the entire bay but additional inputs of organic matter from bacterial and terrestrial vascular plants contribute to particulate organic matter in the innermost bay. Ishida et al. (2005) found that the $\delta^{13}$C of organic matter in suspended particulates and sediment in Hakata Bay and its inflowing rivers was likely to be low in the upstream and midstream, and high in the estuary and bay.

Lake Shinji and Lake Nakaumi can be viewed as two different areas within a large brackish system, but they also maintain semi-permanent differences in salinity and productivity. The differences in the salinity regime for each lake is controlled by geological, astronomical, climatic and hydrological conditions.

In addition the salinity of the Honjo area in Lake Nakaumi is a result of human engineering with constructed dikes mostly enclosing this water body. Food webs within these different aquatic ecosystems will be altered by changes in the salinity and flow regimes in these brackish lake systems. In 2009 human engineering has begun major modification of the Honjo area with removal of some of the dikes. Stable isotope signatures could be useful tools for monitoring both short-term and long-term changes in brackish water ecosystems. Further stable isotope studies in Lakes Shinji and Nakaumi will provide important information on both human and natural changes in the environment and the aquatic food web structure.

## ACKNOWLEDGMENTS

This study was conducted by grants-in-aid KAKENHI 16710048 from the Ministry of Education, Culture, Sports, Science and Technology, Japan and in part supported by the funds from the Research Project Promotion Institute, Shimane University. The authors wish to thank Kenji Toda for identifying invertebrates, Hidenobu Kunii and Shuji Otani for identifying water plants, Koji Seto for taking sediments in deeper sites, Keiko Funaki for sorting organisms collected, and Majie Fan for measuring the samples.

## REFERENCES

CANUEL, E. A., J. E. CLOERN, D. B. RINGELBERG, J. B. GUCKERT & G. H. RAU, 1995. Molecular and isotopic tracers used to examine sources of organic matter and its incorporation into the food webs of San Francisco Bay. Limnology and Oceanography, **40**(1): 67-81.

CHANTON, J. & F. G. LEWIS, 2002. Examination of coupling between primary and secondary production in a river-dominated estuary: Apalachicola Bay, Florida, U.S.A. Limnology and Oceanography, **47**(3): 683-697.

CIFUENTES, L. A., J. H. SHARP & M. L. FOGEL, 1988. Stable carbon and nitrogen isotope biogeochemistry in the Delaware estuary. Limnology and Oceanography, **33**(5): 1102-1115.

CLOERN, J. E., E. A. CANUEL & D. HARRIS, 2002. Stable carbon and nitrogen isotope composition of aquatic and terrestrial plants of the San Francisco Bay estuarine system. Limnology and Oceanography, **47**(3): 713-729.

DARNAUDE, A. M., C. SALEN-PICARD, N. V. C. POLUNIN & M. L. HARMELIN-VIVIEN, 2004. Trophodynamic linkage between river runoff and coastal fishery yield elucidated by stable isotope data in the Gulf of Lions (NW Mediterranean). Oecologia, **138**: 325-332.

FENTON, G. E., 1996. Diet and predators of *Tenagomysis tasmaniae* Fenton, *Anisomysis mixta australis* (Zimmer) and *Paramesopodopsis rufa* Fenton from south-eastern Tasmania (Crustacea: Mysidacea). Hydrobiologia, **323**: 31-44.

FRY, B., 2006. Stable isotope ecology: 1-308. (Springer, New York).

GUEST, M. A., R. M. CONNOLLY & N. R. LONERAGAN, 2004. Carbon movement and assimilation by invertebrates in estuarine habitats at a scale of meters. Marine Ecology Progress Series, **278**: 27-34.

ISHIDA, T., T. ARIMA, Y. INAGAKI, K. IDEMITSU, H. KAWAMURA, M. NAKASHIMA & N. MATSUOKA, 2005. Carbon isotopic compositions of suspended particulate and sedimentary organic matters collected in summer at Hakata Bay, Japan. Journal of Japan Society on Water Environment, **28**: 405-410. [In Japanese with English Abstract.]

JASCHINSKI, S., D. C. BREPOHL & U. SOMMER, 2008. Carbon sources and trophic structure in an eelgrass *Zostera marina* bed, based on stable isotope and fatty acid analyses. Marine Ecology Progress Series, **358**: 103-114.

KANAYA, G., S. TAKAGI, E. NOBATA & E. KIKUCHI, 2007. Spatial dietary shift of macrozoobenthos in a brackish lagoon revealed by carbon and nitrogen stable isotope ratios. Marne Ecology Progress Series, **345**: 117-127.

KIBIRIGE, I., R. PERISSINOTTO & C. NOZAIS, 2003. Grazing rates and feeding preferences of the mysid shrimp *Gastrosaccus brevifissura* in a temporarily open estuary in South Africa. Marine Ecology-Progress Series, **251**: 201-210.

KOIKE, H., T. NAKAJIMA & N. NAKAI, 1989. Stable carbon isotopic and gut content analyses of a tidal flat food web. Benthos Research, **37**: 1-10. [In Japanese with English Abstract.]

MACKO, S. A., W. Y. LEE & P. L. PARKER, 1982. Nitrogen and carbon isotope fractionation by two species of marine amphipods: laboratory and field studies. Journal of Experimental Biology and Ecology, **63**: 145-149.

MATSUO, H., H. ARIYAMA, T. IKEMOTO, K. OMORI & I. TAKEUCHI, 2009. Analysis of food web structure in an artificial tidal flat in Osaka Bay using stable isotopes of carbon and nitrogen. Journal of Japan Society on Water Environment, **32**: 99-104. [In Japanese with English Abstract.]

MCCUTCHAN, J. H., W. M. LEWIS, C. KENDALL & C. C. MCGRATH, 2003. Variation in trophic shift for stable isotope ratios of carbon, nitrogen, and sulfur. Oikos, **102**: 378-390.

NYUNJA, J., M. NTIBA, J. ONYARI, K. MAVUTI, K. SOETAERT & S. BOUILLON, 2009. Carbon sources supporting a diverse fish community in a tropical coastal ecosystem (Gazi Bay, Kenya). Estuarine Coastal and Shelf Science, **83**: 333-341.

PETERSON, B. J., R. W. HOWARTH & R. H. GARRITT, 1985. Multiple stable isotopes used to trace the flow of organic matter in estuarine food webs. Science, **227**: 1361-1363.

SAKURAI, I., S. YANAI, K. ITO & T. KANETA, 2007. Quantitative evaluation of a food chain that originates from leaf litter in a river mouth. Scientific Reports of Hokkaido Fisheries Experiment Station, **72**: 37-45. [In Japanese with English Abstract.]

TAKAI, N., 2005. Carbon and nitrogen stable isotope ratios indicative of organic matter flow in the ecosystem of the Seto Inland Sea, Japan. Japanese Journal of Ecology, **55**: 269-285. [In Japanese.]

TAKAI, N., A. HOSHIKA, K. IMAMURA, A. YOROZU, T. TANIMOTO & Y. MISHIMA, 2001. Distribution of carbon and nitrogen stable isotope ratios in macroalgae in Hiroshima Bay. Japanese Journal of Ecology, **51**: 177-191. [In Japanese.]

THORNTON, S. F. & J. MCMANUS, 1994. Application of organic carbon and nitrogen stable isotope and C/N ratios as source indicators of organic matter provenance in estuarine systems: evidence from the Tay Estuary, Scotland. Estuarine, Coastal and Shelf Science, **38**: 219-233.

VANDERKLIFT, M. A. & S. PONSARD, 2003. Sources of variation in consumer-diet $\delta^{15}$N enrichment: a meta-analysis. Oecologia, **136**: 169-182.

YAMADA, Y. & T. YOSHIOKA, 1999. Stable isotope analysis in aquatic ecosystems. Japanese Journal of Ecology, **49**: 39-45. [In Japanese with English Abstract.]

YOKOYAMA, H., 2008. Food sources of consumers in temperate estuaries and coastal waters: achievements and potential problems of isotopic studies. Japanese Journal of Ecology, **58**: 23-36. [In Japanese.]

YOKOYAMA, H. & Y. ISHIHI, 2007. Variation in food sources of the macrobenthos along a land — sea transect: a stable isotope study. Marine Ecology Progress Series, **346**: 127-141.

First received 16 December 2009.
Final version accepted 3 March 2010.